W9-AHP-907
SB
951.3
1972b

9-5

Systemic Fungicides

Systemic Fungicides

Edited by R. W. Marsh O.B.E.
with the assistance of R. J. W. Byrde
and D. Woodcock

A Halsted Press Book

SB
951.3
M36
1972b

John Wiley & Sons
New York

CATALOGUED

Longman Group Limited
London
Associated companies, branches and representatives throughout the world

© Longman Group Limited 1972

All rights reserved. No part of this publication may be reproduced, stored in a retrieval system, or transmitted in any form or by any means, electronic, mechanical, photocopying, recording, or otherwise, without the prior permission of the Copyright owner.

Published in the U.S.A. by
Halsted Press, a Division
of John Wiley & Sons, Inc.,
New York.

First published 1972

ISBN 0 470 57250 7
Library of Congress Catalog Card No. 72-4058

Printed in Great Britain by
William Clowes & Sons, Limited, London, Beccles and Colchester

Contents

		Preface	ix
		List of contributors	xi
R. L. Wain and G. A. Carter	**1**	**Nomenclature and definitions**	1
		Mechanisms of chemotherapeutant activity	3
R. L. Wain and G. A. Carter	**2**	**Historical aspects**	6
		Early investigations	6
		Development from mid-1930s	8
		The assessment of therapeutic and systemic activity	10
		The search for new chemotherapeutants	23
D. Woodcock	**3**	**Structure–activity relationships**	34
		Organophosphorus compounds	35
		Antibiotics	42
		Carboxylic acid anilides	54
		Heterocyclic compounds	64
		Aromatic compounds	79
D. Woodcock	**4**	**Toxicological considerations**	86
		Organophosphorus compounds	87
		Antibiotics	88
		Carboxylic acid anilides	89
		Heterocyclic compounds	89

S. H Crowdy	5	**Translocation**	92
		Entry into the free space within the tissues	93
		Apoplastic movement	100
		Symplastic movement	102
		Translocation of systemic fungicides	109
		Conclusion	113
A. E. Dimond	6	**Effects on physiology of the host and on the host/pathogen interactions**	116
		Introduction	116
		Effects of systemic fungicides on the host	117
		Production of fungitoxicants in plants	118
		Stomatal closure as a physical barrier to host entry	119
		Systemic compounds that alter growth of the host	120
		Treatments that block pathogenic processes	126
		Future prospects	131
A. Kaars Sijpesteijn	7	**Effects on fungal pathogens**	
		Introduction	132
		Effect of individual compounds on the fungus	135
		Factors determining activity: mode of action; selectivity and resistance; metabolic conversion	150
J. Dekker	8	**Resistance**	156
		Introduction	156
		Occurrence of acquired resistance	157
		Origin of resistance	161
		Mechanism of resistance	164
		Emergence of resistance in the field	170
		Avoidance of fungicide resistance	171
E. Evans	9	**Methods of application**	175
		Biological activity and systemicity	175
		Seed treatment	176
		Soil treatment	178
		Leaf and stem treatments	179

Post-harvest treatments 183
Timing systemic fungicide treatment 183

D. H. Brooks **10 Results in practice – I. Cereals** 186
Introduction 186
Barley and wheat 186
Rice 204
Maize 205

D. M. Spencer **10 Results in practice – II. Glasshouse crops** 206
Introduction 206
Disease control in the glasshouse 208
Use of selected systemic fungicides 209
Methods of application in glasshouses 218
Resistance 220
Non-fungicidal effects of benzimidazole compounds 221
Biological control 222
Present trends: future needs 223

R. B. Maude **10 Results in practice – III. Vegetable crops** 225
Introduction 225
Seed-borne diseases 225
Soil-borne diseases 228
Leaf diseases 232
Post-harvest diseases 235

R. J. W. Byrde **10 Results in practice – IV. Fruit crops** 237
Introduction 237
Temperate fruit 238
Sub-tropical fruit 249
Tropical fruit 252
Conclusions 254

References 255

Index 311

Preface

Critics will doubtless say that this book was out of date before it was published, thus confirming the rapid progress now being made in the study of systemic fungicides. This accelerated progress started around 1968 and it will be noted that of the 1174 references listed on pp. 255–309, 591 are papers published since 1967.

Because research in this field is developing so rapidly we felt that this was not the time to produce a definitive account of systemic fungicides. Instead we have aimed at making a relatively brief survey of the position now reached, including an outline of the basic principles of systemic fungicidal action, a summing-up of the present state of knowledge in the various branches of the subject, examples of practical applications and a look at possible future developments.

Experimental work on systemic fungicides is now worldwide, but the major centres of research have been in the Netherlands, England and the United States. We have been specially fortunate in securing the authoritative contributions of Professor Dekker and Dr Kaars Sijpesteijn from the Netherlands, of Professor Wain, Professor Crowdy, Dr Woodcock, Dr Byrde, Dr Brooks, Dr Spencer, Dr Carter, Mr Maude and Dr Evans from this country and of Dr Dimond from USA. Dr Dimond's death on 4 February 1972 was a serious loss to plant pathology: his contribution to *Systemic Fungicides* was his last completed manuscript.

This book is aimed at research workers in the field of fungicides, manufacturers of crop protection products, students of agriculture and horticulture and all concerned practically with fungicides and their use.

R. W. Marsh

Long Ashton Research Station, 1972

Note added at Proof

At the time of going to press, the development of triarimol has been suspended. This action was taken by the manufacturers after undesirable toxicological effects were observed in one species of laboratory animal following long-term administration at high dietary levels.

Acknowledgements

We are grateful to the following for permission to reproduce copyright material:

Society of Chemical Industry and authors for Table I and Table II on p. 44 of *Pesticide Science* **2** by Jank & Grossman (1971); British Crop Protection Council for Tables I and II on pp. 564 and 565 of *Proc. 5th Br. Insectic. Fungic. Conf.* **2** by Pommer & Kradel (1969) and Tables 4 and 5 on pp. 460 and 461 of *Proc. 6th Br. Insectic. Fungic. Conf.* **2** by Ten Haken & Dunn (1971); Cambridge University Press for Tables 3 and 4 on p. 469 of *Ann. appl. Biol.* **57** by Pluijers & Kaars Sijpesteijn (1966) and Table 3 on p. 479 of *Ann. appl. Biol.* **61** by Kaars Sijpesteijn *et al.* (1968); Gordon and Breach for Tables 4 and 5 of the *Proc. of the 2nd Inter. cong. of Pest. Chem.* (1971) by Clifford *et al.*; Koninklijke Nederlandse Chemische Vereniging and authors for part of Table V on p. 816 of *Rec. Trav. Chim.* **79**, 807–822 (1960) by Van den Bos *et al.*; Nederlandse Planteziektenkundige Vereniging for Tables 3 and 4 on p. 136 and Table 5 on p. 137 of *Neth. J. Pl. Path.* **74** by Tempel *et al.* (1968); Springer Verlag for Fig. 1 on p. 134 of *Residue Reviews* **25** by Kado & Yoshinaga (1970), Table I on p. 95 of *Residue Reviews* **25** by Misato (1969), Table V on p. 103 of *Residue Reviews* **25** by Misato (1969), Table I in column 8 of *U.S. 3,249,499* by Von Schmeling *et al.* (1966) and Table IV in column 10 of *U.S. 3,249,499* by Von Schmeling *et al.* (1966); Faculteit Landbouwwetenschappen, University of Ghent, for Tables 1 and 2 on p. 737 of *Meded. Rijksfaculteit Landbouw-weten-schaffen, Gent* **32** by Pommer & Kradel (1967) also Table 3 on p. 738 and Table 2 on p. 1219 of *Meded. Landbhoogesch Opzoek Stns, Gent* **27** by Dekker (1962).

We have been unable to trace the following authors and would appreciate any information that would enable us to do so:

D. T. Misato for Polyoxins table on p. 17 and Table 1 on p. 15 of: *Jap. Pest. Inform.* **1** by Misato and the authors for Table 1 on p. 701 of *Agr. Biol. Chem.* **34** by Maeda *et al.* (1970).

For permission to redraw diagrams we are grateful to Academic Press Inc., to Annual Reviews Inc., and to Blackwell Scientific Publications Ltd.

List of contributors

D. H. Brooks Ph.D. Head of Plant Pathology Section, Plant Protection Ltd, Jealott's Hill Research Station, Bracknell, Berks.

R. J. W. Byrde B.Sc., Ph.D. Head of Plant Pathology Section, Long Ashton Research Station; Research Fellow, University of Bristol

G. A. Carter Ph.D. Senior Scientific Officer (Plant Pathologist), Agricultural Research Council Growth Substances and Systemic Fungicides Unit, Wye College (University of London), nr Ashford, Kent

S. H. Crowdy B.Sc., Ph.D., A.R.C.S. Professor of Botany, University of Southampton

A. E. Dimond (the late) Ph.D. Chief, Department of Plant Pathology and Botany, Agricultural Experiment Station, New Haven, Connecticut, USA

J. Dekker Ph.D. Professor; Director of Laboratorium voor Fytopathologie, Wageningen, The Netherlands

E. Evans Ph.D., D.Sc. Senior Plant Pathologist, Fisons Pest Control Ltd, Chesterford Park Research Station, nr Saffron Walden, Essex

A. Kaars Sijpesteijn Ph.D. Head of Department of Biochemistry and

Microbiology, Institute for Organic Chemistry TNO, Utrecht, The Netherlands

R. B. Maude B.A., M.Sc. Plant Pathologist, National Vegetable Research Station, Wellesbourne, Warwick

D. M. Spencer Ph.D. Head of Mycology and Bacteriology Department, Glasshouse Crops Research Institute, Rustington, Littlehampton, Sussex

R. L. Wain C.B.E., D.Sc., F.R.I.C., F.R.S. Professor of Agricultural Chemistry, Wye College, University of London; Honorary Director, Agricultural Research Council Growth Substances and Systemic Fungicides Unit

D. Woodcock M.Sc., Ph.D., D.Sc., F.R.I.C. Head of Organic Chemistry Section, Long Ashton Research Station; Reader, University of Bristol

1

Nomenclature and definitions

by R. L. Wain and G. A. Carter

Accepting the broad definition that a **fungicide** is an agent that kills or inhibits the development of fungus spores or mycelium, the fungicides used on plants may be classified as protectant, systemic and eradicant on the basis of their uptake by, and mobility within, plant tissues.

The use of a **protectant** fungicide is an example of **prophylaxis**; it is a treatment intended to prevent or protect against infection. Protectant fungicides, which may be applied to seeds, soil or the plant surface, cannot penetrate into plant tissues in effective amounts. They must, therefore, act outside the plant prior to infection by the pathogen. By contrast, the use of a **systemic fungicide**, which is taken up by the plant, is one form of **therapy**, i.e. the cure of an established infection, where penetration of the host cuticle is used as the criterion of infection. A **therapeutant** is an agent that inhibits the development of a disease syndrome in a plant when applied subsequent to invasion by a pathogen. Therapy can be achieved by physical means, such as the hot-water treatment of smut-infected grain, but it is more usually brought about by chemical means; it is then termed **chemotherapy** and the active agent is a **chemotherapeutant**.

A chemical that can penetrate the plant cuticle and move through cell membranes is a potential therapeutant for any plant disease. A few fungal pathogens, notably the powdery mildews (Erysiphaceae), after penetrating their host, develop externally on the surface of the plant; non-penetrating fungicides can often be used as therapeutants for diseases caused by pathogens of this type.

Any compound capable of being freely translocated after penetrating the plant is termed a **systemic compound** or just a **systemic**. **Translocation** can be defined as the movement of a compound within the plant body to tissues remote from the site of application. Such a compound has advantages in use over one that remains localised in tissues adjacent to the site of application, i.e. a compound working **topically.**

Topical therapeutants are more often referred to as **eradicant fungicides** or **eradicants**. Typically these are fungitoxic chemicals which, when applied at an infection site, are capable of limited penetration leading to the elimination of an established infection.

When non-systemic compounds are applied to a plant and there is no surface redistribution, the areas missed in the initial application remain unprotected, as does any subsequent growth. The surface deposits are also subject to weathering. Repeated applications are therefore necessary to ensure continuous protection. Furthermore it is difficult to protect inaccessible organs, such as roots, with non-systemic compounds.

Because they are able to move freely from applications made at readily available sites to distant, untreated tissues, systemic compounds suffer none of these disadvantages. It may be possible to delay the application of systemics until disease symptoms are apparent, thereby eliminating the need for routine and often wasteful pre-symptom applications. One application of a systemic compound cannot, however, be expected to give continuous protection to the whole plant because translocation may be limited to certain tissues and the compound will be subjected to dilution and possibly metabolic breakdown as the plant grows (see Chapters 5 and 7).

Although applications of systemic fungicides are often said to 'protect' new growth from infection, their action is usually therapeutic in that it occurs after penetration of the host even if no visual symptoms of disease arise. Systemic protectant activity could arise if a compound applied, for example, through the roots became translocated and exuded on to the plant surface. The germination of fungal spores alighting on this surface might then be inhibited in the pre-penetrative phase. The antifungal phenolics exuded by varieties of onions resistant to 'smudge' (*Colletotrichum circinans*) are thought to act in this manner (Walker and Stahmann, 1955) and fungitoxic materials have been washed from intact leaf surfaces (Topps and Wain, 1957). As yet, no externally-applied systemic compound has been shown to exert such a protectant action on the leaf surface although both griseofulvin (Stokes, 1954) and cycloheximide (Wallen and Millar, 1957) have been detected in the guttation droplets of wheat seedlings following application to the roots. Microscopic examination of wheat leaves inoculated with powdery mildew and taken from seedlings treated with three systemic anti-mildew compounds showed that inhibition of mildew development became apparent only after penetration had started (Dekker and van der Hoek-Scheuer, 1964).

A high degree of mobility of a compound within the host plant is not always necessary for disease control; this property is more important, for example, when applying a root treatment against a leaf-attacking pathogen than when trying to control a pathogen causing a foot-rot or a vascular wilt. Chapman (1951) showed that the effectiveness of various compounds applied through the roots differed against the twig-invading Dutch elm disease pathogen *(Ceratostomella ulmi)* and the root-invading *Fusarium* wilt pathogen of tomato; the differences correlated with the mobility of these compounds in cellulose.

The direction in which the systemic chemical moves is also important. Movement is usually upwards in the xylem but for the protection of roots using foliar sprays, basipetal movement down the stem is required and this has proved very difficult to achieve.

MECHANISMS OF CHEMOTHERAPEUTIC ACTIVITY

A chemotherapeutant can modify, and be modified by, the tissues of the plant to which it is applied. It does not necessarily, therefore, reduce disease by direct fungitoxic action against the pathogen and so need not itself be fungicidal. The postulated mechanisms of therapeutic activity are very varied but three broad categories have been distinguished; direct toxic action against the pathogen, inactivation of toxins produced by the pathogen and enhancement of resistance of the host.

Direct activity This requires that an effective level of a fungitoxicant accumulates at the infection site. The compound applied need not itself be fungitoxic *in vitro* as it might be converted to an active compound by metabolism within the plant.

To inhibit a fungal pathogen inside the plant a fungitoxic chemical must be effective within the tissues – and these present a very different environment from that of the standard laboratory tests for *in vitro* fungitoxicity. Certain **systemic fungitoxicants**, e.g. rimocidin and pimaricin (Oort and Dekker, 1960), can be detected by bioassay in extracts made from treated plants yet show no therapeutic activity in those plants. Antagonism by cell constituents (Gottlieb, 1957) and irreversible binding to adsorption sites can render a powerful *in vitro* fungitoxicant inactive *in vivo*.

A direct-action therapeutant must show selective activity – it must be toxic to the pathogen but innocuous to the host plant (Dimond, 1963b). In practice this has proved difficult to achieve because the basic biochemistry of fungi and higher plants has much in common. Indeed, the mechanism by which a fungicide exerts its toxic action may well determine whether or not it is a potential chemotherapeutant (Kaars Sijpesteijn, 1970).

Some workers in the field of chemotherapy of both plant and animal diseases have measured selectivity in terms of a therapeutic index, this being calculated by dividing the minimum curative dose by the maximum tolerated dose. It is not possible, however, to assign a single index to each therapeutant for, as pointed out by Dimond (1962), some plant tissues, e.g. leaves, are more susceptible to toxic damage than are others, e.g. seeds.

The presence of a reactive toxophore in the molecule of a fungitoxicant may render it both poorly translocated and phytotoxic. A derivative of such a molecule, however, may possess improved translocation and therapeutic activity (Dimond and Davis, 1953; van Raalte *et al.*, 1955; Hamilton *et al.*, 1956; Pluijgers, 1959) although *in vitro* fungitoxicity is often lost. Such compounds have been referred to as **masked fungicides** (Kaars Sijpesteijn, 1961b) the intention being that the original fungitoxic molecule be released within the plant, preferably only at infection sites.

Inevitably most chemotherapeutants will be metabolised within plant tissues and where the metabolites are inactive this will result in a steady loss of effectiveness. This subject is dealt with in Chapter 7 and has also been reviewed by Kaars Sijpesteijn and van der Kerk (1965), Wain and Carter (1967) and Kaars Sijpesteijn (1969).

While the term **systemic fungicide** is often employed to describe therapeutants with a direct toxic action, some pathologists use the term synonymously with **systemic therapeutant**, thereby employing it in a wider sense. This usage has been criticised (see e.g. Horsfall, 1956) on the grounds that an indirect-action therapeutant is not itself a fungicide: if this view is accepted, the term 'systemic fungicide' should be restricted to direct-action systemic therapeutants.

Toxin inactivation This topic is reviewed by the late Dr Dimond in Chapter 6. There is evidence that toxins secreted by the pathogen are at least partially responsible for inducing the disease symptoms in the host (Braun and Pringle, 1959; Ludwig, 1960; Wheeler and Luke, 1963; Deverall, 1964; Pringle and Scheffer, 1964). Certain chemotherapeutants are thought to antagonise these toxins, thereby reducing host symptoms without appreciably affecting the growth of the pathogen. The therapeutic action of 8-quinolinol salts against Dutch elm disease, for example, is thought to be due to toxin inactivation (Horsfall and Zentmyer, 1942).

In some ways related in their action to these fungal toxins are the extracellular hydrolytic enzymes, secreted by many fungi and known to induce disease symptoms, notably in the vascular wilts (Wood, 1960). Grossmann (1962d) showed that several *in vitro* inhibitors of fungal pectinases can alleviate symptoms in tomato cuttings infected with the *Fusarium* wilt organism.

Enhancement of resistance of the host plant It has frequently been suggested that a chemotherapeutant operates indirectly by altering the metabolism of the host plant in such a way as to render it more resistant to disease. Widely differing mechanisms by which this increased resistance can be brought about have been suggested (see Chapter 6 and Dimond, 1963a, b, 1965; Grossmann, 1968b). These include alterations to (*a*) the host surface; (*b*) the nature of pectic substances (Edgington *et al.*, 1961); (*c*) wood morphology (Beckman, 1958); (*d*) carbohydrate levels (Horsfall and Dimond, 1957); and (*e*) phenolic constituents (Kaars Sijpesteijn and Pluijgers, 1962; Holowczak *et al.*, 1962). Several 'masked fungicides' appear to exert their therapeutic activity indirectly rather than by liberation of the free fungicidal molecule (Kaars Sijpesteijn, 1961b; van Andel, 1962b; Dekhuijzen, 1964).

A treatment that reduces disease without actually enhancing the disease resistance of the host has been demonstrated by Király *et al.* (1962). This operates by modifying the plant's growth so as to shorten the susceptible phase of development.

Although a classification of fungicides in terms of where, when and how they act in controlling plant disease is both feasible and useful, uptake, translocation and mode of action can be influenced by many factors as will be apparent in later chapters. Therefore, such a system of classification must inevitably be less rigid and precise than one based on the chemical structure of the fungicidal molecule. It must also be remembered that a compound which is a successful systemic therapeutant for one disease frequently fails to be so when a different host or pathogen is involved.

2

Historical aspects
by R. L. Wain and G. A. Carter

EARLY INVESTIGATIONS

Since this topic has been discussed in the reviews of Müller (1926), Roach (1939), Stoddard and Dimond (1949) and Horsfall (1945, 1956), only a brief outline will be given here.

The idea of introducing substances into plants goes back at least to the twelfth century when solid materials such as spices, medicines and colouring matters were inserted under the bark or into a borehole in a tree with the aim of imparting new odours, tastes or colours to the developing fruit. Such experiments were carried out spasmodically for several hundred years and in the fifteenth century Leonardo da Vinci describes the use of injections of arsenic into fruit trees to render the fruit poisonous. In 1602 an anonymous writer describes methods for improving the flavour and colour of fruits by inserting spices and dyestuffs into the pith; he also mentions the idea of killing 'wormes', i.e. larvae of wood-boring beetles, by inserting a mixture of pepper, laurel, incense and wine into boreholes. Over 150 years later Wilson records the use of liquid mercury inserted into boreholes in branches of trees to combat insect pests. This early work was, inevitably, empirical – there was at that time little knowledge of the uptake and movement of compounds in plants, although pioneer work on movement of sap was performed by Hales in 1726.

During the eighteenth century, when the study of plant physiology developed rapidly, studies were made on the movement of injected dyestuffs and mineral salts. These experiments revealed that certain compounds could be transported readily within the plant but great differences in the extent of uptake and translocation were observed. A further impetus to these studies was provided by the work of Liebig and other agricultural chemists in their pioneer investigations on the mineral nutrition of plants which led to the recognition of nutrient deficiency diseases. In attempts to cure such diseases, and especially chlorosis resulting from iron deficiency, injections of mineral salts were made. This work was pursued by workers in France, Germany, Russia and the USA, culminating in the studies in this country by Roach (1934, 1938, 1939).

Mokrzecki (1903) observed that certain nutrient solutions when injected appeared to alleviate attacks by pests and diseases. Slightly earlier Ray (1901) had injected various liquids, some of biological origin, into leaves via capillary tubes with the specific intention of rendering the leaves immune from disease. In 1906 Bolley reported that injections of copper sulphate, iron sulphate and formaldehyde reduced attacks of *Taphrina* on fruit trees and ten years later Norton reduced the incidence of *Septoria lycopersici* on tomato plants by injecting copper sulphate solution.

At the turn of the century in Russia and Italy, poisonous substances, particularly potassium cyanide, were injected into plants in the hope of combating insect pests. Work with cyanide was continued in the USA where its movement and persistence in the plant were studied in some detail.

Early this century Brooks and his co-workers began investigations on the silver leaf disease of plum incited by *Stereum purpureum* and Brooks and Bailey (1919) found that several dyes and disinfectants, when injected into diseased trees, allowed some recovery. Later Brooks and Storey (1923) reported the very high *in vitro* fungitoxicity of 8-quinolinol sulphate towards *S. purpureum* and this compound also proved effective against silver leaf when injected into the host plant (Roach, 1939).

In the USA Rumbold (1920a, b) made extensive studies on the injection of sweet chestnut trees as a method of curing blight, caused by *Endothia parasitica*. She found that injections of lithium salts could check disease development but the effect was only transient. Later Scherer (1927) and Jacobs (1928) reported the efficacy of injecting a solution of thymol for the control of both fungal and bacterial pathogens.

Throughout most of the early studies on therapy of disease, injection directly into the host plant was the preferred method of application. However, attempts were made to eradicate established infections with surface-applied compounds; Viala (1893) reported that copper sulphate had an eradicant action against anthracnose *(Elsinöe ampelina)* of the vine and Bolley (1891) controlled scab *(Actinomyces scabies)* on potato tubers with mercuric chloride. Later it was shown that zinc chloride could effectively destroy cankers of fireblight *(Erwinia amylovora)* on pear trees (Day, 1928; McClintock, 1931).

Attempts, too, were made to eradicate pathogens, such as the loose smuts and certain *Helminthosporium* species, carried internally within seeds. Geuther (1895) reported the effectiveness of formaldehyde, Hiltner (1915) of mercuric chloride and Riehm (1914) of organomercurials; during the early 1900s formaldehyde was used extensively in the treatment of cereal seed.

Root applications for the control of foliar pathogens were also employed by some early workers. In 1903 Massee applied copper sulphate to the roots of cucumber plants and reduced the incidence of *Cercospora melonis*, whilst in 1913 Spinks found that the application of salts of lithium to wheat and barley growing in soil or water culture effectively prevented the development of powdery mildew *(Erysiphe graminis)*.

A landmark in the study of chemotherapy of plant disease was the publication in 1926 of Müller's monograph *Die innere Therapie der Pflanzen*. In it he summarised much earlier work and described his own injection experiments designed to control insect pests and fungal pathogens. However, despite the publication of this work and the scattered reports of successful chemotherapy of plant disease, little interest in the subject was aroused at that time. Disease control was based on the use of successful protectant fungicides, such as Bordeaux mixture and lime sulphur, and the development of resistant varieties by selection and breeding. The problem of selectively inhibiting one organism within the living tissues of another was thought to be insuperable.

DEVELOPMENT FROM MID-1930s

Interest in chemotherapy was revived in the mid 1930s and it developed rapidly during the next decade. There were several reasons for this. Firstly, there were limitations both in the use of protectant fungicides, which offered little hope of curing infected plants or of controlling vascular diseases, soil-borne pathogens and viruses, and in the use of resistant varieties which frequently lost their resistance rapidly as new races of the pathogen arose. Secondly, many new organic chemicals were now available and these offered more scope than the inorganic salts used by most early workers. Thirdly, the chemotherapy of human diseases, pioneered by Ehrlich, had now been shown to be both possible and strikingly effective.

In 1935 Dogmagk showed that 'Prontosil' could cure bacterial infections in the bloodstream and in the same year Trefouël showed that the active compound, produced from Prontosil within the human body, was *p*-aminobenzene sulphonamide. This chemical became the parent of a whole range of sulphonamide or 'sulpha' drugs. Three years later Hassebrauk protected wheat seedlings from attack by rust fungi by treating them at the roots with *p*-aminobenzene sulphonamide.

A second breakthrough in medical chemotherapy came with the dawn of the 'antibiotic era' when in 1940 Florey and Chain isolated penicillin, which proved outstanding in controlling certain bacterial infections in man. A search for further antibiotics commenced and chloramphenicol, aureomycin and streptomycin were soon discovered. In 1944 Brown and Boyle achieved a cure of crown gall on several species with applications of crude penicillin and in 1947 Anderson and Nienow showed streptomycin to be translocated in plants.

The second world war aided the development of potential chemotherapeutants by stimulating the commercial production of antibiotics. It also stimulated Schrader's work on organic phosphates as poison gases; subsequently these compounds, some of which are freely mobile within plants and non-phytotoxic, were developed into highly successful systemic insecticides.

The development of chemotherapy continued to expand and many hundreds of compounds were examined for therapeutic activity. Progress has been reviewed by many workers, including those listed below:

1946 Zentmyer *et al.*
1949 Dimond *et al.*; Stoddard and Dimond
1951 Horsfall and Dimond (a, b); Horsfall *et al.*
1952 Anderson and Gottlieb; Brian (a, b); Dimond *et al.*; Sharvelle; Wain.
1953 Dimond; Wain
1954 Brian; Leben and Keitt
1956 Horsfall; van der Kerk; Rudd Jones
1958 Byrde and Ainsworth; Ford *et al.*; Zaumeyer
1959 Dimond; Dimond and Horsfall; Goodman; Howard and Horsfall; Livingston and Hilborn; Wain.
1960 Brian; Oort and van Andel
1961 Cremlyn; Horsfall; Kaars Sijpesteijn (a, b); van der Kerk; Pridham; Wain *et al.*
1962 Dimond; Goodman (a, b); Rhodes
1963 Dekker (a); Dimond (a, b); Wain
1965 Dimond; Wain
1967 Goodman
1968 Grossmann (b)
1969 Byrde
1970 Byrde; Erwin; Fawcett and Spencer

THE ASSESSMENT OF THERAPEUTIC AND SYSTEMIC ACTIVITY

As investigations into chemotherapy increased in number, varying methods were employed to detect therapeutic and systemic activity. Rapid laboratory and glasshouse tests were devised for screening large numbers of test compounds, different methods of application and different host/pathogen combinations being used. The choice of test organism was usually made either because of its suitability for routine use or because it was an important pathogen not readily controlled by protectant fungicides.

Seed-borne pathogens

Studies on the eradication of internally-borne seed pathogens were continued, favoured test organisms being the *Helminthosporium* diseases of cereal seeds, which are carried in the outer tissues of the seed, and *Ascochyta pisi*, which is carried deeper in the tissues of pea seeds. The *Helminthosporium* species could be controlled with organomercurials partly because of their volatility (Leben and Arny, 1952; Arny and Leben, 1954). The antibiotics antimycin A, helixin B (Leben and Arny, 1952; Leben *et al.*, 1953, 1954; Lockwood *et al.*, 1954) and mycothricin (Rangaswami, 1956) were also found to be effective.

Volatility may also assist in the eradication of *A. pisi* infections with chloropicrin (Kennedy, 1961) and pyridinethiol N-oxide (Kaars Sijpesteijn *et al.*, 1958; Dekker *et al.*, 1958), both of which are effective against this disease. Antibiotics such as GS-1 (Ark and Dekker, 1958), XG (=fungistatin) (Wallen and Skolko, 1950, 1951), cycloheximide (Wallen and Skolko, 1951), pimaricin and rimocidin (Dekker, 1955, 1957; Oort and Dekker, 1960) were also found effective against this pathogen. Another important group of internally-borne seed pathogens are the loose smuts and these were shown to be controlled by a treatment with 'Spergon' (tetrachloro-*p*-benzoquinone) (Porter, 1956) and streptomycin (Paulus and Starr, 1951).

Root pathogens

As root-attacking pathogens are difficult to control with protective fungicides, many studies were made on the therapy of diseases caused by soil-borne species of *Fusarium, Phytophthora* and *Rhizoctonia*, using chemicals applied as a soil drench. Although some control was often obtained it was difficult to ascertain how much of the effect was due to chemotherapy and how much to direct contact action against the inoculum in the soil.

Stoddard (1952) treated sand-grown plants of *Matthiola* with 8-quinolinol sulphate, lifted them, washed the roots, and replanted them in infected soil. The plants remained healthy, suggesting that the compound was operating within the plant. Previously Stoddard (1951b) had used this technique to demonstrate the therapeutic action of disodium ethylenebisdithiocarbamate against red core (*Phytophthora fragariae*) of strawberry. Zentmyer and Gilpatrick (1960) root-treated avocado seedlings already infected with *Phytophthora cinnamomi* and obtained evidence of a curative effect with 'Dexon' (sodium *p*-dimethyl-aminobenzenediazo sulphonate).

For many crops the preferred method for controlling root diseases would be by the application of a foliar spray containing a chemical which could be absorbed and translocated down to the roots. Successful control, however, has rarely been achieved even under experimental conditions. Zentmyer (1954) sprayed avocado seedlings growing in soil infested with *P. cinnamomi* with 8-quinolinol benzoate and noted a decrease in wilting and an increase in new root growth.

Vascular pathogens

These pathogens remain deep within the plant tissues. They offer one of the greatest challenges to chemotherapy and much work has been done with them, in particular with Dutch elm disease (caused by *Ceratostomella ulmi*) and *Fusarium* wilt of tomato (caused by *Fusarium oxysporum* f. *lycopersici*).

In the early 1900s Rumbold attempted to control chestnut blight by injecting chemicals but with little success. She did, however, suggest that toxins produced by the fungus were important in the etiology of the disease and that these might be antidoted with chemicals. Howard (1941) and Howard and Caroselli (1941), working at the Rhode Island Agricultural Experimental Station, studied the bleeding canker disease of maples (caused by *Phytophthora cactorum*) and found evidence that fungal toxins were important in inducing disease symptoms. They separated the toxin and demonstrated not only that it could be antagonised by diaminoazobenzene dihydrochloride but also that this compound, when injected into infected trees, allowed a recovery.

Dutch elm disease appeared in Europe after the first World War and spread rapidly. Several attempts were made to control it by chemotherapy and both Boudru (1935) and Fron (1936) reported success with 8-quinolinol salts. The disease was introduced into the USA in 1933 and was established there by 1940. In efforts to deal with this situation, research was initiated at the Connecticut Agricultural Experiment Station into various control methods, including chemotherapy. Much of the work done there has been described by Zentmyer *et al.* (1946) and Dimond *et al.* (1949).

Initially over 100 compounds were examined for their toxicity towards the pathogen *in vitro* and the most promising were examined in the field by injecting them into infected elm seedlings. Many compounds were ineffective or phytotoxic but *p*-nitrophenol, hydroquinone and 8-quinolinol sulphate or benzoate showed some promise: 8-quinolinol benzoate, which gave the most lasting effects, was tested in field trials and it was found that soil injection around the feeding roots was a superior method of application to the trunk injections used previously.

Following the earlier work of Howard on bleeding canker, studies were made which established that *C. ulmi* also produced toxins capable of causing at least part of the disease syndrome. Horsfall and Zentmyer (1942) suggested that the activity of 8-quinolinol salts against Dutch elm disease was brought about by toxin inactivation and they pointed out that both this compound and diaminoazobenzene contain basic nitrogen atoms in the molecule. They later tested malachite green and urea as therapeutants with some success; both compounds antidoted toxins from cultures of *C. ulmi*. In 1950 Feldman *et al.* showed that above pH 6 hydroxyl ions irreversibly inactivated the toxins of *C. ulmi in vitro* and that placing applications of basic compounds, including lime, around the roots of infected elms mitigated the disease symptoms (Feldman and Caroselli, 1951). This work led to the introduction of 'Carolate' as a treatment for Dutch elm disease, this being a mixture of low magnesia lime, urea, salicylic acid and an azo dye.

Hoffman (1951, 1952, 1953) used *in vitro* screening against fungal toxins in his attempts to find a therapeutant for oak wilt, caused by *Ceratocystis fagacearum*.

To assay potential vascular wilt therapeutants, a technique using *Fusarium* wilt of tomatoes was developed at Connecticut (Dimond *et al.*, 1952). Unlike *C. ulmi* which invades through the stem, this pathogen invades through the roots, so a method was devised to prevent root-applied compounds from coming into direct contact with the inoculum. Tomato plants were grown in sand and the test compounds applied as a drench; the treated plants were then lifted, their roots washed and deliberately damaged. The injured roots were then dipped into a bud cell suspension of the pathogen and the plants repotted in fresh sand. Subsequent disease development was assessed on the extent of vascular discoloration. Compounds which were effective when applied to sand were later applied to soil-grown plants and it was found that many were inactive. Keyworth and Dimond (1952) showed that chemical as well as mechanical injury to the roots altered their metabolism and reduced their susceptibility to attack by *Fusarium* wilt.

Compounds effective against tomato wilt were also tried against *Fusarium* wilt of carnations and *Verticillium* wilt of eggplants. A high correlation was obtained between the activity shown by compounds towards these three root-invading pathogens but effective compounds were often inactive against *C. ulmi*. Chapman (1951) attempted to correlate the relative effectiveness of compounds against tomato wilt or Dutch elm disease with their mobility in the plant as measured by their diffusibility on cellulose.

In investigations on the wilt diseases of tomato and carnation (Stoddard and Dimond, 1951; Dimond and Chapman, 1951; Dimond and Davis, 1952, 1953; Davis and Dimond, 1952), several promising chemotherapeutants were discovered, notably 4-chloro-3,5-dimethylphenoxyethanol, 2-norcamphene methanol, *n*-octadecyltrimethylammonium pentachlorophenate and salts of 2-carboxymethyl mercaptobenzothiazole.

Attempts to select potential therapeutants on the basis of their high fungitoxicity, high water solubility and low phytotoxicity failed and Davis (1952) found no correlation between the *in vitro* fungitoxicity and chemotherapeutic activity of test compounds. Several of the most effective chemotherapeutants were not fungitoxic but were found to have profound effects on the physiology of the host plant; these appeared to be similar to those caused by root injury. At high concentrations both 4-chloro-3,5-dimethylphenoxyethanol and 2-carboxymethyl-mercaptobenzothiazole salts produced formative effects in treated plants and Davis and Dimond (1953) found that certain growth regulators provided enhanced resistance to *Fusarium* wilt. By contrast Waggoner and Dimond (1952) had found that treatment of tomato plants with maleic hydrazide increased their susceptibility to wilt.

Since these early investigations, studies on the chemotherapy of both *Fusarium* wilt of tomato and Dutch elm disease have continued. Davis *et al.* (1954) found several unsubstituted heterocycles to be effective against *Fusarium* wilt; their fungitoxicity was very low but the appearance of foliar symptoms suggested that they affected the physiology of the host.

Corden and Dimond (1957, 1959) found that the therapeutic activity of certain naphthalene acetic acids was correlated with their ability to inhibit root elongation. In 1960 it was shown by Corden and Edgington that α-naphthylacetic acid (NAA) did not increase the wilt resistance of calcium-deficient tomato plants and studies on the nature of the pectins in treated plants (Edgington and Dimond, 1959; Edgington *et al.*, 1961) produced evidence that NAA enhanced wilt resistance by rendering the pectins of the host more resistant to attack by fungal enzymes.

Extra-cellular pectolytic enzymes are thought to play a major role in inducing the *Fusarium* wilt syndrome (Wood, 1960), and Grossmann (1962a, b, c; 1968a) examined both synthetic chemicals and plant extracts for their ability to antagonise fungal pectolytic enzymes *in vitro*. He tested the most promising materials on the plant and found that rufianic acid (1,4-dihydroxyanthraquinone-2-sulphonic acid) reduced wilting although it was less successful in preventing vascular browning or invasion of the tissues by the fungus. Earlier Grossmann (1957) had found that pre-infection treatments with several dimethyldithiocarbamates reduced wilt, as did applications of chloronitrobenzenes (Grossmann, 1958). Despite the high fungitoxicity of these compounds no fungitoxic principle was found in the bleeding sap of treated plants. In 1961 the same worker found applications of certain sulphonamides to be partially effective against *Fusarium* wilt.

Beckman and Howard (1957) observed that thiazolyl mercaptoacetates were effective against Dutch elm disease only if applied prior to infection and early in the season and that they checked the production of large-vesseled spring wood in treated elm trees. *C. ulmi* penetrates plant cell walls slowly and it was suggested that the modified wood development induced by these compounds acts as a mechanical barrier. Since then evidence has been obtained that 2,3,6-trichlorophenylacetic acid (Smalley, 1962), halogenated benzoic acids (Geary and Kuntz, 1962) and an aminotrichlorophenylacetic acid (Edgington, 1963) exert a control on the development of Dutch elm disease by a similar mechanism.

Foliar pathogens

Although foliar pathogens are more amenable to control by protectant sprays than are root or vascular pathogens, they have been widely used in studies on chemotherapy. The test method most frequently used is to apply a compound to the roots of a healthy plant growing in soil, sand or water culture and, after a suitable period, to inoculate the foliage and assess the severity of disease development.

In early studies on the therapeutic activity of aryloxyalkanecarboxylic acids Crowdy and Wain (1950, 1951) used *Botrytis fabae*, causal agent of 'chocolate spot' of broad bean, as a test organism. Root applications of certain of these acids, notably 2,4,6-trichlorophenoxyacetic acid, reduced symptoms of chocolate spot as shown by a reduction in lesion size rather than lesion numbers. Studies on the activity of structurally-related aryloxy- and arylthio-alkanecarboxylic acids were continued at the Agricultural Research Council Plant Growth Substance and Systemic Fungicide Unit at Wye College (Fawcett *et al.*, 1955, 1957). The systemic activity of 2,4,6-trichlorophenoxyacetic acid against chocolate spot, when applied to the roots of sand-grown broad bean plants, was confirmed and other acids were also found to exert a systemic action, notably 3-phenoxybutyric and 5-phenoxycaproic acids. Although these acids proved to be moderately fungitoxic *in vitro* it was thought that other factors were involved in their activity as therapeutants. *Botrytis fabae* on broad bean has been used for assessing the activity of root-applied (aryloxythio)trichloromethanes (Fawcett *et al.*, 1958), hydroxynitroalkanes (Bates *et al.*, 1963), inorganic salts (Carter and Wain, 1964) and N-carboxymethyl dithiocarbamic acid derivatives (Heyns *et al.*, 1966).

This host-parasite combination was also used by Crowdy and Davies (1952) who examined various compounds previously reported to be effective therapeutants for wilt diseases. Activity was assessed by measuring the mean lesion diameter; the *in vitro* fungitoxicity of these compounds was also determined. They found that whilst 8-quinolinol derivatives, Helione A, salicylic acid, certain metallic dinicotine salicylates, *p*-aminobenzoic acid and several ureas and thioureas all significantly reduced susceptibility to *B. fabae*, there was no clear correlation between fungitoxicity and systemic activity.

Byrde *et al.* (1953) found that applications in water culture of chlorinated β-naphthyloxyacetic acids and β-naphthols enhanced the resistance of broad beans to chocolate spot. Later Napier *et al.* (1957) reported that root applications of the protectant fungicide 'captan' [N-(trichloromethylthio)-cyclohex-4-ene-1,2-dicarboximide] also gave protection against chocolate spot and that when this fungicide was applied to the dorsal surface of leaves it protected not only the ventral surface of the treated leaf but untreated leaves above and below. Rombouts and Kaars Sijpesteijn (1958) painted pyridine-thiol-N-oxide and its carboxymethyl derivative on to the upper surface of broad bean leaves and obtained protection against *B. fabae* on the untreated surface of the leaflet and also in the twin leaflet of the treated one; lower leaves were also protected to a lesser degree. Root applications of these compounds were not effective.

Another foliar pathogen used in early studies on chemotherapy was *Alternaria solani*, causal agent of *Alternaria* or early blight of tomato. As early as 1903 Massee had shown that root applications of copper sulphate were active against this pathogen. Later McNew and Sundholm (1949) reported that when one leaf of a tomato plant was immersed in a suspension of certain substituted pyrazoles, untreated leaves showed an enhanced resistance to *A. solani*; 1-aryl-3,5-dimethyl-4-nitrosopyrazoles were the most effective.

Stubbs (1952) developed a method for the root application of compounds against *A. solani*, using the method of McCallan and Wellman

(1943) for inoculation and disease assessment. When applied to the roots of sand-grown tomato plants griseofulvin gave almost complete control; lower levels of activity were shown by 3,5-dimethyl-4-nitrosopyrazoles and 2,4,6-trichlorophenoxyacetic acid. Applying the test compounds to leaves of tomato plants followed by the inoculation of untreated leaves gave variable results. In the same year Dimond and Davis showed that soil applications of the alkali salts of 2-carboxymethyl-mercaptobenzo-thiazole enhanced the resistance of tomato plants to *A. solani*.

Applying test chemotherapeutants to the roots of tomato plants growing in sand was the method adopted by Fawcett *et al.* (1955, 1957) in their investigations on aryloxyalkanecarboxylic acids, several of which significantly reduced the incidence of *Alternaria* blight. Later (1958) the same workers found certain (aryloxythio)trichloromethanes to be systemically active against *A. solani* as were some inorganic salts (Carter and Wain, 1964).

In 1958 Grossmann applied 1,3,5-trichloro-2,4,6-trinitrobenzene to the roots of tomato plants and reduced the incidence of *A. solani*. Later, in 1961, he found that pre-infection applications of certain sulphonamide drugs including 'Prontalbin' (*p*-aminophenyl sulphonamide) and 'Albucid' (acetyl sulphanilamide), were also partially effective. In the same year Ristich and Cohen showed that soil drenches of 2-pyridinethiol-N-oxide were effective against this disease. About the same time Tsao *et al.* (1960) used *A. solani* as a test organism in their examination of the culture filtrates from soil actinomycetes for systemic antifungal activity. Culture filtrates which were active *in vitro* against conidia of *Alternaria solani* were tested further using excised tomato leaves. A culture filtrate from *Streptomyces griseus*, coded B-74, was found to be active in these tests and this material showed systemic activity when applied to intact tomato plants.

A Research Group for Internal Therapy of Plants was established in the Netherlands and here a test for systemic activity was developed using *Cladosporium cucumerinum*, the causal agent of cucumber scab (van Raalte *et al.*, 1955). In this test seven-day-old cucumber seedlings are placed in an aqueous solution of the test compound for two days before being inoculated with conidia of *C. cucumerinum*. They are then kept under high humidity for four to seven days, the severity of disease symptoms being assessed on an arbitrary scale. Alternatively the compound is applied by dipping one cotyledon into an aqueous acetone solution of the compound and then rinsing the treated area prior to inoculating the whole seedling. This test system was used in the intensive investigations on the systemic activity of derivatives of dithiocarbamic acid (van Raalte *et al.*, 1955; Pluijgers, 1959) in which the marked therapeutic activity of S-carboxymethyl-N,N-dimethyldithiocarbamate was discovered. Later investigations showed that DL-serine and DL-threonine (van Andel, 1958), pyridinethiol-N-oxide (Rombouts and Kaars Sijpesteijn, 1958), fluorophenylalanine (van Andel, 1962), thioacetamide and phenylthioureas (Kaars Sijpesteijn and Pluijgers, 1962) and thiosemicarbazides (Pluijgers and Kaars Sijpesteijn, 1966) are also effective in this test system. Earlier (1956) Rich had found that foliar applications of captan to young cucumber plants could give some protection against scab in leaves developed subsequently.

Rusts Rust fungi have proved one of the most devastating groups of foliar pathogens, attacks of black or stem rust (*Puccinia graminis tritici*) in the wheat lands of Canada and the USA being particularly serious. Many investigations have been made into the chemical control of rust diseases (Dickson, 1959) particularly since the early 1920s when the rust resistance of specially bred wheat cultivars broke down with the emergence of new physiologic races of the pathogen. Various formulations of inorganic sulphur were found to be effective but uneconomic, as were the copper, mercury and dithiocarbamate compounds examined. Other protectant fungicides tested were ineffective and although the antibiotic cycloheximide showed some promise it was erratic in its effect and often phytotoxic.

The failure to achieve economic control of rusts with protectant fungicides led to many investigations on the chemotherapy of these diseases, particularly using leaf or brown rust, *Puccinia recondita (P. triticina)* and stem rust (*P. graminis*) of wheat, *Uromyces appendiculatus* (*U. phaseolicola)* on *Phaseolus vulgaris* and *U. fabae* on broad bean. In most studies either pre-inoculation root applications have been made to plants growing in sand, soil or water culture or the test compounds have been applied as eradicant sprays to infected leaves.

In 1936 Sempio made extensive investigations into the activity of root-applied compounds upon the resistance of *Phaseolus* to *U. appendiculatus* and wheat to *P. striiformis* (*P. glumarum*). Cadmium and nickel salts proved active as did certain alkaloids. Previously Gigante (1935) had found that applications of sodium borate to soil in glasshouse and field plots increased the resistance of wheat to both *P. striiformis* and *P. recondita*. Subsequently, applications of boron compounds were shown to reduce attacks of *P. graminis* on wheat (Hart and Allison, 1939) and *Melampsora lini* on flax (Heggeness, 1942).

Gassner and Hassebrauk (1936) tested 174 compounds for their ability to control brown rust of wheat and barley (*P. hordei*) when applied to the soil surface, rust development being assessed on an arbitrary scale based on the number of sori present. Several compounds including acridine, bromonaphthols, picric acid and sodium *p*-toluenesulphochloramide were effective; the last two compounds were also active when the plants were sown in pre-treated soil, although the chloramide was phytotoxic. It was also found that dipping the leaves of susceptible varieties of wheat into solutions of sulphides could enhance their resistance towards *P. recondita* and *P. graminis* as shown by changes in reaction type. Later Hassebrauk (1938) showed that *o*- and *p*-toluenesulphonamide, recently developed for pharmaceutical use, gave complete control of *P. striiformis, P. graminis* and *P. recondita* on wheat, *P. hordei* on barley *P. secalina* on rye and *P. coronata* on oats, although both compounds were more phytotoxic than the moderately effective picric acid. When applied in the field, however, *p*-toluenesulphonamide failed to control yellow or brown rust of wheat and proved phytotoxic (Hassebrauk, 1940).

Hart and Allison (1939) found that root applications of borax, picric acid and both *o*- and *p*-toluenesulphonamides controlled stem rust of wheat but all were phytotoxic; the sulphonamides were more effective in a sandy soil than in a loam.

Hassebrauk (1951) examined other sulphonamide drugs for their systemic activity against cereal rusts, and particularly *P. recondita*. Whilst many of them were inactive, others were highly effective when applied to roots; seed treatment, however, did not enhance the resistance of the developing seedlings to rust and phytotoxicity was again evident with many of the active compounds. Hassebrauk (1952) later showed that 'Ladogal' (sodium salt of *p*-aminobenzylsulphonoxymethylamide-N-*d*-glucoside sulphonic acid), previously shown to be active against *P. recondita,* was also effective against *P. graminis* and *P. striiformis*. The compound remained active in the soil for four weeks after application but as the humus content increased, activity decreased. Infections of *P. recondita* were eradicated by leaf sprays.

Crowdy and his co-workers made studies on the uptake and translocation of certain sulphonamides fed to the roots of seedlings of oat, wheat and broad bean growing in water culture (see Rudd Jones, 1956). Treated plants were inoculated with rust and a relationship was established between the high mobility of sulphanilamide in the plant and its systemic activity as compared with the ineffectiveness of the poorly translocated sulphaguanidine. Another sulphonamide drug, sulphadiazine, was found to be effective against brown rust of wheat by Yamada (1955).

Kent (1941a, b) applied salts of lithium to soil-grown wheat plants and reduced their susceptibility to leaf rust and Heggeness (1942) found applications of borax enhanced the resistance of flax to *Melampsora lini*. Other investigators observed that root applications of 2,4-D (von Witsch and Kasperlik, 1953), ferrous sulphate (Forsyth, 1957) and maleic hydrazide (Samborski *et al*., 1960) were partially effective in controlling cereal rusts.

Straib (1941) applied sprays to rust-infected cereal, antirrhinum and flax plants and found picric acid, sulphonamides and acridine to be partially effective but phytotoxic. Mitchell *et al.* (1950) applied several sulphanilamides to wheat as eradicants for stem rust; the sodium salt of sulphadiazine was the most effective and formulation with a wetter or keeping the sprayed plants in humid conditions was shown to enhance activity. Hotson (1952, 1953) also examined sulphonamides as eradicants of wheat stem rust. He applied sprays eight days after planting the seed and the rust susceptibility of the treated seedlings was assessed in terms of their reaction type. Several of the compounds were active; some, e.g. sulphanilamide, gave only a temporary inhibition of rust whilst others, e.g. sulphadiazine, inhibited development throughout the experimental period.

Also in 1953 Livingston examined 179 compounds for their eradicant activity against leaf and stem rust of wheat by spraying plants five days after inoculation. Promising compounds were then applied in the field with knapsack sprayers or by aeroplane. Sulphanilamide was found to retard leaf rust development but its low solubility prevented its effective application in aerial sprays and sulphanilic acid proved to be more effective. Salts of sulphamic acid were also found to be active in these experiments, the calcium salt being the least phytotoxic. Acosta and Livingston (1955) confirmed that both sodium sulphanilate and calcium sulphamate reduce stem rust on wheat, barley and oats. Sodium sulphanilate was not phytotoxic but the sulphamate markedly reduced the viability of the grain produced from treated crops; its phytotoxicity, however, was found to depend on the stage of growth when it was applied. Further evidence of the effectiveness of sulphanilic acid as an eradicant of *P. graminis* was provided by Campos and Borlaug (1956).

Compounds other than sulphonamides have also been reported as eradicants for rust diseases. Yarwood (1947, 1948) observed that lime sulphur and hydrogen sulphide or hydrogen cyanide inhibit the development of infections of *U. appendiculatus* on leaves of *Phaseolus* and Ibrahim (1951) found that sprays of 2,4-D retarded the development of *P. graminis* on oats. In a survey by Hotson (1953) picric acid, nitrobenzoic acids, saccharin and methylene blue were found to be effective eradicants of wheat stem rust and, in a similar survey, Livingston (1953) found phenylhydrazine to be effective in the glasshouse although its performance in the field was disappointing. N-phenyl-N′-3-sulpholanyl hydrazine, however, was found to be a useful therapeutant for rust on *Phaseolus* by Evans and Saggers (1962) and Jaworski and Hoffman (1963) reported various phenylhydrazones to be effective eradicants of wheat leaf rust. Livingston (1953), Silverman and Hart (1954) and Wallen (1955, 1958) tested cycloheximide against both stem and leaf rust of wheat. It was found to show promise both in the glasshouse and in the field although it often proved phytotoxic. Certain salts of cycloheximide, especially the acetate and semicarbazone (Hacker and Vaughn, 1957a, b), gave good and prolonged control of wheat stem rust in the field.

Many other antibiotics have been examined for their eradicant activity against rusts but most proved ineffective. Limited activity has been shown for anisomycin (Zaumeyer and Wester, 1956), F-17 (Preston *et al.*, 1956), oligomycin (Zaumeyer, 1957), P-9 (Davis *et al.*, 1960; Hagborg *et al.*, 1961) and phleomycin (Smale *et al.*, 1961; Purdy, 1964).

Keil *et al.* (1958) applied eradicant sprays in glasshouse tests using leaf rust of rye (*P. secalina*) and several nickel salt-amine complexes were the best compounds tested. Nickel amine complexes applied against wheat leaf rust in the field were also found to be effective by Peturson *et al.* (1958) and Forsyth and Peturson (1959a, b) showed that inorganic nickel salts could also eradicate wheat leaf rust in the field without causing phytotoxicity although the levels necessary for rust eradication proved to be too toxic for use on oats or sunflower. In 1960 these workers showed that nickel salts controlled both stem and leaf rust of wheat in the field and that the timing of their application was less critical than that of the protectant fungicide zineb. A year later, Keil and Frohlich took out a US patent for the use of nickel compounds as rust eradicants.

Powdery mildews Powdery mildew diseases, caused by members of the Erysiphaceae, are found on almost all species of crop plant. Some of the more important are those occurring on vine (*Uncinula necator*), apple (*Podosphaera leucotricha*), cucumber (*Erysiphe cichoracearum* or *Sphaerotheca fuliginea*) and cereals (*Erysiphe graminis*). The cucumber and cereal pathogens have been used extensively in studies on chemotherapy.

Elemental sulphur applied alone in dust, colloidal dispersion or vapour form, or combined in the form of lime sulphur, was early recognised as a cheap and effective control for powdery mildews but it tended to be phytotoxic. It has been partially replaced with organic fungicides such as dinocap.

Since the mycelium of powdery mildew fungi develops externally upon the leaf surface, foliar sprays do not have to penetrate the plant tissues to be effective eradicants and for this reason most protectant mildew fungicides also show some eradicant activity. Uptake by the sprayed leaves and redistribution within the plant would undoubtedly enhance disease control but only a few studies have been made (e.g. Sharples, 1963) in which the effect of foliar sprays on the disease resistance of unsprayed leaves has been assessed.

Roach (1939, 1942) showed that injections of sodium thiosulphate could protect apple trees against attacks of *P. leucotricha* and since the early work of Spinks (1913) root applications of inorganic salts have frequently been examined for systemic anti-mildew activity. Salts of copper (Olsen, 1939), cadmium (Sempio, 1938; Tomlinson and Webb, 1960), zinc (Tomlinson and Webb, 1958, 1959, 1960), magnesium (Colquhoun, 1940) and boron (Eaton, 1930) have all been shown to possess systemic activity and there have been repeated reports of the systemic anti-mildew activity of both silicates (Germar, 1934; Wagner, 1940; Minabe, 1951; Grosse-Brauckmann, 1957, 1958) and salts of lithium (Spinks, 1913; Wortley, 1936; Kent, 1941a, b; Vidali and Ciferri, 1951; D'Armini, 1953 and Yarwood, 1959). In 1964 Carter and Wain examined a large number of inorganic salts for their systemic activity against *E. graminis* when applied to the roots of wheat seedlings growing in sand. Most salts were either phytotoxic or ineffective but the marked activity of lithium salts against mildews was again confirmed.

Heyns *et al.* (1966) made root applications of N-carboxymethyldithiocarbamates to seedlings of wheat, pea, cucumber and apple and found that certain of these compounds showed systemic antimildew activity. Previously Darpoux *et al.* (1958a) found that the manganese salt of pyridinethiol-1-oxide was systemically active against barley mildew and Koopmans (1960) showed that root applications of several substituted triazoles were also effective against this disease; 1-bis(dimethylamido)-phosphoryl-3-phenyl-5-aminotriazole-1,2,4 has been marketed as a systemically active mildew fungicide by N. V. Philips Duphar under the trade name 'Wepsyn'.

Sphaerotheca fuliginea on cucumber has been used by Dekker in his investigations at Wageningen into plant disease chemotherapy. In 1961 he investigated an earlier report that procaine penicillin G was systemically active against cucumber mildew and he showed that the procaine (2-diethylaminoethyl *p*-aminobenzoate) moiety was responsible for this activity. Procaine hydrochloride, applied to the roots of cucumber plants grown in water culture, protected the leaves against infection and was detected chromatographically in leaf sap. The compound was shown to be effective against several powdery mildews but not other plant pathogens and it was inactive *in vitro*. Microscopic examination of inoculated leaves from treated plants revealed no effect upon the germination of mildew conidia and inhibition of the fungus was only apparent when it was penetrating the host. The hydrolytic products of procaine, *p*-aminobenzoic acid and diethylaminoethanol, proved inactive but several analogues of procaine showed some activity although less than procaine itself. Procaine proved to be disappointing in the field and in further studies with many procaine analogues and derivatives (Niemann, 1964;

Niemann and Dekker, 1966a), none proved better than the parent compound. Following an observation that 3-methyl-N,N-diethyl-*p*-phenylenediamine is active against cucumber mildew, the same workers examined a series of *p*-phenylenediamines. Marked differences in the systemic activity of structurally-related compounds were found (Niemann, 1964; Niemann and Dekker, 1966b).

In other studies Dekker used cucumber leaf discs, inoculated with *S. fuliginea* and floated upon the test solution, as a preliminary method for detecting chemotherapeutic activity. Kinetin (6-furfurylaminopurine) was found to be effective in this test but it was inactive when applied to the roots of cucumber plants (Dekker, 1963b). Methionine too proved active in the leaf disc test but not following root application (Dekker, 1969a). The activity of kinetin led to the examination of a number of purines and pyrimidines; five of these proved effective in leaf disc tests but only one, 6-azauracil, was active when applied to roots (Dekker, 1962). This compound was later tried in the orchard against *P. leucotricha* on apple trees. It was effective at low concentrations and gave better control of the disease than some non-systemic fungicides but it damaged the new growth of treated trees (Oort and Dekker, 1964; Dekker and Roosje, 1968).

Root applications of barbiturates have given control of powdery mildew of both marrow (Zaracovitis, 1965) and cucumber (Tempel and Kaars Sijpesteijn, 1967), and in studies on the systemic activity of antibiotics both griseofulvin (Rhodes *et al.*, 1957) and cycloheximide derivatives (Jones and Swartwout, 1961a, b) have shown systemic activity against powdery mildews following root application.

Some investigators have sown treated seed and looked for an enhanced disease resistance in the developing seedling. Bouchereau and Atkins (1950) reported the effectiveness of seed treatment with copper salts against *Ophiobolus miyabeanus* whilst Rodigin and Krasnova (1959) found copper salts effective against leaf rust of wheat when used as a seed dressing, as was boric acid and zinc sulphate. Treatment of rice seed with 2-methyl-1,4-naphthaquinone, 2,4-D, boric acid or copper salts was reported by Akai (1955) as effective in reducing the incidence of *O. miyabeanus* and Volger (1959) found that treatment of the seeds of *Pinus sylvestris* with TMTD rendered the developing seedlings more resistant towards species of *Pythium* and *Rhizoctonia*. There is no evidence in these studies whether the applied chemical entered the seed and moved directly into the tissues of the seedling or whether it was leached from the seed and taken up by the seedling roots.

THE SEARCH FOR NEW CHEMOTHERAPEUTANTS

Since the 1930s many hundreds of compounds have been screened for chemotherapeutic activity and some of the more active are listed in Table 2.1.

In most studies, made with the aim of discovering compounds of practical use in the field, activity was measured directly by application to the growing plant. Some workers, however, have used more rapid and economic techniques such as floating inoculated leaf discs on solutions of the test compounds and van Raalte (1954) devised a rapid test, using excised petioles, for detecting the translocation of antifungal chemicals. This test has been used in a modified form by Spencer (1957) and Darpoux *et al.* (1958b). Initially, compounds were either tested at random or selected on the basis of high water solubility and fungitoxicity but it soon became apparent that there is rarely a high correlation between *in vitro* fungitoxicity and therapeutic activity.

The ability of test compounds to antagonise fungal toxins *in vitro* has been employed as a screening technique but with only limited success possibly because the role of some toxins, produced in culture, in the development of the disease syndrome in the infected plant is not clearly established (Dimond and Waggoner, 1953). Furthermore, toxin antagonists frequently reduce only some of the disease symptoms (e.g. Feldman and Caroselli, 1951) and evidence has been obtained that a single pathogen can produce more than one toxin (Brooks and Brenchley, 1931; Dimond *et al.*, 1949).

The observation that several successful chemotherapeutants are non-fungitoxic *in vitro* yet systemic, led to the concept of indirect activity, the compound exerting its effect by influencing the metabolism of the host plant. Plant growth regulators, and in particular 2,4-D (Ibrahim, 1951; Davis and Dimond, 1953, 1956; von Witsch and Kasperlik, 1953; Natti and Szkolnik, 1954; Akai, 1955; Waggoner, 1956) were examined as therapeutants. It was found that some of these compounds enhanced the resistance of plants to certain diseases and increased their susceptibility towards others. In an attempt to explain this, Horsfall and Dimond (1957) postulated that the protection afforded by growth regulators was related to carbohydrate levels in the treated plant, some pathogens favouring tissues with a high sugar content and others favouring those with low levels.

Crowdy and Wain (1950, 1951) chose to work with aryloxyalkanecarboxylic acids because these compounds are structurally related to synthetic growth regulators known to be readily translocated in plants (Wain, 1951). By selecting compounds which lacked the structural requirements

TABLE 2.1 **Compounds reported as showing chemotherapeutic activity**

Compound	*Host/pathogen*	*Reference(s)*
Acridine	Rust diseases	Hassebrauk (1938), Straib (1941)
Alkaloids	*Phaseolus/Uromyces appendiculatus*	Sempio (1936)
Alkylmercapto-tetrahydro-pyrimidines	*Phaseolus/U. appendiculatus*	Nickell *et al.* (1961)
Amino acids	Various	see van Andel (1966b)
p-Aminobenzoic acid	Broad bean/*Botrytis fabae*	Crowdy and Davies (1952)
β-Aminoethyl-aryl ketones	*Phaseolus/U. appendiculatus*	Pellegrini *et al.* (1965)
	Vine/*Plasmopara viticola*	
Aminotrichlorophenyl acetic acid	Elm/*Ceratostomella ulmi*	Edgington (1963)
Anthraquinones	Tomato/*Fusarium oxysporum*	Fuchs (1964)
Antibiotics (cycloheximide, griseofulvin, etc.)	Various	see Zaumeyer (1958), Goodman (1959), Dekker (1963a, b)
Aryl-3,5-dimethyl-1,4-nitrosopyrazoles	Tomato/*Alternaria solani*	McNew and Sundholm (1949), Stubbs (1952)
	Broad bean /*B. fabae*	Crowdy and Wain (1951)
Aryloxyalkanecarboxylic acids	Broad bean/*B. fabae*	Crowdy and Wain (1951), Stubbs (1952), Fawcett *et al.* (1955, 1957)
	Tomato/*A. solani*	
(Aryloxythio)trichloromethanes	Tomato/*A. solani*	Fawcett *et al.* (1958)
6-Azauracil	Cucumber/*Sphaerotheca fuliginea*	Dekker (1962), Oort and Dekker (1964), Dekker and Oort (1964)
Azo dyes	Maple/*Phytophthora cactorum*	Howard and Caroselli (1941), Caroselli and Howard (1942)
	Broad bean/*B. fabae*	Crowdy and Davies (1952)
Barbiturates	Cucumber, Marrow/*S. fuliginea*	Zaracovitis (1965), Tempel and Kaars Sijpesteijn (1967)

Benzoic acid	Elm/*C. ulmi*	Zentmyer (1942), Zentmyer and Horsfall (1943), Zentmyer *et al.* (1946)
	Oak/*Ceratocystis fagacearum*	Phelps *et al.* (1966)
Bromonaphthols	Wheat/*Puccinia recondita*	Hassebrauk (1938)
Captan	Various	see text
4-chloro-3,5-dimethylphenoxy-ethanol	Tomato/*F. oxysporum*	Dimond and Chapman (1951)
Chlorogenic acid	Apple/*Venturia inaequalis*	Kirkham (1957)
Chloronitrobenzenes	Tomato/Various	Grossmann (1958)
Coumaric acid, esters, salts	Apple/*V. inaequalis*	Kirkham and Flood (1963), Kirkham and Hunter (1964, 1965)
Dibenzothiophene	*Pinus banksiana*/*Rhizoctonia solani*	Vaartaja (1955)
Dithiocarbamates	Various	see text
2,4-D	Various	see text
Gibberellins	Potato/*Phytophthora infestans*	Weindlmayr (1963)
	Wheat/*Tilletia foetida*	Király *et al.* (1962)
Hydroquinone	Elm/*C. ulmi*	Zentmyer and Horsfall (1943), Zentmyer *et al.* (1946)
IAA	Tomato/*Verticillium albo-atrum*	Davis and Dimond (1953, 1956), Sinha and Wood (1967)
	Oak/*C. fagacearum*	Phelps *et al.* (1966)
Inorganic salts	Various	see Carter and Wain (1964)
Maleic hydrazide	Wheat/*P. recondita*	Samborski *et al.* (1960)
2-Mercaptobenzothiazole salts	Tomato/*F. oxysporum, A. solani*	Dimond and Davis (1952)
	Elm/*C. ulmi*	Dimond and Davis (1952), Tarjan and Howard (1953), Beckman and Howard (1957), Beckman (1958)
	Oak/*C. fagacearum*	Phelps *et al.* (1966)
2-Methyl-1,4-naphthaquinone	Rice/*Ophiobolus miyabeanus*	Akai (1955)
Naphthyl- and naphthoxy-acetic acids	Broad bean/*B. fabae*	Byrde *et al.* (1953)
	Tomato/*F. oxysporum*	Corden and Dimond (1957)
	Tomato/*V. albo-atrum*	Davis and Dimond (1953, 1956)

TABLE 2.1 *contd.*

Compound	*Host/pathogen*	*Reference(s)*
Nitrobenzoic acids	Wheat/*Puccinia graminis*	Hotson (1953)
Nitrophenols	Elm/*C. ulmi*	Zentmyer *et al.* (1946)
	Phaseolus/Erysiphe polygoni	El-Zayat *et al.* (1968)
N-octadecyltrimethyl ammonium pentachlorophenate	Tomato/*F. oxysporum*	Dimond *et al.* (1952), Davis and Dimond (1952)
Norcamphene methanol	Carnation/*Fusarium dianthi*	Stoddard (1951a)
Organomercurials	Various	see text
Pentachlorophenol	Rice/*O. miyabeanus*	Akai and Kato (1962)
Phenylhydrazines	Wheat/*P. recondita, P. graminis*	Livingston (1953)
	Phaseolus/U. appendiculatus	Evans and Saggers (1962)
Phenylhydrazones	Wheat/*P. recondita*	Jaworski and Hoffman (1963)
3-Phenylsydnones	Wheat/*P. recondita*	Davis *et al.* (1959)
	Phaseolus/U. appendiculatus	
Phenylthioureas	Apple/*V. inaequalis*	Kuć *et al.* (1957)
	Cucumber/*Cladosporium cucumerinum*	Kaars Sijpesteijn and Pluijgers (1962)
Picric acid	Rust diseases	Hassebrauk (1938), Hart and Allison (1939), Hotson (1953), Straib (1941)
Plant growth retardants	Tomato/*V. albo-atrum*	Sinha and Wood (1964, 1967)
	Phaseolus/Sclerotium rolfsii	Tahohri *et al.* (1965)
	Blackcurrant/various	Smith and Corke (1966)
Procaine and derivatives	Powdery mildew diseases	Dekker (1961a, b), Niemann (1964), Niemann and Dekker (1966a)
Sodium propionate	Wheat/*P. graminis*	Futrell and Berry (1964)

Pyridinethiol-N-oxide	Pea/*Ascochyta pisi*	Kaars Sijpesteijn *et al.* (1958), Dekker *et al.* (1958)
	Phaseolus/Colletotrichum lindemuthianum	
	Barley/*E. graminis*	Darpoux *et al.* (1958a)
	Cucumber/*Colletotrichum lagenarium*	Ristich and Cohen (1961)
	Broad bean/*B. fabae*	
	Peach/*Cytospora cincta*	Helton and Rohrbach (1967)
Pyrogallol	Elm/*C. ulmi*	Zentmyer *et al.* (1946)
8-Quinolinol salts	Various	see text
Rufianic acid	Tomato/*F. oxysporum*	Grossmann (1962a, c, 1963)
Salicylic acid	Broad bean/*B. fabae*	Crowdy and Davies (1952)
	Elm/*C. ulmi*	Tarjan and Howard (1953)
Sulphanilamides, sulphonamides and related compounds	Various, mainly rust diseases	see text
Tetrachlorobenzoic acids	Elm/*C. ulmi*	Beckman (1959)
	Oak/*C. fagacearum*	Geary and Kuntz (1962)
Thioacetamide	Cucumber/*C. cucumerinum*	Kaars Sijpesteijn and Pluijgers (1962)
Thiosemicarbazides	Wheat/*P. graminis*	Fuchs and Bauermeister (1958)
	Cucumber/*C. cucumerinum*	Pluijgers and Kaars Sijpesteijn (1966)
TIBA	Tomato/*V. albo-atrum*	Davis and Dimond (1953, 1956)
	Oak/*C. fagacearum*	Phelps *et al.* (1966)
Triazoles	Barley/*Erysiphe graminis*	Koopmans (1960)
	Apple/*P. leucotricha*	
Sodium trichlorophenoxy-acetate	Peach/*Cytospora cincta*	Helton and Rohrbach (1967)
Salts of trichlorophenyl acetic acid	Elm/*C. ulmi*	Smalley (1962)
Unsubstituted heterocycles	Tomato/*F. oxysporum*	Davis *et al.* (1954)
Ureas and thioureas	Broad bean/*B. fabae*	Crowdy and Davies (1952)

for growth-regulating activity, they attempted to find chemicals that were readily translocated, non-phytotoxic yet effective against the pathogen.

Most of the compounds tested for chemotherapeutic activity were either inactive or too phytotoxic though some showed promise in glasshouse tests. A few of these were examined in the field but generally they proved disappointing. Many compounds have been examined only by one worker in a single test system but others have been examined more extensively and these will now be briefly classified (see also Pluijgers, 1959).

Inorganic salts

Much early work was done using inorganic salts, particularly those of the fungitoxic metals such as copper and mercury. From these studies salts of lithium emerged as showing some promise but in the field they proved relatively ineffective and phytotoxic. Carter and Wain (1964) made extensive studies on the systemic activity of inorganic salts in glasshouse tests but, whilst the activity of lithium salts was confirmed, most compounds proved poorly translocated and often phytotoxic. The most promising application of inorganics in the field appears to be the use of nickel salts as rust eradicants.

Sulphonamides

Sulphonamides and related compounds have been examined extensively and their potential use in controlling plant disease was discussed by Rudd Jones (1956). Although they have shown therapeutic activity against facultative parasites (Dimond *et al.*, 1952; Grossmann, 1957, 1961) they are more effective against obligate parasites, particularly rusts. Against these diseases they have proved to be very effective in the glasshouse and moderately effective in the field although they are frequently phytotoxic. Their phytotoxicity can be reduced by applying *p*-aminobenzoic acid and as this also antagonises their therapeutic activity against rusts, they may act directly against the pathogen as antimetabolites. The activity of sulphonamides against rusts is fungistatic rather than fungicidal and when treatment ceases then the disease re-establishes itself. Work on their uptake and metabolism by plants has shown that these compounds can be acetylated in plant roots and that their mobility within the plant varies greatly with changes in chemical structure.

Organomercurial fungicides

Commercial protectant fungicides have generally shown little or no therapeutic activity but some organomercurials have shown limited eradicant activity against both seed-borne (Leben and Arny, 1952; Arny and Leben, 1954; Porter, 1956) and foliar (Hamilton, 1931, 1948; Howard and Sorrell, 1943; Moore, 1966, 1967) pathogens.

Captan

There have been reports of therapeutic activity for captan, not only as an eradicant of seed (Maude, 1966) and foliar (Rich, 1956; Albert and Groves, 1966) pathogens but as showing systemic activity against foliar pathogens after root application (Stoddard, 1954; Napier *et al.*, 1957; Rosen, 1959). Somers and Richmond (1962), however, detected only a very slight movement of root-applied captan into the shoots of broad bean plants. Fawcett *et al.* (1958) examined other compounds containing the trichloromethyl group and found some of them to be systemic; captan itself was not active in their tests.

Dithiocarbamates

The dithiocarbamates are one of the most successful groups of protectant fungicides and there have been reports of therapeutic activity for ferbam (Taylor, 1953), nabam (Stoddard, 1951a, b), zineb (Taylor, 1953; Rich, 1956) and thiram (Grossmann, 1957; Volger, 1959).

In 1954 van Raalte, using his excised petiole test for demonstrating translocation, showed that the free dimethyldithiocarbamate ion was not readily translocated but that tetramethylthiuram monosulphide (TMTM) was; TMTM was known to liberate the free dimethyldithiocarbamate ion only slowly and it was suggested that the remainder of the TMTM molecule was acting as a protective carrier for the fungitoxic ion. Investigations were therefore made into the systemic activity of compounds with various potential carrier groups attached to the dimethyldithiocarbamate ion (van Raalte *et al.*, 1955) and this led to the discovery of systemic activity in the S-carboxymethyl compound whose *in vitro* fungitoxicity, however, was low. Further studies with related compounds (Pluijgers, 1959) did not lead to the discovery of a more effective chemotherapeutant and later work on the metabolism of these compounds within plants (see Dekhuijzen, 1964) has indicated that the S-carboxymethyl compound, which is an active plant growth regulator, works through its effect upon the metabolism of the host plant rather than by liberation of dimethyldithiocarbamate ions.

8-Quinolinol salts

8-Quinolinol and its salts are highly antifungal water-soluble compounds which were used extensively in early studies on chemotherapy, particularly the therapy of vascular wilt pathogens. They have shown activity against silver leaf of plum (Roach, 1934; Grosjean, 1951; Bennett, 1962) and Dutch elm disease (Boudru, 1935; Fron, 1936; Horsfall and Zentmyer, 1942; Zentmyer and Horsfall, 1943; Zentmyer *et al.*, 1946; Stoddard, 1946; Dimond, 1947; Dimond *et al.*, 1949) as well as oak wilt (Hoffman, 1952, 1953) and the *Fusarium* wilts of geranium (Stoddard, 1957), tomato (Dimond, *et al.*, 1952), carnation (Stoddard, 1951a; Stoddard and Dimond, 1951) and guava (Jain, 1956). Activity has also been reported against *P. cinnamomi* on avocado (Zentmyer, 1954), *Rhizoctonia* on *Matthiola* (Stoddard, 1952), *B. fabae* on broad bean (Crowdy and Davies, 1952) and *Cytospora cincta* on peach and plum (Helton and Rohrbach, 1967). These compounds have been used commercially for the control of Dutch elm disease and carnation wilt. Against the former, 8-quinolinol benzoate gives moderate protection for a year if applied prior to infection but when applied post-infection it does not effect a cure although disease symptoms may be suppressed.

Microbial antibiotics

These compounds have played a major role in the development of chemotherapy. Their use in plant disease control has been reviewed by Brian (1952b), Leben and Keitt (1954), Pridham (1961), Zaumeyer (1958), Goodman (1959, 1962a, b, 1967), Thirumalachar (1968) and Dekker (1963a, 1969b). It has been shown that many antibiotics are taken up by and translocated within plants (see Crowdy and Pramer, 1955; Goodman, 1962b; Wain and Carter, 1967) and they are therefore of great interest as potential systemic therapeutants.

The first commercially available antibiotics were antibacterial and by 1952 streptomycin was being used in the field for the control of halo blight *(Pseudomonas phaseolicola)* on French bean and fireblight *(Erwinia amylovora)* on pear. Streptomycin, often in combination with oxytetracycline, has been widely used since then for the control of bacterial diseases of plants. Although essentially an antibacterial compound, streptomycin has shown systemic activity against certain fungal pathogens such as hop downy mildew *(Pseudoperonospora humuli)* (Horner and Maier, 1957; Horner, 1963). Other non-fungitoxic antibiotics, notably phleomycin (Smale *et al.*, 1961; Purdy, 1964), have also shown systemic activity against fungal diseases. Antifungal antibiotics have also been discovered and two of these, cycloheximide (Ford *et al.*, 1958) and griseofulvin (Brian, 1960; Rhodes, 1962), have been extensively tested.

Cycloheximide, discovered in 1946, is very active *in vitro* against a wide range of fungi and it was released for use in crop protection in 1954–55, being marketed under the name 'Actidione'. It has limited systemic activity and has proved an effective eradicant against *Coccomyces hiemalis* on cherry (Petersen and Cation, 1950; McClure and Cation, 1951; Hamilton and Szkolnik, 1953, 1958; Klos and Fronek, 1964), cereal rusts (Livingston, 1953; Silverman and Hart, 1954; Wallen, 1955, 1958), tea rust (Hocking and White, 1965), blister rust on pine (Moss *et al.*, 1960; Moss, 1961), *Cytospora cincta* on peach and plum (Helton and Rohrbach, 1967) and wilt of oak (Phelps *et al.*, 1957; 1966). It has also been found effective as an eradicant of seed-borne pathogens (Wallen and Skolko, 1951; Krüger, 1960) and as a soil drench against powdery mildew of rose (Jones and Swartwout, 1961a, b). Its usefulness, however, is limited by its high phytotoxicity. Certain derivatives, such as the acetate, semicarbazone and oxime have been examined for improved translocation and lower phytotoxicity; the oxime has been found to be effective against *C. hiemalis* (Hamilton *et al.*, 1956) and the semicarbazone against wheat stem rust (Hacker and Vaughn, 1957a, b). It has been shown that foliar injury is reduced when chlorophyllin is added to the spray material (Ark and Thompson, 1958; Wilson and Ark, 1958).

Griseofulvin, isolated in the early 1940s is active *in vitro* against a wide range of fungi (Napier *et al.*, 1956). Since it is readily translocated in plants (Crowdy *et al.*, 1955, 1956) and has low phytotoxicity, it has offered much promise as a chemotherapeutant. It has been shown to exert systemic activity against *B. cinerea* (Brian *et al.*, 1951), *Alternaria solani* (Brian *et al.*, 1951; Stubbs, 1952), *Botrytis tulipae* (Aytoun, 1956), *Oidium chrysanthi* (Rhodes *et al.*, 1957), *Pseudoperonospora humuli* (Horner and Maier, 1957) and *Sclerotinia fructigena* (Byrde, 1959). It has been used commercially in Japan for the control of *Sclerotinia mali* and *Mycosphaerella melonis*. Over 300 derivatives of griseofulvin have been synthesised (Crosse *et al.*, 1964), some of which have been examined for their systemic activity (Crowdy *et al.*, 1959a, b). None of them, however, has been developed as a commercial fungicide for crop protection.

Since 1950 many polyene macrolide antibiotics have been discovered; these are highly antifungal *in vitro* but although some of them have been shown to be taken up by plants they are not very effective as therapeutants, apparently because they are antagonised by plant metabolites (Oort and Dekker, 1960). Some have shown activity as eradicants of seed infections, e.g. pimaricin and rimocidin (Dekker, 1955, 1957; Oort and Dekker, 1960), helixin (endomycin) (Leben and Arny, 1952; Leben and Keitt, 1952; Leben *et al.*, 1953, 1954; Krüger, 1960) and filipin (Krüger, 1960).

Polypeptide antibiotics, although often possessing antifungal activity, appear generally to be poorly translocated and phytotoxic. Antimycin (Leben *et al.*, 1954; Lockwood *et al.*, 1954) and mycothricin (Rangaswami,

1956) have shown some activity when applied to infected seeds and a streptothricin complex is active against powdery mildews (Sharples, 1963). Phytoactin has also shown some promise, especially against diseases of trees (Phelps *et al.*, 1957, 1966; Moss, 1961; Helton and Rohrbach, 1967).

Oligomycin and anisomycin are two antifungal antibiotics that have shown limited activity against rust and anthracnose on *Phaseolus vulgaris* (Zaumeyer, 1957). Blasticidin-S, a purine antibiotic discovered in Japan in 1958, has a very selective antifungal activity but has proved very effective against the rice blast pathogen *(Pyricularia oryzae)*. It shows therapeutic activity and is now used commercially for the control of this disease (Misato *et al.*, 1959).

In some investigations, culture filtrates of micro-organisms have been examined for activity against plant pathogenic fungi. Pridham *et al.* (1956) in an examination of the filtrates from *Streptomyces* cultures discovered the F-17 antibiotic complex which has shown systemic activity against rust fungi (Preston *et al.*, 1956; Mitchell *et al.*, 1959). In a similar study Davis *et al.* (1960) discovered the P-9 antibiotic which is also systemically active against rusts and effective in the field (Hagborg *et al.*, 1961). Rhodes *et al.* (1961) whilst screening soil actinomycetes for activity against the apple scab pathogen *(Venturia inaequalis)* discovered venturicidin, and Tsao *et al.* (1960) also working with soil actinomycetes, tested filtrates for systemic activity and discovered B-74, systemically active against *Alternaria solani*.

Compounds from higher plants

Constituents from higher plants have been examined for chemotherapeutic activity. Particular attention has been paid to amino acids (see review by van Andel, 1966b). Naturally-occurring L-amino acids, other than methionine, have shown little activity and D-isomers have usually proved more effective. Certain analogues of natural amino acids show useful activity; some of these, e.g. L-threo-β-phenylserine, apparently operate through an effect on the host (van Andel, 1966a), whilst others, e.g. canavanine, ethionine and fluorophenylalanine, are fungitoxic antimetabolites which exert a direct action against the pathogen.

Plants contain a wide range of antifungal compounds (see Fawcett and Spencer, 1969; Wain, 1969). Many of these, and in particular those of a phenolic nature, have been implicated in the natural resistance which is shown by plants towards most fungi (Walker and Stahmann, 1955; Farkas and Király, 1962; Kuć, 1963; Byrde, 1963; Goodman *et al.*, 1967). A good example of such compounds is 'wyerone', isolated from broad bean tissue at Wye. Antifungal activity was demonstrated in broad bean tissue by Spencer *et al.* (1957) and shown (Wain *et al.*, 1961) to be due to the presence of a phenolic compound, a hydrophilic component and a highly-antifungal lipophilic substance. This lipophilic material has since proved to be an acetylenic keto-ester (Fawcett *et al.*, 1965, 1968). It shows a wide spectrum of activity but it is less effective against pathogens of broad bean (Fawcett *et al.*, 1969) and may play an important role in protecting the broad bean plant against other, non-pathogenic, species.

In addition to the preformed antifungal compounds of the wyerone type, there has been much interest in antifungal compounds (phyto-alexins) which are produced by plant tissues as a response to infection. This work has developed rapidly since the findings of Müller *et al.* (1939) (see for example Cruickshank, 1963; Fawcett and Spencer, 1969).

Naturally-occurring antifungal compounds, whether preformed in healthy tissues or phytoalexins, are of considerable interest as potential chemotherapeutants. Some, such as the phenolic compounds and quinones, are structurally similar to synthetic compounds already known to be active. Other classes of naturally-occurring antifungal compounds, however, have received little attention though some of these compounds have been examined for systemic activity at Wye. The results are summarised in the review of Fawcett and Spencer (1970).

Studies in chemotherapy prior to the mid-1960s failed to produce a fully effective compound suitable for use in the field because even the best compounds were insufficiently active, too phytotoxic or too costly. These studies did, however, reveal the potentialities and problems in the practical application of chemotherapeutants.

It soon became apparent that most chemotherapeutants have a narrow spectrum of activity. Furthermore, the activity of such a compound against a particular disease has been found to depend upon the method, site and time of application. These topics are discussed in later chapters of this book. The complex interaction of factors which determines the systemic activity of structurally-related compounds is well demonstrated by such studies as those of Crowdy *et al.* (1959a, b) with griseofulvin analogues.

3

Structure–activity relationships

by D. Woodcock

The objects of chemical structure–biological activity studies are to discover active basic structures, to establish which functional groups are essential for, or promote high activity, to locate optimal homologues and to complement mode of action studies. Although such work has often been done in a desultory fashion in the past, it is clear that physico-chemical parameters such as pK_a, partition coefficient and integrated absorption intensity of the hydrogen bond, are being increasingly investigated and are often correlated with fungicidal activity. This has led to limited predictions from partition coefficients using the Hansch relationship, but such forecasts are usually only in a limited context and so far have often remained unpublicised.

In the past decade, the early stages of research and development of systemic fungicides inexorably followed the systemic insecticide lead and explored organophosphorus compounds for activity. Since then a wide variety of organic molecules have been scrutinised for systemic activity, and many interesting data have been published for a large number of pathogens.

In the following account, systemic activity has been interpreted in the broadest possible sense and compounds showing only limited movement, such as in trans-laminar activity, have also been mentioned.

ORGANOPHOSPHORUS COMPOUNDS

Substituted 1,2,4-triazoles

It was perhaps inevitable that the quest for phosphorus-containing compounds with insecticidal and acaricidal properties should involve heterocyclic nuclei. The compound synthesised by van den Bos *et al.* (1960) from 3-amino-1,2,4-triazole and bis(dimethylamido) phosphoryl chloride, however, not only possessed good insecticidal and acaricidal activity, but was also found to be effective against powdery mildew of barley (caused by *Erysiphe graminis* var. *hordei*). This interesting spectrum of biological activity led to the synthesis of some forty more substituted 3-amino-1,2,4-triazoles containing a phosphoryl group and structural investigations by van den Bos (1960) later established the structure I for these compounds.

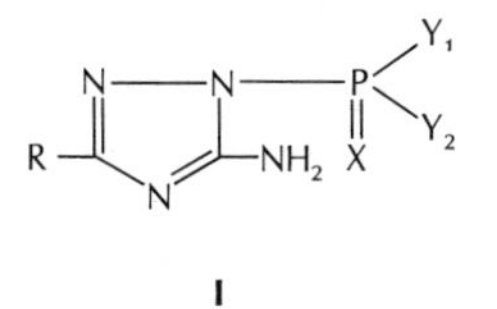

I

From the biological results obtained it was clear that the primary requirement for fungitoxicity was the presence of a bis(dimethylamido) phosphoryl group in the 2-position, a thiophosphoryl or other phosphorus-containing substituents being detrimental (Table 3.1). The nature of the substituent in the 5-position was also important, and in the *n*-alkyl series activity declined after the *n*-amyl member. Introduction of a methylene group between a substituent phenyl group and the triazole ring also caused a reduction in activity, as did substitution in the phenyl nucleus. It was noticeable that the insecticidal and acaricidal activity, as well as the phytotoxicity and mammalian toxicity, paralleled fungitoxicity, and it was significant that several compounds without contact or protectant activity nevertheless showed distinct systemic effects on insects, mites and powdery mildews.

Two compounds (I: $R = C_5H_{11}$ or phenyl) were chosen for further trials and the latter, claimed as the first systemic fungicide, has reached commercial acceptance with the common name triamiphos. Although it has been used for the control of powdery mildew on ornamentals such as roses, on which curative effects have also been reported by Betgem (1960), attempts to control apple powdery mildew by administering it to the trunks and roots have generally failed because of phytotoxicity.

TABLE 3.1 **Biological activities of triamiphos analogues (from van den Bos *et al.*, 1960)**

Substituents in formula I							
R	X	Y_1	Y_2	*Fungitoxicity*†	*Insecticidal activity*‡	*Acaricidal activity*‡	*Phytotoxicity*§
H	O	NMe_2	NMe_2	4	4	4	0
C_5H_{11}	O	NMe_2	NMe_2	5	2	3	2
$C_{11}H_{23}$	O	NMe_2	NMe_2	1	0	0	4
Phenyl	O	NMe_2	NMe_2	4	4	5	0
p-Chlorophenyl	O	NMe_2	NMe_2	3	3	4	1
p-Methoxyphenyl	O	NMe_2	NMe_2	2	1	0	0
Benzyl	O	NMe_2	NMe_2	1	–	4	0
H	S	NMe_2	NMe_2	0	1	1	1
C_5H_{11}	S	NMe_2	NMe_2	2	1	0	3
Phenyl	O	NHMe	NHMe	0	0	0	0
Benzyl	O	NHMe	NHMe	0	0	0	1
H	O	OEt	OEt	0	0	0	1
C_5H_{11}	O	OEt	OEt	0	–	4	0
Phenyl	O	OEt	OEt	0	–	–	4
C_5H_{11}	S	OEt	OEt	0	1	1	2
Phenyl	S	OEt	OEt	0	1	1	1

† O = <1/100 unit of dinocap, 5 = >3 units
‡ MLD O = >1000 ppm, 5 = 10 ppm
§ O = no damage, 5 = leaves dead

Further analogues have been synthesised (van den Bos *et al.*, 1961, 1966) and the structure-fungicidal activity relationships of the two series (II) and (III) are worth examining.

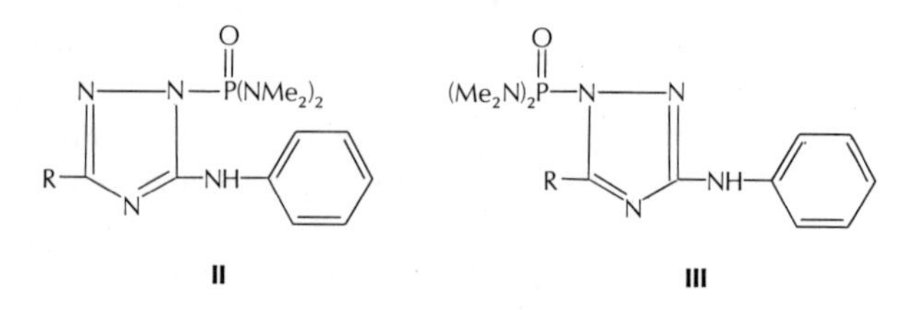

Fungicidal activity against *Sphaerotheca fuliginea* on cucumber seedlings and *Podosphaera leucotricha* on apple seedlings was examined by Tempel *et al.* (1968) using two tests. In the spray test, leaves were sprayed with aqueous suspensions of the compounds, whilst in the systemic test seedlings were grown in nutrient solution and then transferred to a fresh solution containing the test chemical and kept there for 48 h before inoculation. In another test using apple seedlings, designed to discover whether translocation took place after the leaf application, the test compounds were applied to upper and lower surfaces of one half of the leaf, and the infection on both halves assessed after the incubation period. The most striking feature of this work was the contrast between the activity of compounds with the phosphoryl group in position 1 (N_1 series) and those of the N_2 series: in both tests, with both pathogens, the N_1 members were far superior to their N_2 isomers. Within the active series there was some divergence in the results of the leaf application and systemic tests, and the alkyl chain length appeared to be critical. On cucumber seedlings only the lower members were active in the systemic test, whereas in the leaf spray test the effect was less well defined, except that activity virtually disappeared with the 2-hexyl member (Table 3.2).

TABLE 3.2 **The activity of 1- and 2-[bis(dimethylamido)phosphoryl]-3-alkyl-5-anilino-1,2,4-triazoles against *Sphaerotheca fuliginea* on cucumber seedlings. Infection in percentages (Tempel *et al.*, 1968)**

	N_1 series		*N_2 series*	
R =	*Leaf spray test, 1000 ppm*	*Systemic test, 100 ppm*	*Leaf spray test, 1000 ppm*	*Systemic test, 100 ppm*
H	30	0	100	100
CH_3	27	0	65	8
C_2H_5	4	7	100	39
i-C_3H_7	0	73	94	40
C_3H_7	29	61	94	100
C_4H_9	0	94	100	100
C_5H_{11}	6	82	—	—
C_6H_{13}	92	100	100	100
Triamiphos	81	86	—	—
Control			100	100

With apple seedlings infection was minimal with the *n*-amyl member (II: $R = C_5H_{11}$) in the leaf spray test, and with the unsubstituted compound (I: R = H) in the systemic test (Table 3.3). Although no translocation from a

TABLE 3.3 **The activity of 1- and 2-[bis(dimethylamido)phosphoryl]-3-alkyl-5-anilino-1,2,4-triazoles against *Podosphaera leucotricha* on apple seedlings. Infection in percentages (Tempel *et al.*, 1968)**

R =	N_1 series		N_2 series	
	Leaf spray test, 300 ppm	*Systemic test, 30 ppm*	*Leaf spray test, 300 ppm*	*Systemic test, 30 ppm*
H	31	0	79	56
CH_3	18	13	17	56
C_2H_5	19	30	24	58
$i\text{-}C_3H_7$	15	20	41	71
C_3H_7	8	53	36	70
C_4H_9	3	86	57	94
C_5H_{11}	0	69	—	—
C_6H_{13}	7	58	59	77
Triamiphos	14	11		
Control			100	100

TABLE 3.4 **The leaf systemic activity of 1- and 2-[bis(dimethylamido)-phosphoryl]-3-alkyl-5-anilino-1,2,4-triazoles against *Podosphaera leucotricha* on apple seedlings. Infection in percentages (Tempel *et al.*, 1968)**

R =	N_1 series		N_2 series	
	Treated half-leaf 300 ppm	*Untreated half-leaf*	*Treated half-leaf 300 ppm*	*Untreated half-leaf*
H	0	40	44	100
CH_3	0	24	22	76
C_2H_5	0	16	22	82
$i\text{-}C_3H_7$	0	16	—	—
C_3H_7	2	18	46	80
C_4H_9	0	38	38	80
C_5H_{11}	0	44	52	96
C_6H_{13}	0	100		
Control			100	100

treated to an untreated leaf took place, lateral movement was demonstrated, and it was clear that the alkyl group again played a critical role, maximum activity being shown by the ethyl, isopropyl and propyl analogues (Table 3.4). It is possible that this is a reflection of optimum partition characteristics for movement in the leaf surface, rather than a distance factor as suggested by Tempel *et al.* (1968).

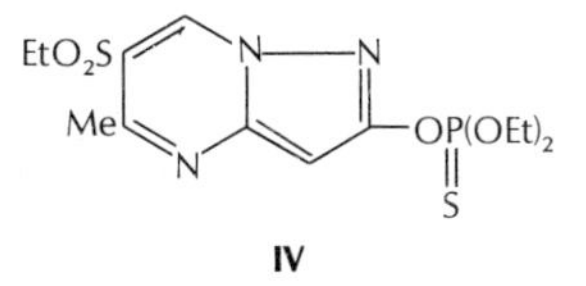

IV

More recently, a new group of heterocyclic organophosphorus compounds, involving phosphoric esters of pyrazolopyrimidines, have been introduced. 2-(O,O-Diethyl thionophosphoryl)-5-methyl-6-ethoxycarbonylpyrazolo-(1,5,a)-pyrimidine (IV) is absorbed by the foliage and shoots and is translocated in the plant. Root uptake is poor and precludes any possible use by soil application or as a seed dressing. Nevertheless trials have shown that the material is non-phytotoxic to a wide range of crops and excellent control of the powdery mildews *Sphaerotheca humili*, *S. fuliginea* and *Erysiphe graminis* has been obtained. Against powdery mildew of apple control comparable to that given by binapacryl and dinocap has been obtained, and noticeable improvements in vigour, colour of foliage, crop quality and yield have often resulted (Hay, 1971).

Dialkylphosphorothiolates and related compounds

Among the fungal diseases of rice, paddy blast caused by *Pyricularia oryzae* is of major economic importance. Control in the past has been with organomercury compounds such as phenyl mercuric acetate and iodide and N-phenyl mercuric *p*-toluene sulphonanilide, which are being currently criticised on the grounds of hazard to human health. More recently some antibiotics have been used, and pentachlorobenzyl alcohol has also been shown to be particularly specific *in vivo*, although without action *in vitro* (Sumi *et al.*, 1968).

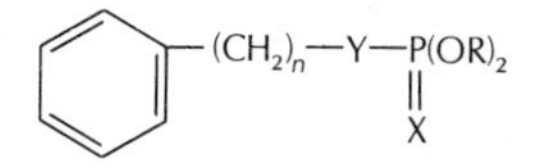

V

The benzyl group also features in a group of substituted phosphoric esters (V) recently developed in Japan (Kado *et al.*, 1965) and found to possess excellent systemic properties against rice blast in the field (Yoshinaga *et al.*, 1965). O,O-Dialkyl-S-benzyl phosphorothiolates (V: Y=S, X=O, R=alkyl) were found to be more effective than the corresponding O-benzyl phosphates (V: Y=X=O, R=alkyl) and phosphorothionates (V: Y=O, X=S, R=alkyl), and the S-benzyl phosphorodithioates (V: Y=X=S, R=alkyl), Fig. 3.1.

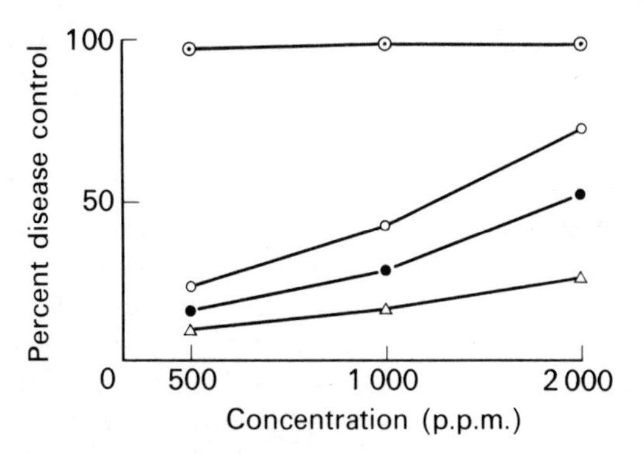

Fig. 3.1. Relation between the percent control of rice blast on potted plants and the concentrations of chemicals (Kado *et al.*, 1965): ⊙ = O,O-diethyl S-benzyl phosphorothiolate; ○ = O,O-diethyl S-benzyl phosphorodithioate; ● = O,O-diethyl O-benzyl phosphate; and △ = O,O-diethyl O-benzyl phosphorothionate (Kado and Yoshinaga, 1969, p. 134)

Among a limited number of alkyl groups which were surveyed, greatest activity was shown by ethyl, propyl and butyl members. Two S-benzyl-phosphorothiolates have reached commercial development – Kitazin (O,O-diethyl-S-benzyl phosphorothiolate; V: $n=1$, Y=S, X=O, R=C_2H_5) and Kitazin P – the isopropyl analogue; and two related compounds have also been developed – Hinozan (O-ethyl-S,S-diphenyl phosphorodithiolate; VI) and Inazin (VII: O-ethyl-S-benzylphenyl phosphonothiolate) (Kado and Yoshinaga, 1969).

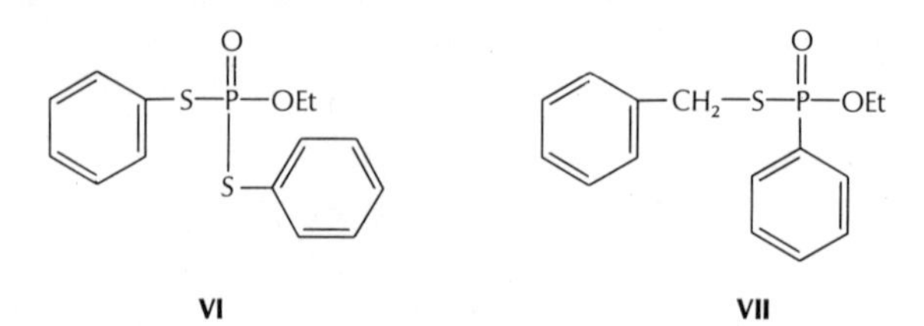

Adsorption of Kitazin into the rice plant by leaf application and subsequent translocation within the plant, was confirmed by autoradiography using the ^{32}P- or ^{35}S-labelled material. The compound is fairly water-soluble (500 ppm) and it was also readily absorbed through the root system and rapidly translocated in the transpiration stream to the site of pathogenic infection. This made possible Kitazin P granules (containing 17 per cent active ingredient), formulated for direct application in the paddy fields, where fungicidal effect is maintained for at least three weeks. This is far superior to foliar application, and there is a welcome bonus in the remarkable increase in rice yield, and a shortening of the stalks which reduces the tendency for lodging.

Mode of action studies have been carried out by Misato and his associates. They found that Kitazin at its fungistatic concentration of 55 ppm inhibited the incorporation of ^{14}C-glucosamine into the cell wall fraction in *P. oryzae*, without affecting the mycelial respiration or interfering with protein and nucleic acid synthesis, and they suggested that the site of action of the fungicide might be the cell wall synthesis system (Kakiki *et al.*, 1969). Further confirmatory investigations were carried out using a number of analogues (Maeda *et al.*, 1970). Of the ten compounds tested, the thiolates (V: X = O, Y = S) both strongly inhibited mycelial growth *in vitro* and gave excellent disease control on potted rice seedlings (Table 3.5). By contrast, the phosphates (V: X = Y = O), though strong inhibitors of mycelial growth, showed poor blast control and this was thought to be due to breakdown of the compounds – a possibility which seems to be

TABLE 3.5 **Fungitoxicity of Kitazin and its analogues (Maeda et *al.*, 1970)**

Substituents in formula V R =	X	Y	*n*	*Inhibition of mycelial growth*†	*Blast control*‡ *(pot test)*
Ethyl	O	S	1	A	A
Isopropyl	O	S	1	A	A
n-Butyl	O	S	3	A	A
Isopropyl	O	O	1	B	D
n-Propyl	O	O	3	A	B
n-Butyl	O	O	1	A	D
Isopropyl	S	O	3	C	D
Ethyl	S	O	1	C	D
n-Butyl	S	S	1	C	D
Isopropyl	S	S	1	C	D
Control				C	D

† A = inhibition ratio > 80%; B = inhibition ratio 60–80%; C = no effect
‡ A = protective value > 80%; D = no effect

supported by decreasing *in vitro* activity with prolonged incubation. The complete absence of any effect on mycelial growth *in vitro* or any suggestion of disease control *in vivo* reported earlier for the thionates (V: X = S, Y = O) and dithioates (V: X = Y = S) was confirmed. It was also significant that thiolates and phosphates, which inhibit mycelial growth *in vitro,* also inhibited ^{14}C-glucosamine incorporation into the cell wall fraction of *P. oryzae*, whereas the non-fungitoxic thionates and dithioates did not do so. This positive correlation strongly indicated that the Kitazin site of action is related to fungal cell wall synthesis, and the accumulation of UDP-acetylglucosamine strongly supports the view that the fungicide affects a stage beyond the formation of the amino sugar derivative. Maeda *et al.* (1970) suggest that all the evidence so far obtained points to penetration of the cytoplasmic membrane as being the site of action of these compounds.

ANTIBIOTICS

The translocation of some antibiotics in higher plants and their systemic antifungal activity was first demonstrated some twenty years ago, and the review of the literature up to 1962 by Dekker (1963a) is to be commended.

Griseofulvin

Possibly the most interesting is griseofulvin, a neutral antibiotic first isolated from the mycelium of *Penicillium griseofulvum* by Oxford *et al.* (1939) and more recently recognised as a metabolic product of many *Penicillium* spp. Translocation was observed in lettuce and tomatoes, uptake from aqueous solution through the root system being rapid, with a slower (passive) movement in the shoots in the transpiration stream. Translocation was demonstrated initially by bioassay (Brian *et al.*, 1951) but later the antibiotic was isolated from the tissues of treated plants and identified (Crowdy *et al.*, 1955).

Griseofulvin does not prevent germination of fungal spores, but it has a remarkable effect on fungal development, a concentration as small as 10 μg/ml causing stunting of hyphae, excessive and abnormal branching, hyphal distortion and loss of apical dominance on germinating conidia of *Botrytis allii*. Not all fungi are as sensitive and it was originally thought that griseofulvin exerted this characteristic effect by interfering with the synthesis of cell wall chitin, since fungi having cellulose based cell walls are unaffected. However there are a few species of fungi with chitinous cell walls that show no response to griseofulvin (Napier *et al.*, 1956), and it has been suggested that nucleic acids might be the primary target (McNall,

1960). This is supported by recent work using the highly griseofulvin-sensitive animal pathogen *Microsporium gypseum* (El-Nakeeb and Lampen, 1964a, b; El-Nakeeb *et al.*, 1965, 1965a, b). Whatever its *modus operandi* it is relatively non-phytotoxic, but unfortunately slowly degraded in plant tissue and more quickly in soil (Crowdy *et al.*, 1956). This liability makes its antifungal activity of limited practical use in agriculture and horticulture, though it is quite effective against a number of mildews (Brian, 1960). These include the powdery mildew of roses (caused by *Sphaerotheca pannosa*), *Pseudoperonospora cubensis* on cucumber and *Oidium chrysanthemi* on chrysanthemums. It has also controlled some *Botrytis* diseases including grey mould on lettuce, tulip fire and the turf diseases caused by *Fusarium nivale* and *Sclerotinia homeocarpa*. However, although griseofulvin affords some control of these plant pathogens it is no better than, and often inferior to other established fungicides in general use.

The chemistry and absolute configuration of griseofulvin (VIII: R = Cl), which has asymmetric centres at positions 2 and 6′, have been well established (Grove, 1963) and a considerable amount of interesting

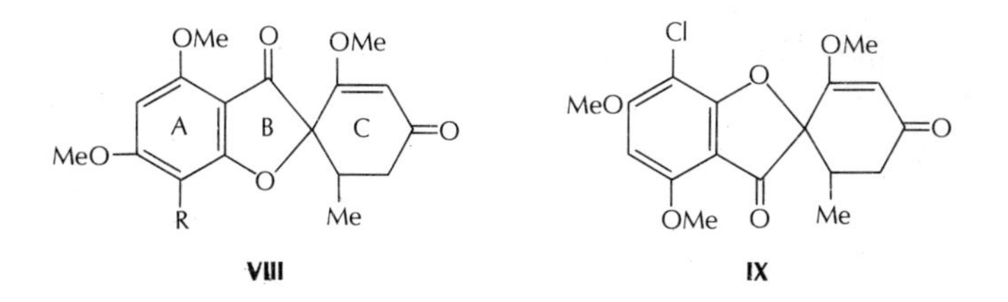

structure-activity data is available. The stereochemistry is critical for antifungal activity and the racemic form has only half of the activity of (+)-griseofulvin (VIII: R = Cl) itself. The diastereoisomer (IX), known as epi-(+)-griseofulvin, and all the transformation products with this same stereochemistry are inactive. However, some activity is still present after the elimination of the asymmetric 6′-centre, for dehydrogriseofulvin (X) and the synthetic grisenone (XI) are weakly active, though grisan-3,4′-dione (XII), without ring A substituents, is not.

When *Penicillium nigricans* and *P. griseofulvum* are grown in chloride-deficient, and also to some extent in normal media, the dechloro analogue (VIII: R = H) is produced. Addition of potassium bromide to the chloride-deficient medium leads to the production of 7-bromo-7-dechloro-griseofulvin (VIII: R = Br) – the first bromine-containing fungal metabolic product to be characterised. Both analogues are less active than griseofulvin itself, but the (+)-fluoro analogue (VIII: R = F), synthesised by Taub *et al.* (1962), was just as active in the disc plate assay using *Botrytis allii*. A

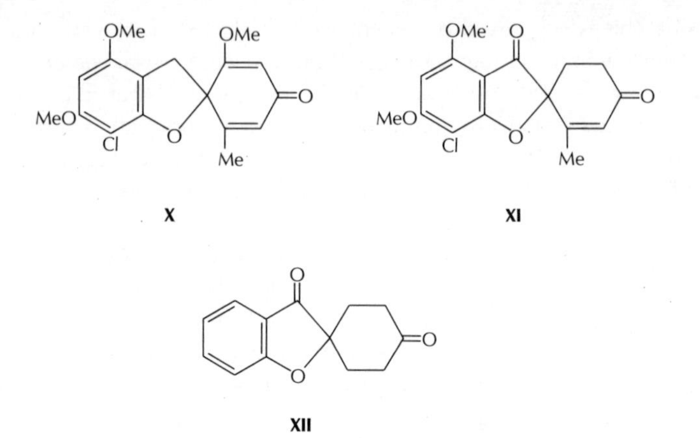

free 5-position appears to be essential for high activity, however, for the substitution of a halogen in position 5, instead of position 7, led to reduced activity (Gerecke *et al.*, 1962). Moreover although the 5,7-dichloro compound, geodin (XIII: R = Cl) showed some fungistatic activity, as do many dienones, the typical curling effects on fungal hyphae were absent. A basic requirement would also appear to be the tricyclic system (XIV), synthetic compounds with only a five-membered ring C, such as XV, being

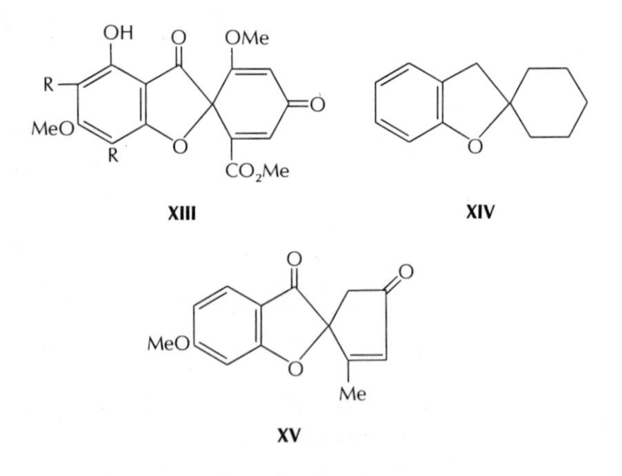

inactive. Compounds more active than griseofulvin can be obtained, however, by increasing the oil–water partition coefficient and homologues such as XVI (R = OPr^n or OBu^n) had 20–50 times the activity of griseofulvin *in vitro* (Crowdy *et al.*, 1959). On the other hand, a polar group not necessarily in the 4′-position in ring C seems to be desirable for activity since the

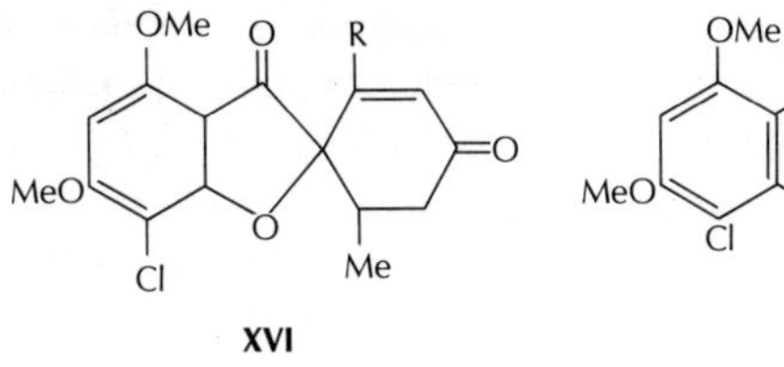

XVI

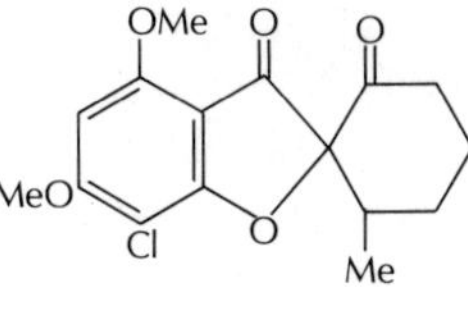

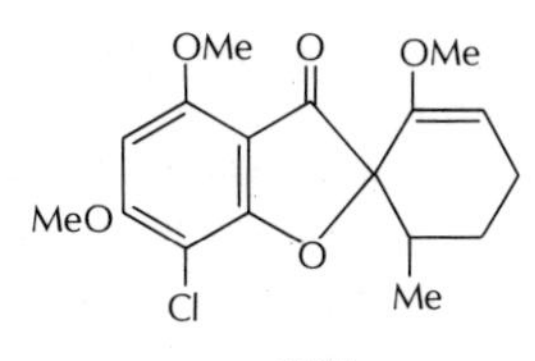

XVIII

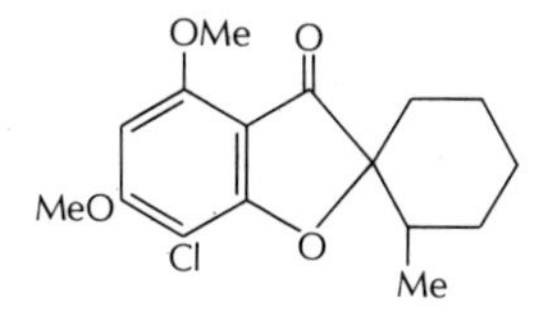

XIX

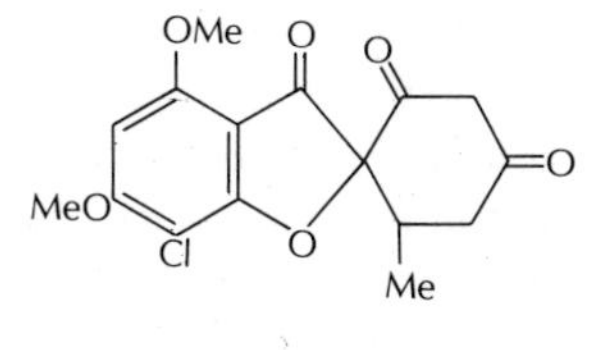

XX

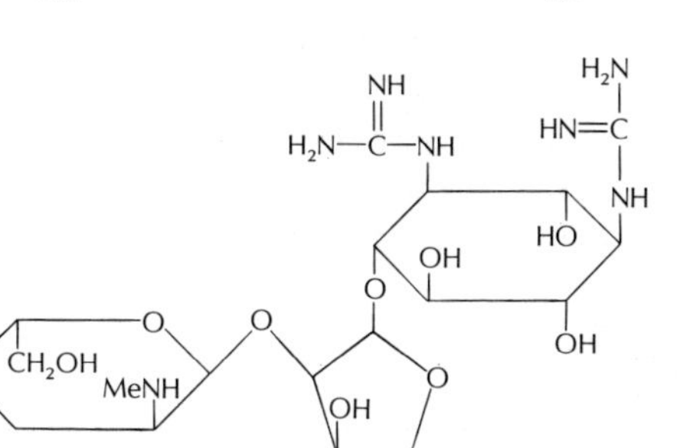

XXI

3-2′-ketone (XVII) was active, whereas compounds XVIII and XIX were not. The inactivity of the acidic trione (XX) and the basic griseofulvamine (XVI: $R = NH_2$) also seems highly significant.

Streptomycin

Other antibiotics have been examined for possible translocation and accumulation in the tissues of higher plants. Thus Mitchell *et al.* (1954) found that of nine antibiotics tested for their ability to protect bean plants against the halo-blight organism, *Pseudomonas medicagensis* var. *phaseolicola* (Burk, Stapp and Kotte), only streptomycin sulphate gave complete protection, although dihydrostreptomycin sulphate and terramycin hydrochloride showed slight activity. This ready translocation of streptomycin (XXI) confirmed previous reports of uptake and translocation of the antibiotic by higher plants (Anderson and Nienow, 1947; Winter and Willeke, 1951) and further confirmation was provided by Pramer (1953) who identified the antibiotic chromatographically in tissue extracts of cucumber leaves after root uptake.

Whilst streptomycin is active against a large number of bacteria, however, it has relatively little effect on anaerobic bacteria, fungi, protozoa and viruses. Thus whilst it has been effectively used as a plant protective agent against such bacterial diseases as fire blight on apples, pears and some ornamental shrubs of the rose family (caused by *Erwinia amylovora*) and against walnut blight (caused by *Xanthomonas juglandis*), its proven systemic properties appear to be of little avail in the control of fungal pathogens. It has however been used to some extent to control early infections of hop downy mildew (caused by *Pseudoperonospora humuli*) and actual translocation in the hop plant was first demonstrated by Maier (1960). It has also some effect on Phycomycetes, such as those causing late blight of tomatoes and potatoes and, even as little as 4 μg/ml of nutrient solution is sufficient to check *P. infestans* completely on tomatoes (Müller *et al.*, 1954; Crosse *et al.*, 1960). The use of streptomycin as a seed disinfectant has also been extensively explored with considerable success and the only disadvantage of such treatments is the possibility of encouraging streptomycin-resistant pathogens.

Cycloheximide

Also produced by *Streptomyces griseus* is the antifungal antibiotic cycloheximide – possibly more commonly recognised as 'Actidione' – a trade name of the Upjohn Co. It is obtained as a by-product in streptomycin manufacture and is remarkably active against a wide range of fungi and yeasts, but inactive against bacteria. Its use against plant diseases has been rather limited by its phytotoxicity. Cycloheximide, the common name for β-[2-(3,5-dimethyl-2-oxocyclohexyl)-2-hydroxyethyl] glutarimide (XXII), has four asymmetric centres at C-2, C-4, C-6 and C-2′, and a few of the other possible stereoisomers have also been found to be produced by *Streptomyces* spp., including isocycloheximide (Lemin and Ford, 1960), naramycin B (Okuda *et al.*, 1958) and two 'cycloheximide diastereoisomers' (Rao and Cullen, 1960). Isocycloheximide may also be obtained by isomerisation of cycloheximide using deactivated alumina, and it seems

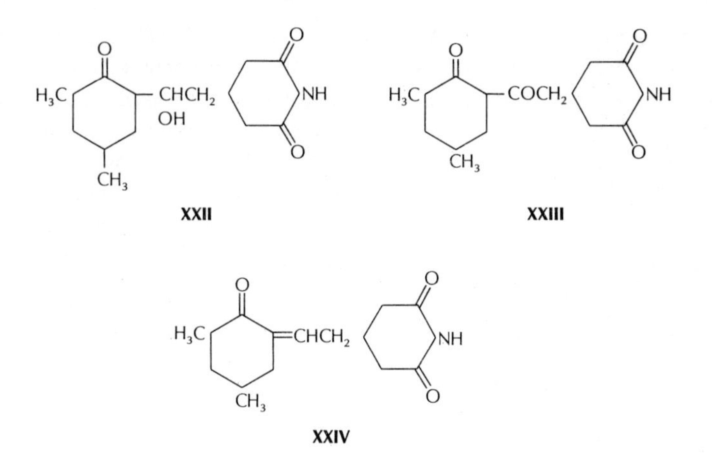

likely that neither the carbon atom carrying the secondary alcoholic group nor the C-4 asymmetric centre are involved. Moreover, since there are two isomeric dehydrocycloheximides (XXIII), inversion must be at C-2 or C-6, and because both cycloheximide and isocycloheximide yield the same anhydrocycloheximide (XXIV), it would seem that cycloheximide is 2,6-*trans* whilst isocycloheximide is 2,6-*cis*. Both naramycin B and isocycloheximide appear to be less active than cycloheximide against yeasts, the former only 32 per cent as active against *Saccharomyces sake* (Okuda *et al.*, 1958) and the latter about 30 per cent as active against *S. pastorianus* (Lemin and Ford, 1960). Anhydrocycloheximide, on the other hand, was just as active, but the related compounds dehydrocycloheximide (XXV) and dihydrocycloheximide (XXVI) were much less so

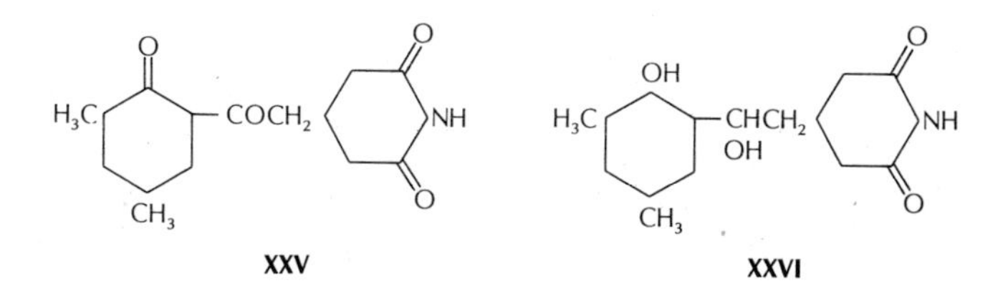

(Lee and Wilkie, 1965). It is possible however that the lower activity of these last two compounds may be due to a decrease in the amount entering the cell since they are both insoluble in water, in contrast to cycloheximide and anhydrocycloheximide. Another dehydrocycloheximide, inactone (XXVII) formed by *Streptomyces griseus* also shows no antifungal activity (Paul and Tchelitcheff, 1955).

A number of cycloheximide derivatives have been prepared and assayed against the fungus *Cytospora cincta*. The acetate and aceto-acetate were the least fungitoxic, while the oxime, carbazone, thiosemi-carbazone and methylhydrazone, although active, were less so than the parent antibiotic (Harvey and Helton, 1962). Whilst these analogues are less fungitoxic *per se*, they are often less phytotoxic and more readily translocated, so much so that use has been made of these attributes in particular applications. Lemin and Magee (1957) using ^{14}C-cycloheximide

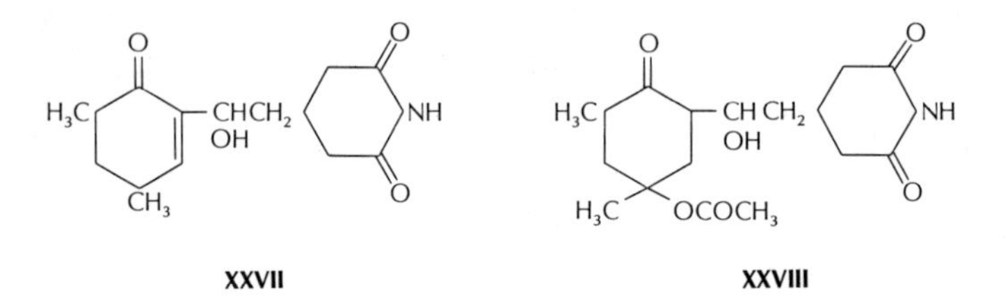

acetate, have shown that it is taken up by the roots of tomato plants, and thus renders the leaves highly antifungal, though whether because of activity *per se* or because of cycloheximide release is not certain. The oxime shows systemic activity against the cherry leaf spot fungus *Coecomyces hiemalis* Higgins, even protecting new shoots (Hamilton *et al.*, 1956), and the semicarbazone confers persistent protection against stem rust in wheat (Anderson and Rowell, 1960). Other compounds closely related to cycloheximide include the antibiotic E-73 (XXVIII) produced by *Streptomyces albulus* which has an acetoxy group at C-4 (Rao, 1960), and the streptovitacins formed by a grain of *S. griseus* (Eble *et al.*, 1959) which are cycloheximide with the cyclohexyl ring hydroxylated in positions 5, 4 and 3 for streptovitacins A, B and C_2 respectively, together with D – which is ring hydroxylated at an unknown position. These compounds are much less active than cycloheximide and have been studied mainly because of their high anti-tumour activity.

Kerridge (1958), working with *Saccharomyces carlsbergensis*, first suggested that cycloheximide inhibits both protein and DNA synthesis, and Siegel and Sisler (1963, 1964, 1965) have done much to probe and confirm this. Using a cell-free system of *S. pastorianus* and ^{14}C-labelled leucine, they demonstrated inhibition of this amino acid into protein and suggested that the antibiotic prevents the transfer of aminoacyl-soluble RNA to the ribosomes and formation of the peptide bond. This has been confirmed by both Ennis and Lubin (1964) and by Wettstein *et al.* (1964).

Antibiotics in Japan *Blasticidin* Nowhere is the agricultural usage of antibiotics increasing faster than in Japan (Table 3.6). The first to be de-

TABLE 3.6 **Agricultural antibiotics used commercially in Japan (Misato)**

Antibiotic	*Disease*	*Amount used in 1968* *Weight (ton)*	*Value (1000 yen)*
Blasticidin S	Rice blast		
Dust		7752	403 092
Wettable powder		48	36 000
Solution		374	280 500
Kasugamycin	Rice blast		
Dust		19 360	1 153 928
Wettable powder		51	41 820
Solution		322	257 600
Polyoxin	Rice sheath blight Fungal diseases of fruits		
Dust		3822	286 650
Wettable powder		131	327 500
Solution		148	156 880
Cellocidin + delan	Rice bacterial leaf blight	1	2 500
Chloramphenicol + copper	Rice bacterial leaf blight	21	73 500
Streptomycin	Bacterial diseases of fruits and vegetables	198	259 470
Cycloheximide	Onion downy mildew Shoot blight of Japanese larch	2	4 000
Griseofulvin	Apple blossom blight *Fusarium* wilt of melon	3	6 900

veloped was blasticidin S (XXIX) which is produced by *Streptomyces griseochromogenes* (Fukunaga *et al.*, 1955), the structure of which was established by Otake *et al.* (1965). It is active against both bacteria and

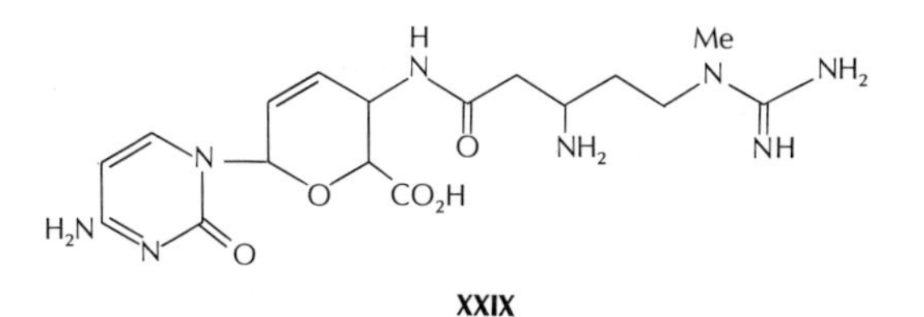

XXIX

fungi, and is particularly selective in its antifungal activity. Thus it inhibits growth of *Pyricularia oryzae* at concentrations as low as 5–10 μg/ml, and this has made the control of paddy blast on rice possible. The antibiotic is up to 100 times as effective as organomercury fungicides (Table 3.7) though concentrations of 20–40 ug/ml may cause chlorotic spots on rice plants. Among a number of derivatives tested, the monobenzyl-aminobenzyl sulphonate appeared to be least phytotoxic to the rice plant

TABLE 3.7 **Effect of blasticidin S against leaf blast (Misato, 1969)**

		No. of disease spots/pot†	
Fungicide	*Concentration a.i. (mcg/ml)*	*Protective effect (spray four days before inoculation)*	*Therapeutic effect (spray one day after inoculation)*
Blasticidin S	2	96	19·3
(wettable powder)	5	64	3·7
	10	100	1·5
	20	76	0·2
	50	42	0
	100	11	0
	200	10	0
Phenylmercuric acetate	10‡	48	10·0
(emulsifiable concentrate)	20‡	28	7·8
Untreated check	—	86	76·3

† Three pots per treatment, 10 seedlings per pot, four leaves per seedling
‡ As Hg

(Asakawa *et al.*, 1963) and this salt is the active ingredient in commercial blasticidin S preparations. Its mode of action against *P. oryzae* is the inhibition of protein synthesis, but no structure-activity information appears to be available.

Polyoxins Another group of antibiotics which also contain the pyrimidine nucleus are the polyoxins (XXX), components of the complex produced by *Streptomyces cacaoi* var. *ascensis* (Suzuki *et al.*, 1965; Isono *et al.*, 1967). These have similar physical and chemical properties but show a marked specificity in their antifungal activity (Table 3.8). Thus of the twelve polyoxins (A–L) characterised so far, only C and I show little activity against phytopathogens. Of the remainder, polyoxin A is effective against brown spot of rice (caused by *Helminthosporium oryzae*) B and G against pear black spot fungus and apple cork spot, whilst D is most effective for rice sheath blight (caused by *Pellicularia sasakii*). Most attention has been devoted to polyoxin D, particularly in mode of action studies (Endo and

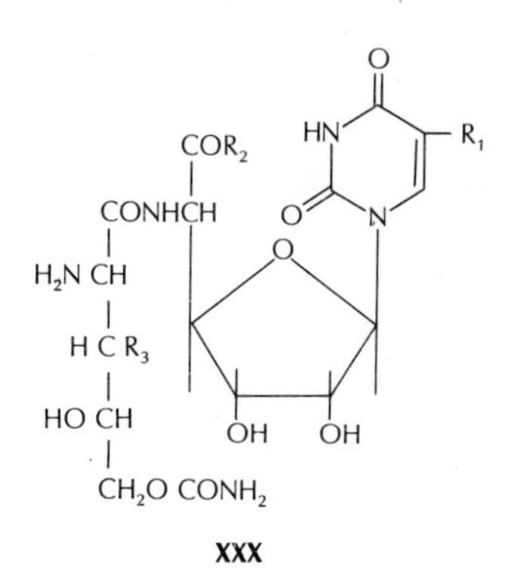

XXX

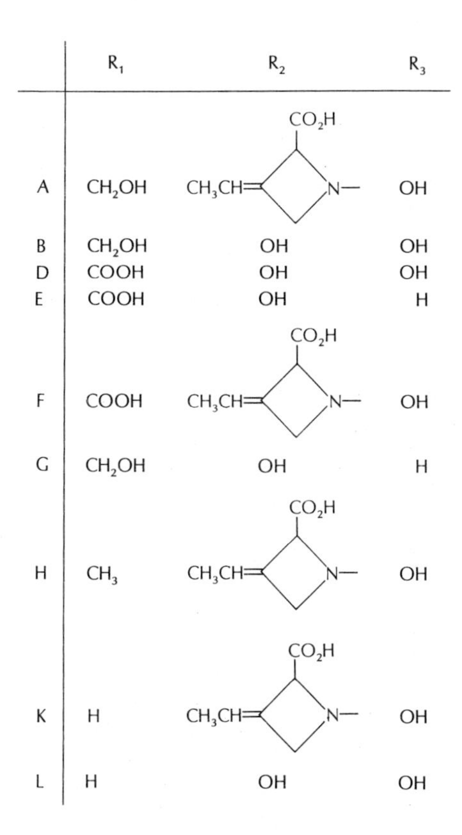

	R_1	R_2	R_3
A	CH_2OH	CO_2H; CH_3CH= (azetidine ring) N—	OH
B	CH_2OH	OH	OH
D	COOH	OH	OH
E	COOH	OH	H
F	COOH	CO_2H; CH_3CH= (azetidine ring) N—	OH
G	CH_2OH	OH	H
H	CH_3	CO_2H; CH_3CH= (azetidine ring) N—	OH
K	H	CO_2H; CH_3CH= (azetidine ring) N—	OH
L	H	OH	OH

TABLE 3.8 **Antifungal activities of polyoxins (Misato, 1969) M.I.C. (μg/ml)**

Test organism	A	B	C	D	E	F	G	H	I
Pyricularia oryzae	3·12	6·25	>100	3·12	12·5	25	6·25	3·12	>100
Cochliobolus miyabeanus	3·12	3·12	>100	6·25	12·5	6·25	3·12	25	>100
Pellicularia sasakii	12·5	1·56	>100	<1·56	1·56	50	1·56	50	>100
Alternaria kikuchiana	50	12·5	>100	50	50	>100	6·25	12·5	>100
Glomerella cingulata	>100	>100	>100	>100	>100	>100	>100	>100	>100
Physalospora laricina	25	3·12	>100	100	50	>100	6·25	12·5	>100
Cladosporium fulvum	3·12	1·56	>100	100	25	25	3·12	6·25	>100
Fusarium oxysporum	>100	>100	>100	>100	>100	>100	>100	>100	>100

Misato, 1969). It seems clear that this antibiotic selectively inhibits the synthesis of cell wall chitin in *Neurospora crassa* at levels which are comparable to those required for antifungal activity. Kinetic studies in a cell-free system showed that the incorporation of N-acetylglucosamine from UDP-N-acetylglucosamine into chitin was strongly inhibited. The greater sensitivity to the antibiotic of the chitin synthetase system, compared to the mycelium seems to point to some kind of barrier to the passage of the toxicant.

Kasugamycin Two other antibiotics deserve mention, not least because of their obvious potential. Kasugamycin is a water-soluble base obtained from the culture broth of *Streptomyces kasugaensis* (Umezawa *et al.*, 1965), and its structure (XXXI) was established by Suhara *et al.* (1966). It selectively controls *P. oryzae* and *Pseudomonas* spp. and whilst its lowest effective concentration against paddy blast is very similar to that of blasticidin S, it is non-phytotoxic, so much so that heavy infections can be treated safely with high concentrations of kasugamycin for rapid and complete control. Successful paddy blast control for 30 days has also been achieved with the use of seed coated with a 2 per cent wettable powder formulation of the antibiotic.

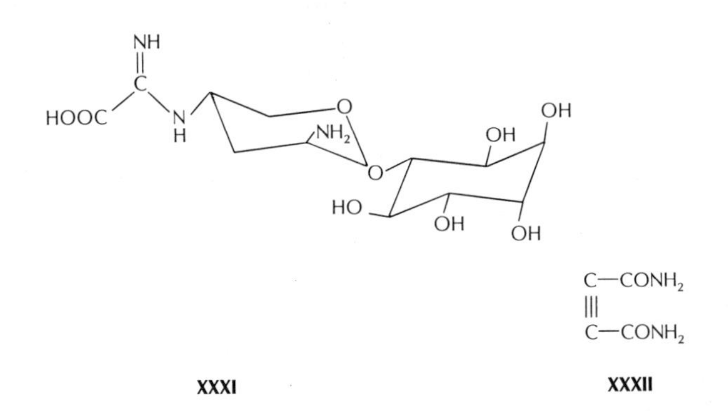

XXXI **XXXII**

Cellocidin Cellocidin (XXXII) an antibiotic produced by *Streptomyces chibaensis* (Suzuki *et al.*, 1958) is unique in its simplicity of structure. This has made synthetic production easy and it is now made from fumaric acid or 1,4-dihydroxybutyne. It has a preventive effect against rice bacterial leaf blight at 100–200 ppm and a study of its effects on the metabolism of *Xanthomonas oryzae* by Okimoto and Misato (1963a, b) suggests that it acts as an inhibitor of the α-ketoglutarate–succinate system.

Polyenes and polyynes Many more antifungal antibiotics too numerous to mention here are known and the number increases yearly. For example, there are the polyene macrolides which consist of a macrocyclic lactonic ring often containing a polyene chromophore. They are all produced by *Streptomyces* spp. and are highly antifungal *in vitro*, with the exception of those without the polyene chromophore, e.g. erythromycin, carbomycin and oleandomycin. A polyene macrolide such as rimocidin (XXXIII) (Bhate *et al.*, 1964), despite its poor aqueous solubility, is nevertheless taken up by

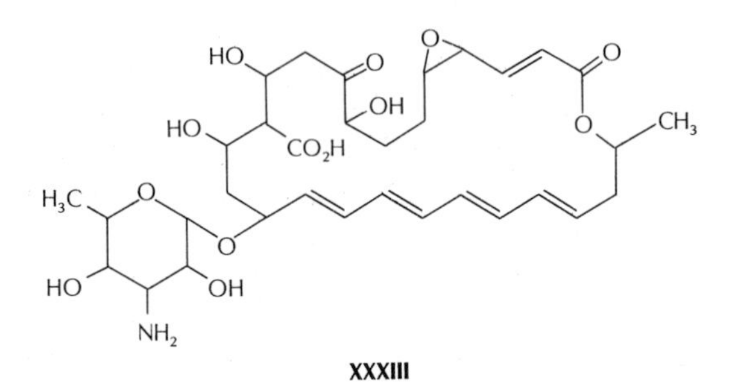

XXXIII

broad bean plant roots and can be subsequently extracted from the leaves. Despite this successful translocation no protection was afforded against *Botrytis fabae*, although a comparable concentration of the antibiotic would have inhibited spore germination *in vitro* (Oort and Dekker, 1960). Use of rimocidin as a seed disinfectant is perhaps more promising, for deep-seated infections of *Ascochyta pisi* in pea seeds have been largely controlled by soaking for 24 h (Dekker, 1957), whilst a significant reduction of sorghum kernel smut (caused by *Spacelotheca sorghi*) has been obtained by similar treatment of the seed (Krüger, 1959).

Whereas polyene macrolides are formed by Streptomycetes and are almost exclusively antifungal, the non-macrolide polyynes and polyenes are often formed by Basidiomycetes and are both antifungal and antibacterial. Many of them are extremely unstable (e.g. half-life of mycomycin (XXXIV) is only 3 h at 27°C) and for this reason alone, interest in their antifungal activity is largely academic.

$$HC{\equiv}C.C{\equiv}C.CH{=}C{=}CH{-}CH{=}CH{-}CH{=}CH{-}CH_2CO_2H$$

XXXIV

CARBOXYLIC ACID ANILIDES

Probably the first compound of this class to be shown to be fungicidal was salicylanilide (XXXV) (Fargher *et al.*, 1930) which has been described as a selective protectant fungicide. Despite a close structural similarity to the

TABLE 3.9 **The systemic fungicidal effect of substituted oxathiins (XXXVI) as measured by their ability to control bean rust (von Schmeling *et al.*, 1966)**

R	*ppm*	*% Control*
Phenyl	12·5	60
	25	95
	50	99
	100	100
2-Methylphenyl	12·5	30
	25	65
	50	90
	100	100
3-Methylphenyl	12·5	30
	25	80
	50	100
	100	100
4-Methylphenyl	125	40
	500	90
	2000	100
2-Chlorophenyl	500	0
	2000	90
4-Chlorophenyl	500	0
	2000	75
4-Ethoxyphenyl	500	10
	2000	75
3-Ethoxyphenyl	125	100
2,4-Dimethylphenyl	125	100
2-Biphenyl	125	90
	500	96
	2000	99
n-Butyl	125	25
	500	100
Cyclohexyl	50	60
	100	97
	200	100
Allyl	500	10
	2000	35
α-Naphthyl	500	0
	2000	35

compounds which follow, systemic properties have never been attributed to it. By contrast the oxathiins (XXXVI) which were first reported by von Schmeling and Kulka (1966) possess highly selective systemic properties acting mainly on Basidiomycetes – a class of fungi which includes such

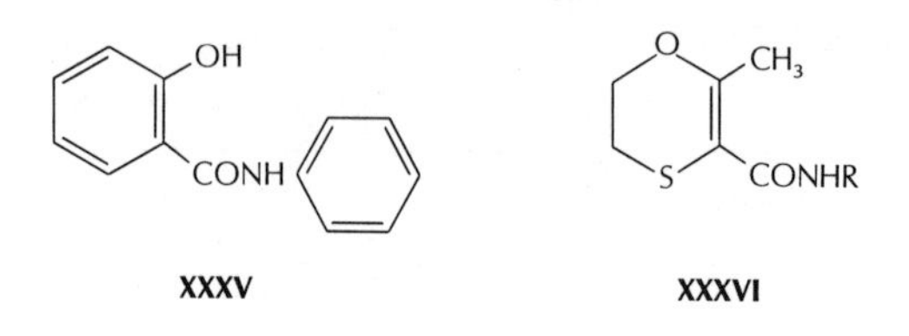

important pathogens as the rusts, smuts and bunts of cereals and the soil fungus *Rhizoctonia solani* Kühn. Evaluation of these compounds in the control of bean rust (caused by *Uromyces phaseoli typica* Arth.) was carried out using a spray application technique and the results are shown in Table 3.9 (von Schmeling *et al.*, 1966). The most active compound was 2,3-dihydro-6-methyl-5-phenylcarbamoyl-1,4-oxathiin (XXXVI: $R = C_6H_5$), since given the common name carboxin. Any substitution in the aromatic ring of chloro, alkyl or alkoxy groups resulted in reduced fungitoxicity, as did its replacement by alkyl, cyclohexyl or other aryl groups. In soil tests against *R. solani* on cotton, the effect of substitution in the phenyl nucleus was less dramatic, and the 5-(2-methylphenyl) carbamoyl analogue was, if anything, slightly more effective against this pathogen than carboxin (Table 3.10).

TABLE 3.10 **Control of *R. solani* on cotton using 20 ppm of substituted oxathiin (XXXVI) in the soil (von Schmeling et *al.*, 1966)**

R	% *Emergence*	% *Stand*
Phenyl	80	80
2-Methylphenyl	88	88
3-Methylphenyl	76	76
2-Chlorophenyl	80	68
4-Chlorophenyl	52	48
Inoculated soil control	56	16
Uninoculated soil control	76	76

The selective toxicity to Basidiomycetes was confirmed by Edgington *et al.*, (1966) who showed that whereas a concentration of 8 ppm or less of carboxin completely prevented the mycelial growth of many Basidiomycetes, other fungal groups such as Phycomycetes, Ascomycetes and Deuteromycetes were only 50 per cent inhibited by concentrations of 32 ppm or higher. Two exceptions to this fungitoxicity pattern were an isolate of *R. solani* which was not as sensitive as other Basidiomycetes, and the Deuteromycete *Verticillium albo-atrum* which was more sensitive.

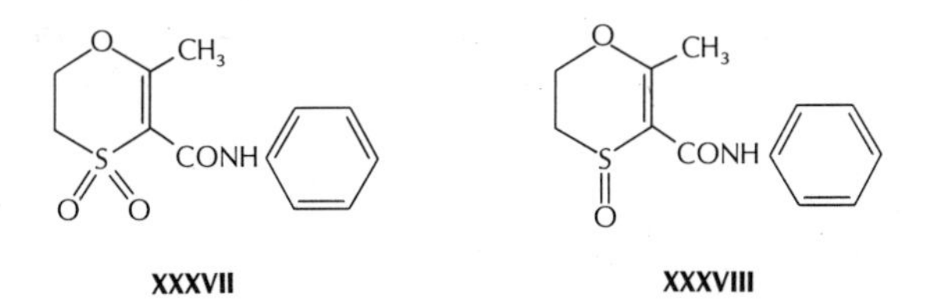

Against bean rust in greenhouse pot tests, carboxin was found to be inferior to the corresponding sulphone, oxycarboxin (XXXVII), possibly because of its oxidation to the less active sulphoxide (XXXVIII). In the field, however, carboxin proved highly effective in the control of loose smut of barley (caused by *Ustilago nuda* (Jens.) Rostr.) whereas oxycarboxin was only moderately successful.

Later work on the uptake of carboxin and oxycarboxin by *R. solani* and *U. maydis*, which are susceptible, and *Fusarium oxysporum* and *Saccharomyces cerevisiae* which are resistant to these chemicals, led Mathre (1968) to suggest that the specificity of the oxathiins may be related to the amount of fungicide taken up and bound to cell organelles.

The stability of carboxin has been examined by Chin *et al.* (1969) who found that oxidation to 2,3-dihydro-6-methyl-5-phenylcarbamoyl-1,4-oxathiin-4-oxide (XXXVIII) took place not only in water and soil, but also in barley, wheat and cotton plants. This change is facilitated by low pH and it is relatively stable above pH 6. There is no evidence so far of any further oxidation to the sulphone and it has been suggested that the energy requirement for this reaction is not available under biological conditions. The sulphoxide has a fungicidal activity against wheat leaf rust (caused by *Puccinia rubigo-vera tritici* (Erikss.) Carleton) *in vitro* at least 5 000 times lower than carboxin, and its production may explain a poor performance of the parent compound in certain conditions. Snel and Edgington (1970) confirmed the uptake of carboxin by bean roots and oxidation to the non-fungitoxic sulphoxide. They also noted fission of the carboxamide linkage in root tissues with the formation of aniline which became bound to plant polymers and also formed highly water-soluble conjugates. The mechanism of action of carboxin in sensitive organisms has been shown to be associated with its ability to inhibit the tricarboxylic acid cycle (Mathre, 1970; Ragsdale and Sisler, 1970). In later work the toxicities of carboxin and eight structurally related compounds (XXXVII–XLIV) were compared with regard to their ability to affect various metabolic pathways in sensitive fungi (Mathre, 1971). Only the fungitoxic compounds (XXXVII, XLII, XLIII and XLIV) strongly inhibited acetate metabolism and RNA synthesis, which seems to indicate a similar mode of action for these compounds, even though two of them are thiazoles (XLII, XLIV). The poor performance of carboxin sulphoxide (XXXVIII) and carboxin sulphone (XXXVII) against *R. solani* and *U. nuda* is interesting especially in the light of previous results,

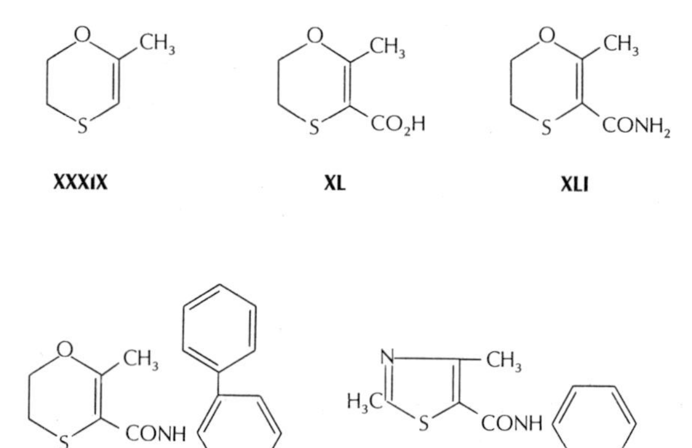

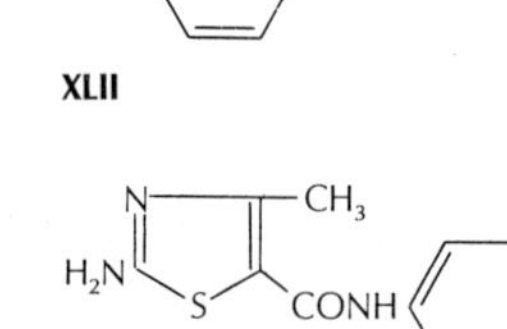

and with the exclusion of these two compounds, the fungitoxicities of these analogues to *R. solani* correlates well with their partition coefficients determined in a water–octanol system. Significantly too, elimination of the aromatic ring in compounds XXXIX, XL and XLI destroyed all activity.

Other substituted benzanilides also have an interesting specificity against Basidiomycetes and neither of the hetero atoms present in carboxin appears to be critical and a methyl group is often present in the 2-position in active compounds. Thus 2,3-dihydro-6-methyl-5-phenylcarbamoyl-4H-pyran (XLV) containing an ortho-methyl group, is also active against rusts, smuts and *Rhizoctonia* spp. (Stingl *et al.*, 1970). The early results with the smut diseases of barley and oats (caused by *U. nuda* and *U. avenae* respectively) were particularly promising, activity comparable to that of carboxin being shown (Jank and Grossmann, 1971) (Tables 3.11 and 3.12). 2,5-Dimethyl-3-phenylcarbamoyl furane (XLVI: R = phenyl) with only a five membered heterocyclic ring and again with a methyl group ortho to the carbamoyl function has recently been announced by Pommer (1971) as a new systemic fungicide. It has a much wider spectrum

TABLE 3.11 **Effect of 2,3-dihydro-6-methyl-5-phenylcarbamoyl-4H-pyran against loose smut of barley (Jank and Grossmann, 1971)**

Fungicide	Application rate, g of active ingredient/ 100 kg of seed	Smutted ears, % Larker Field trial	Smutted ears, % Larker Pot trial	Smutted ears, % Pirol Field trial	Smutted ears, % Pirol Pot trial
Control	—	20·1[a]	18·7[ab]	28·1[a]	23·7[a]
2,3-Dihydro-6-methyl-	6·25	12·9[b]	23·6[a]	10·7[b]	25·6[a]
5-phenylcarbamoyl-	12·5	4·5[cd]	8·3[cd]	6·7[b]	14·8[b]
4H-pyran	25·0	2·3[de]	3·7[de]	0·5[c]	3·2[c]
	50·0	0·4[ef]	1·2[e]	0·3[c]	1·6[cd]
Carboxin	12·5	7·8[bc]	13·5[bc]	7·9[b]	15·4[b]
	50·0	0·3[f]	2·3[e]	0·4[c]	0·3[d]

a–f Treatment means in the same column followed by different letters are statistically different at the 5% level of significance

TABLE 3.12 **Effect of 2,3-dihydro-6-methyl-5-phenylcarbamoyl-4H-pyran against loose smut of oats (Jank and Grossmann, 1971)**

Fungicide	Application rate g of active ingredient/100 kg of seed	Smutted panicles, % Field trial	Smutted panicles, % Pot trial
Control	—	18·2[a]	29·7[a]
2,3-Dihydro-6-methyl-5-phenylcarbamoyl-4H-pyran	6·25	9·1[b]	5·8[b]
	12·5	2·6[c]	1·2[bc]
	25·0	0·2[c]	0·0[c]
	50·0	0·1[c]	0·0[c]
Carboxin	6·25	7·9[b]	1·3[bc]
	50·0	0·1[c]	0·0[c]

a–c Treatment means in the same column followed by different letters are statistically different at the 5% level

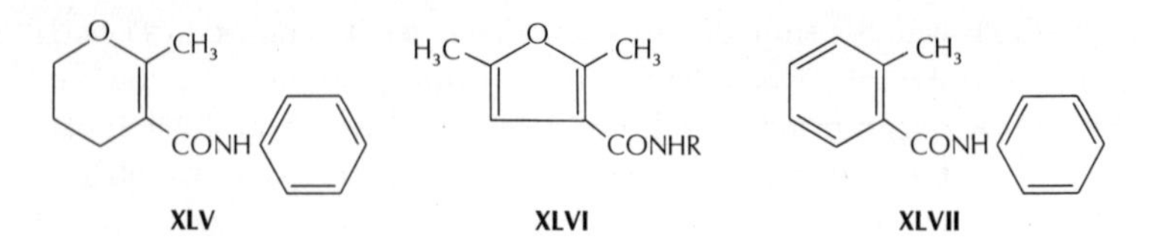

of activity, being particularly effective, not only against loose and stinking smuts, but also for some *Helminthosporium* and *Fusarium* spp. Its systemic effect was demonstrated in seed and soil treatments by the control of bean rust and it is also being currently evaluated as a seed dressing against *U. nuda* in spring barley and against *U. tritici* in winter wheat. Attempts to improve the fungicidal effectiveness by increasing the lipoid solubility of the molecule were unsuccessful. Thus substitution of a methyl group in the phenyl nucleus reduced fungitoxicity by 30–50 per cent, whilst introduction of a chlorine or bromine atom resulted in up to 100 per cent reduction in activity. Interference with methyl groups of the furane ring was even more disastrous, for the compounds 3- or 5-methyl-3-phenylcarbamoyl furane were found to be completely inactive when tested against Basidiomycetes.

The implication from Mathre's results is that the carboxylic acid anilide part of the molecule is a prime requirement for activity and this is supported by the fact that similar activity has been reported for a series of substituted benzanilides which do not involve a heterocyclic ring system. 2-Toluanilide (XLVII) (common name mebenil) without any hetero atoms, represents the best of this series for yellow rust control in cereals (Pommer and Kradel, 1969). The importance of the orientation of the methyl group is clearly shown in Table 3.13, as is the inadequacy of the hydroxyl group in the 2-position. More recently 2-iodobenzanilide has been shown to be more active than mebenil and appears to have a greater margin of crop safety. It is also more persistent than the corresponding chloro and bromo compounds. The species specificity of the active compounds of this group is most marked. The use of 2,5-dimethyl-3-phenylcarbamoyl furane (XLVI: R = phenyl) as a seed dressing for barley, has been limited to some extent because of the sensitivity shown by some varieties under certain climatic conditions. This has been largely overcome by the introduction of the cyclohexyl analogue (XLVI: R = cyclohexyl) which was well tolerated by even the sensitive varieties (Pommer *et al.*, 1971).

A recent study of the systemic activity of another group of carboxanilides by ten Haken and Dunn (1971) has produced even more interesting and valuable structure-activity data. These authors studied the various structural features of oxycarboxin (XXXVII) in relation to its systemic control of broad bean rust (caused by *Uromyces fabae*), using logically modi-

TABLE 3.13 **The fungicidal activities of substituted benzanilides (Pommer and Kradel, 1969)**

	Basidomycetes		*Ascomycetes*		*Phycomycetes*	
Substituent	*R. solani*	Crown rust (oat)	*Aspergillus*	Barley mildew	*P. viticola*	*Pythium* spp.
—	–	–	–	–	–	–
2-Hydroxy	–	–	+	–	+	–
2-Methoxy	–	–	–	–	–	–
2-Methyl	+	+	–	–	–	–
3-Methyl	±	–	–	–	–	–
4-Methyl	–	–	–	–	±	–
2-Trifluoromethyl	+	+	–	–	–	–
2-Chloro	+	+	–	–	–	–
2-Bromo	+	+	–	–	–	–
2-Nitro	+	±	–	–	–	–
2-Amino	±	+	–	–	±	–

\+ Effective; ± Less effective; – Insufficiently or ineffective

fied compounds to assess the contribution of various parts of the molecule. They confirmed the activity of compounds which did not possess hetero atoms and showed that this ring system could in at least one instance be replaced by an open-chain analogue and still retain fungicidal activity. Thus *cis*-crotonanilide (XLVIII) was active, albeit also somewhat phytotoxic, although the corresponding trans-isomer (XLIX) was not.

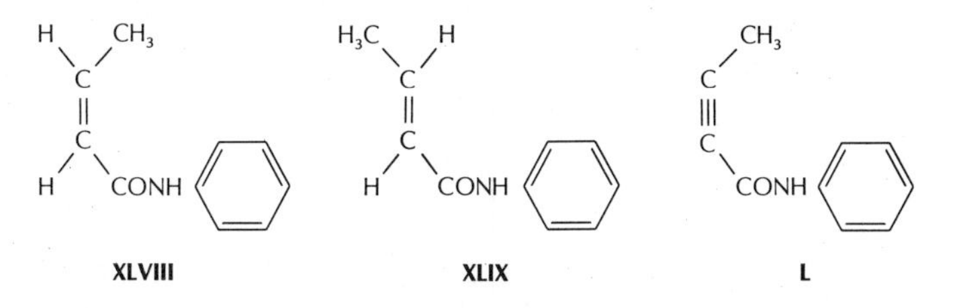

Reduction of the double bond to give *n*-butyranilide, however, destroyed activity as did its replacement by a triple bond (L). ten Haken and Dunn also noted that the double bond may form part of a planar ring system such as benzene, furan, thiazole and oxazole, or of non-planar ring systems like dihydrooxathiin, dihydrofuran and dihydropyran. Their results confirmed the sacrosanctity of a suitable ortho group: replacement by hydrogen, ethyl or phenyl groups in 2,3-dihydro-6-methyl-5-phenylcarbamoyl-4H-pyran (XLV), or hydroxylation in mebenil (XLVII) caused complete loss of activity.

TABLE 3.14 **Importance of the carboxanilido function in the activity of carboxylic acid anilides against *Uromyces fabae* (ten Haken and Dunn, 1971)**

Compound	*Foliar application*		
	Direct	*Systemic*	*Root drench*
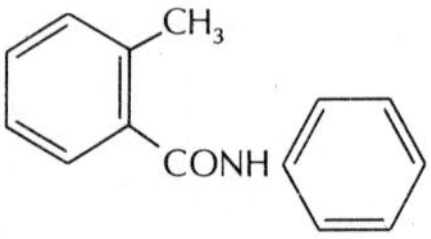 CH_3; CONH	2	2	1
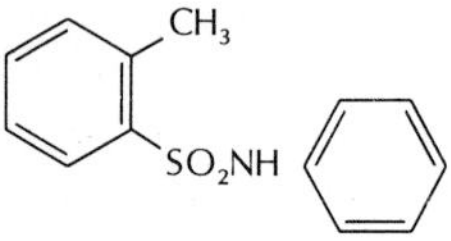 CH_3; SO_2NH	0	0	0
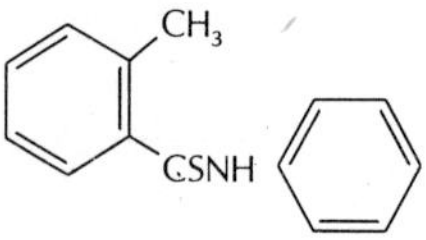 CH_3; CSNH	2	0	0
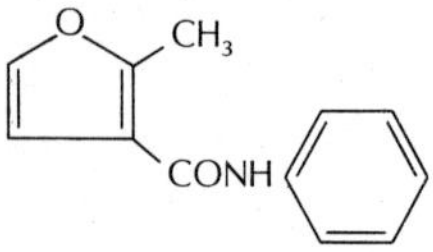 O; CH_3; CONH	2*	2	1
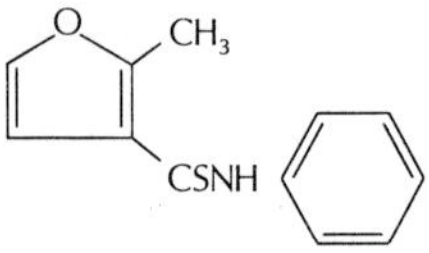 O; CH_3; CSNH	2*	0	0
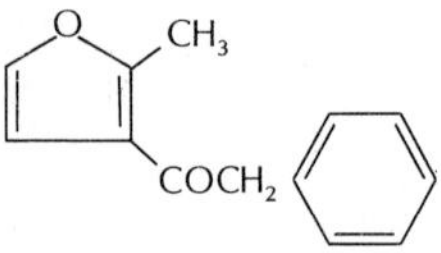 O; CH_3; $COCH_2$	0	0	0
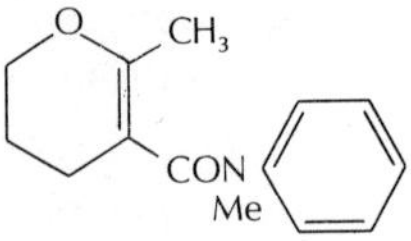 O; CH_3; CON(Me)	1*	0	0
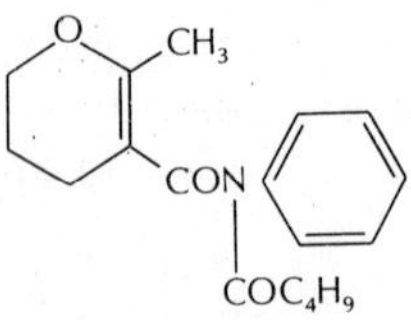 O; CH_3; CON(COC_4H_9)	1	0	0

2 80–100% Reduction in rust symptoms; 1 50–80% Reduction in rust symptoms;
0 Insignificant reduction in rust symptoms; * Slight to moderate necrosis

TABLE 3.15 **Glasshouse evaluation of carboxylic acid anilides against wheat brown rust (ten Haken and Dunn, 1971)**

Compound	*% Rust control*	*Phytotoxicity*
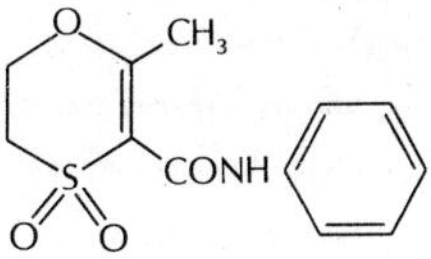 O, CH_3, CONH, S, O, O	100	1
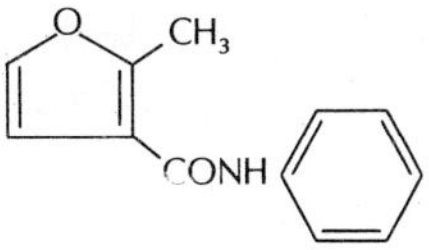 O, CH_3, CONH	50	2
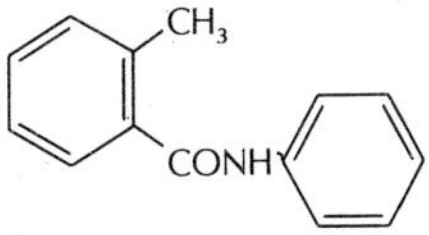 CH_3, CONH	0	1
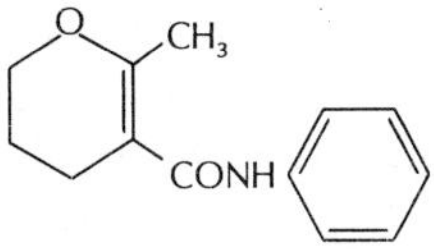 O, CH_3, CONH	100	3
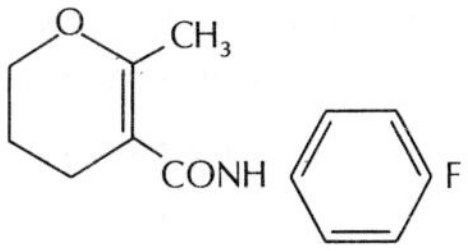 O, CH_3, CONH, F	99	1
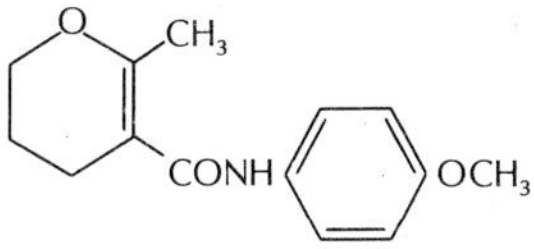 O, CH_3, CONH, OCH_3	36	0
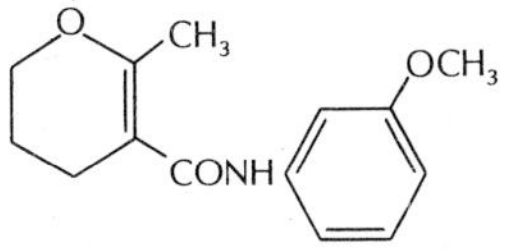 O, CH_3, CONH, OCH_3	100	2
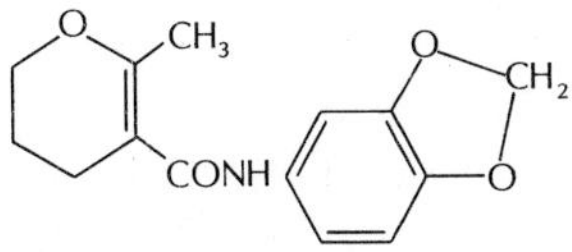 O, CH_3, CONH, O, CH_2, O	97	0

0 No visible phytotoxicity; 1 Slight stunting and chlorosis;
2 Moderate stunting and chlorosis; 3 Severe stunting and chlorosis

The importance of the carboxanilido moiety was also investigated, and in general any interference led to loss of activity (Table 3.14). It is perhaps significant that the function of the —NH— group appears to be more than that of a 'spacer', because replacement by a methylene bridge led to loss of activity, even though the interatomic distances involved were 0·25 nm in both compounds. ten Haken and Dunn also tested some of the compounds against wheat brown rust (caused by *Puccinia recondita*) and found a good correlation with the bean rust results. In this glasshouse evaluation, they found that certain nuclear substituents in the phenylcarbamoyl part of the molecule had considerable influence on both phytotoxicity and fungitoxicity (Table 3.15).

HETEROCYCLIC COMPOUNDS

Benzimidazoles

Although 2-aminobenzimidazole (LI: $R^1 = R^2 = H$) is devoid of fungitoxicity, systemic fungicidal activity was claimed by Klöpping (1960) for various substituted 2-aminobenzimidazoles. This basic work was the prelude to the production of methyl (1-butylcarbamoyl)-2-benzimidazole carbamate (benomyl: LI: $R^1 = CONHC_4H_9$, $R^2 = COOCH_3$), which in addition to eradicant and protectant fungicidal activities and a mite ovicidal action, possesses exceptional systemic fungicidal properties (Delp and Klöpping, 1968). It is active against an impressive list of fungal diseases of fruit and vegetable crops, against *Verticillium albo-atrum* on cotton and against powdery mildew and black spot of roses. However, it has little effect on diseases caused by *Helminthosporium* spp. (Greenaway, 1972) and by Phycomycetes, such as potato late blight, vine downy mildew and various damping-off diseases. In aqueous solution benomyl has been

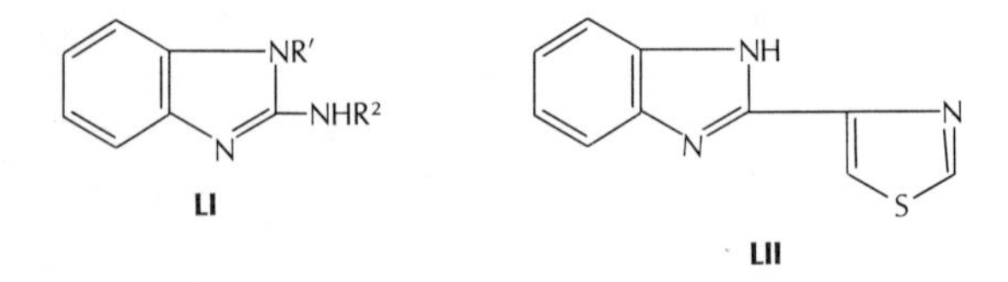

shown to be rapidly hydrolysed to methyl benzimidazol-2-yl-carbamate (MBC) (LI: $R^1 = H$, $R^2 = COOCH_3$) (Clemons and Sisler, 1969), and it is possible that this stable breakdown product is partly or even wholly responsible for the fungitoxicity. Although MBC was 30 times less toxic to *Saccharomyces pastorianus* Han., it was found to be just as toxic to

Neurospora crassa Shear and Dodge and *Rhizoctonia solani* Kühn, as benomyl, and the toxicity to these two organisms must be attributed to the breakdown product. The poorer performance of MBC against the yeast could be merely a reflection of poorer penetration. Other workers have confirmed the suggestion that the active entity in the plant at sites remote from the point of application of benomyl is MBC. Thus, Sims *et al.* (1969) were able to detect it, but not benomyl, in extracts of cotton plants four weeks after treatment, while Peterson and Edgington (1969) found that benomyl was completely transformed to MBC in five days in bean plants. Evidence regarding its mode of action has been provided by Clemons and Sisler (1971) using *N. crassa* and *Ustilago maydis* (DC.) Cda. They showed that though protein and RNA synthesis in *N. crassa* were not inhibited, the action of MBC appeared to involve DNA synthesis or some closely related process such as nuclear or cell division.

It seems highly probable that the many analogous compounds which have been synthesised (Loux, 1961) will by now have been screened for any sign of improved systemic fungitoxicity.

2-(4′-Thiazolyl)-benzimidazole (thiabendazole) (LII), originally introduced as an anthelminthic, has also achieved prominence as a broad-spectrum systemic fungicide effective against many major fungal pathogens (Staron and Allard, 1964; Robinson *et al.*, 1964; Erwin *et al.*, 1968a, b). Translocation in many plants has been reported by bioassays, chromatography and autoradiography, and it seems clear that movement is not only from roots to leaves but also in the less usual reverse direction, and so far there are no indications of any metabolic breakdown *in vivo*. A certain amount of work has been done on its mode of action and like benomyl and MBC it appears to be fungistatic. Allen (1969) showed it to be a potent inhibitor of respiration in isolated mitochondria of *Penicillium atroveretum* G. Smith, though the toxicant level required to inhibit whole-cell respiration was higher than that needed to inhibit growth. Similarly Kaars Sijpesteijn (1970) reported that 100 ppm – nearly 100 times the growth inhibitory concentration – could not inhibit respiration in *Aspergillus niger*.

Another newly introduced benzimidazole is 2-(2′-furyl) benzimidazole (fuberidazole) (LIII). This compound shows considerable promise as a seed treatment against *Fusarium* diseases, such as *F. rivale* of rye and *F. culmorum* on peas (Schuhmann, 1967). Both it and thiabendazole showed good systemic activity, and seed treatments of wheat and barley considerably reduced infections of *Puccinia triticina* Eriksson and *Erysiphe graminis* DC., for several weeks, and also gave extremely effective control of snow mould *(Calonectria rivalis)*, being superior to organomercurials at the same concentration.

The fungitoxicities of benomyl, thiabendazole and fuberidazole *in vitro* have been examined against a wide fungal spectrum and an identical selective antifungal pattern was revealed (Edgington *et al.*, 1971). Thus whilst a similar mode of action for these benzimidazole analogues seems both logical and likely, a recent report from Sisler (1971) suggests that possession of the same basic ring structure may not always imply the same mode of action. Using *N. crassa* and *U. maydis*, he found that whereas

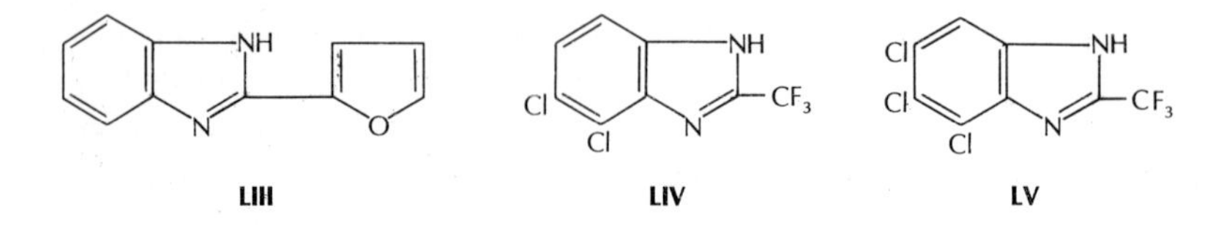

benomyl and thiabendazole prevented cell multiplication, permitted appreciable increases in dry weight, but had little or no effect on respiration, the trifluoromethyl analogues LIV and LV both prevented increase of cell dry weight and strongly stimulated respiration at concentrations below or only slightly above those completely inhibiting growth, though higher concentrations do strongly inhibit respiration. Significantly too, a benomyl-tolerant mutant of *N. crassa* was also tolerant of thiabendazole, but not of compounds LIV and LV, all of which evidence points to a different mechanism of action for the chloro compounds.

Hydroxypyrimidines

Like imidazole, the pyrimidine nucleus is often closely associated with biological activity of one sort or another. An early example was the tetrahydropyrimidines (LVI: R = *n*-alkyl) some of which were highly active as protectant fungicides (Rader *et al.*, 1952), optimum fungistatic activity occurring against the late blights of tomato and celery when $R = C_{17}H_{35}$. These compounds showed no control of powdery mildews and only moderate control of bean rust. By constrast, the 2-*n*-alkylmercapto-1,4,5,6-tetrahydropyrimidine hydrohalides (LVII: R = *n*-alkyl) show an interesting antifungal spectrum *in vivo*, although they appear to be largely inactive *in vitro* (Nickell *et al.*, 1961). Thus they show systemic activity against rusts and powdery mildews and to some extent against *Fusarium* wilt of tomatoes, 8–14 carbon atoms being required in the alkyl group for maximum effect.

More recently, a growing number of antibiotics which include the pyrimidine nucleus have been found to possess systemic fungicidal activity. These include bacimethrin (LVIII) (Tanaka *et al.*, 1962), tubercidin

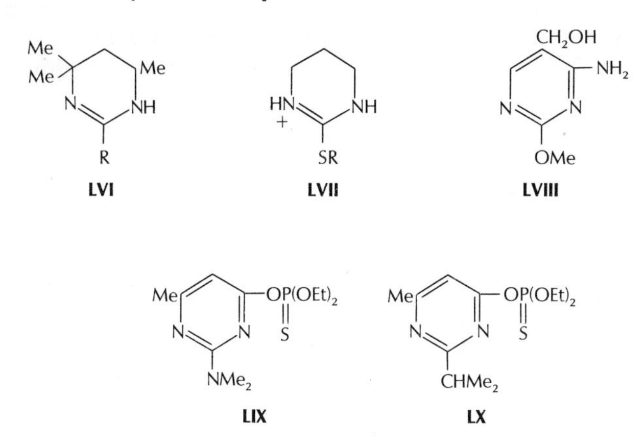

and toyocamycin (Ohkuma, 1961) and pathocidin which is formed by *Streptomyces albus* var. *pathocidus*. The latter which is highly antifungal, is reported to contain 8-azaguanine (Nagatsu *et al.*, 1962).

A series of phosphorylated pyrimidines related to O,O,-diethyl-O-(2-dimethylamino-4-methylpyrimidin-6-yl) phosphorothionate (LIX) developed earlier by I.C.I.† for sheep blowfly control, were found by workers at Jealott's Hill Research Station to have protectant fungicidal activity against certain powdery mildews, although the very closely related diazinon (LX) showed little such activity. This led to further structure-activity work with phosphorylated hydroxypyrimidines,† in which it was found that a wide class of 2-amino-4-hydroxypyrimidines were just as fungicidally active as the derived phosphorothionates. Further synthetic work was therefore concentrated on the synthesis of suitably substituted hydroxypyrimidines, rather than on the relatively more complex organophosphorus derivatives.

Whilst they are generally depicted for convenience as hydroxy compounds, the hydroxypyrimidines exist in solution and in the solid state as the tautomeric 1,4-dihydro-4-oxopyrimidines (LXI). This oxygen function at position 4 makes highly active compounds possible. Its replacement by other groups may result in loss of activity (Table 3.16) and where activity was maintained the compounds were not sufficiently superior to warrant further development. The size of the various alkyl groups can also affect the level or area of activity and the preferred groups at position 6 are methyl and ethyl. The effect of the 5-substituent provides yet another example of gradual change of biological activity with increasing lipophilicity. Fungicidal activity is at a high level when the 5-alkyl group has from two to five carbon atoms, and *n*-alkyl groups were somewhat more

† British Patents Nos., 1 019 227, 1 204 552, 1 205 000, 1 203 026, 1 129 797, 1 129 563 and others

TABLE 3.16 **Activity of substituted pyrimidines against cucumber powdery mildew (Snell, unpublished)**

Substituent R (Formula LXII)	*Activity at 500 ppm (scale 0–3)*		
	Eradicant	*Protectant*	*Systemic*
OH	3	3	3
Cl	0	0	0
NH_2	1	0	0
$NHNH_2$	3	3	3
$NHN{=}CHC_6H_5$	3	3	3
$NHNHCOC_6H_5$	0	3	3
Morpholine	2	0	0
SH	3	3	3
$SCO_2C_2H_5$	3	1	3
$SCH_2CH_2OC_2H_5$	1	1	0

effective than the corresponding branched-chain compounds (Table 3.17). The alkylamino group in position 2 is also useful for the highest fungitoxicity. The influence of various 2-substituted amino groups on the systemic activity against cucumber and barley mildews is shown in Tables 3.18 and 3.19. It will be seen that five of the six compounds presented give 90–100 per cent control of the former pathogen when applied as a root drench at 5 ppm and of these the 2-dimethylamino analogue (LXV: R = NMe_2) was selected for further development with the common name – dimethirimol. It is, for example, very effective against cucurbit powdery mildew, and only slightly less effective against cereal powdery mildews. By contrast the 2-ethylamino analogue (LXV: R = NHEt) – common name

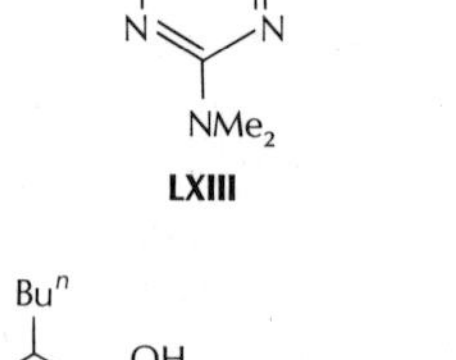

LXI **LXII** **LXIII** **LXIV**

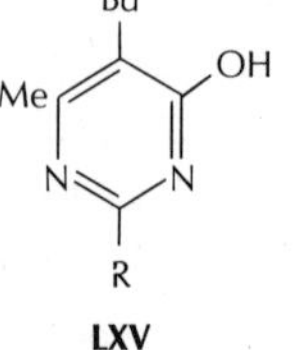

LXV

TABLE 3.17 **Activity of substituted pyrimidines against cucumber powdery mildews (Snell, unpublished)**

Substituent R (Formula LXIV)	*Activity at 500 ppm (scale 0–3)* Eradicant	*Protectant*	*Systemic*
C_2H_5	3	3	3
i-C_3H_7	2	2	0
C_3H_7	3	3	3
$CH_2CH{=}CH_2$	3	3	3
C_4H_9	3	3	3
i-C_4H_9	3	3	3
C_5H_{11}	3	3	3
i-C_5H_{11}	3	3	3
C_6H_{13}	2	2	0
$CH_2C_6H_5$	1	3	3

TABLE 3.18 **Systemic activity of substituted pyrimidines† against cucumber powdery mildew at 5 ppm (Snell, unpublished)**

Substituent R (Formula LXV)	*% Leaf surface disease free*
$NHCH_3$	94
NHC_2H_5	100
$N(CH_3)_2$	92
$N(CH_3)C_2H_5$	99
NHC_3H_7	77
$NHCH_2CH{=}CH_2$	91

† British Patent No. 1182584

ethirimol – was somewhat more effective against cereal mildews. Results obtained in glasshouse testing have in general been confirmed in the field (Brooks, 1970), and both dimethirimol and ethirimol have now achieved widespread commercial acceptance.†

Acylation of 4-hydroxypyrimidines is usually unsuccessful, or may lead to the formation of unstable N-acyl derivatives. Acylation of dimethirimol, however, in which the bulky substituted amino group hinders ring access, invariably occurs on the oxygen atom (Snell, 1968), and several of these O-acyl derivatives also had fungicidal activity (Table 3.20).‡

† British Patent No. 1182584
‡ British Patent No. 1185039

TABLE 3.19 **Systemic activity of substituted pyrimidines† (LXV) against barley powdery mildew at 50 ppm (Snell, unpublished)**

Substituent R *(Formula LXV)*	*Activity (scale 0–36)*
$NHCH_3$	35
NHC_2H_5	36
$NHCH_2CH{=}CH_2$	25
NHC_3H_7i	27
$N(CH_3)C_2H_5$	29

† British Patent No. 1182 584

TABLE 3.20 **Protectant activity of O-acyl dimethirimols against apple powdery mildew (Snell, unpublished)**

Substituent R *(Formula LXVI)*	*% Leaf surface disease free*
$P(O)(OC_2H_5)_2$	70
$P(S)(OC_3H_7)_2$	31
COC_6H_5	94
$COOCH_2C_6H_5$	86

Autoradiographic studies using ^{14}C-labelled compounds showed that following uptake from the soil they are translocated throughout the plant in the xylem. In contrast to their stability in the soil, however, Hemingway and Cavell have shown that they undergo rapid metabolism in the plant

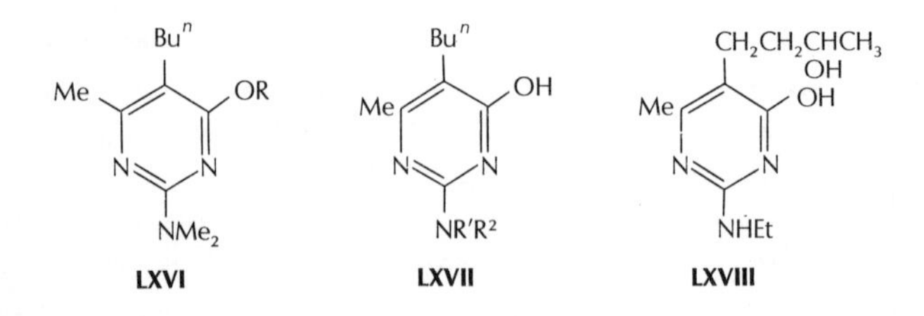

(Slade *et al.*, 1971). Thin-layer chromatographic examination of methanol extracts of cucumber plants which have been treated with ^{14}C-dimethirimol, has shown that dimethirimol is dealkylated not only to the desmethyl derivative (LXVII: $R^1 = H$, $R^2 = CH_3$) which retains fungicidal

activity, but also to the amine (LXVII: $R^1 = R^2 = H$) which is not as active. Both the original fungicide and its primary metabolites react further to form water-soluble conjugates which seem likely to involve glucose, since β-glucosidase brings about at least partial re-conversion. Cavell *et al.* (1971) have demonstrated a very similar metabolic pattern for ethirimol in cereals, dealkylation to the amino compound (LXVII: $R^1 = R^2 = H$) again taking place. Water-soluble metabolites are also formed and these can be hydrolysed under mild conditions to ethirimol, the amine (LXVII: $R^1 = R^2 = H$) and the γ-hydroxybutyl analogue (LXVIII). Some chromatographic evidence for the formation of ethirimol phosphate has also been obtained. Since some of these metabolites and their conjugates are fungitoxic in their own right, the systemic antifungal activity of these two pyrimidine fungicides must be regarded as a composite effect.

Studies on the mode of action of the hydroxypyrimidines have produced some very interesting initial findings. Bent (1970) has described experiments using segments of wheat leaves, in which the fungitoxicity of dimethirimol has been reversed by the addition of certain chemicals, and it is clear that folic acid, riboflavin and to a lesser extent adenine, thiamine and uracil all act as antagonists. Sampson (1969) has extended the fungitoxicity reversal work using cucumber plants grown in water culture and treated first with ethirimol and subsequently with various potential reversal agents, followed by the spore inoculum. The chemicals which caused reversal in varying degrees, of which the most effective were riboflavin, adenine and guanine, were all active in C-1 metabolism, either as co-factors, and products or C-1 donors. This strongly suggests that the fungicide inhibits a key metabolic reaction. However, despite indications of the inhibition of THFA-directed C-1 metabolism by the hydroxypyrimidines in the reversal experiments, no direct evidence could be obtained using key enzymes, isolated from powdery mildew preparations, though limited evidence has been obtained of interference with a number of pyridoxal-dependent enzymes.

Pyridine and pyrimidine alkanes and carbinols

Structure-activity relationships of a new group of pyridine alkanes (LXX) and carbinols (LXIX), several of which show excellent systemic fungicidal activity, have also been reported recently (Brown *et al.*, 1967). The pathogens which have been controlled in greenhouse tests include *Erysiphe polygoni* and those causing dollar spot, brown patch and *Helminthosporium* leaf spot in turf grasses. Maximum activity was shown when the pyridine ring was substituted in the 3-position and similar substitution at positions 2 or 4 led to reduced activity. Against *E. polygoni* the substituted methanols and ethanols (LXIX) were more active than the alkanes (LXX) and this activity was not altered by the substitution of amino, chloro, cyano or methoxy groups on the β-carbon atom, although such changes did cause a reduction in the control of the turf pathogens. Excellent control of *E. polygoni* was also obtained with analogues having two aryl substituents on the α-carbon atom and this activity was further increased by the order being: diphenyl $<$4-chlorodiphenyl$<$bis (2-, 3- or 4-chlorophenyl). The activity of this type of compound against turf pathogens could be increased by replacement of aryl by cycloalkyl groups, and maximum activity was observed when one or two cyclohexyl groups were present.

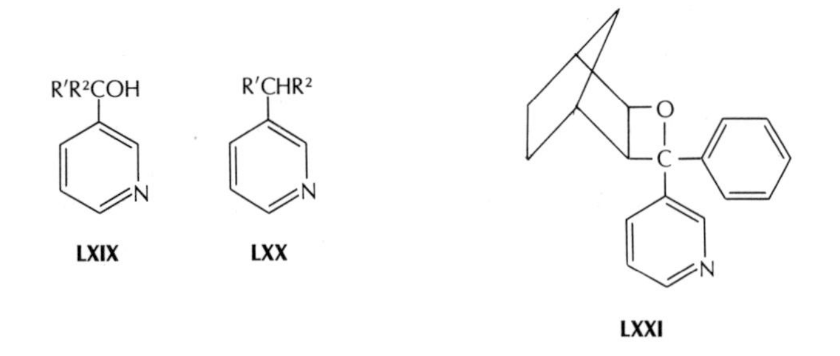

Bis-(*p*-chlorophenyl)-3-pyridine methanol (LXIX: $R^1 = R^2 = p$-chlorophenyl) or parinol was in fact introduced as a new fungicide for the control of powdery mildews, foliar applications at 10 ppm giving $>$95 per cent control of *Erysiphe cichoracearum* on squash, canteloupe and cucumber (Thayer *et al.*, 1967), and $>$90 per cent control of *Podosphaera leucotricha* on apple, *S. pannosa* on roses and *Uncinula necator* on vines was obtained using 25 ppm. Parinol is of considerable interest because it exhibits translaminar activity – a valuable attribute for the control of pathogens such as powdery mildews which when treated with conventional protectant sprays tend to proliferate unhindered on the under surfaces of leaves.

With the related experimental fungicide 4-phenyl-4-(3-pyridyl)-3-oxatricyclo [4.2.1.$0^{2,5}$] nonane (LXXI), no systemic movement of the chemical was observed following soil, stem or leaf applications, but fumigant activity from 100 μg topically applied completely eradicated powdery mildew from the leaf surface of bean plants (Spurr and Chancey, 1970). This is not the first case of fungicide redistribution in the vapour phase, as these authors seem to claim, but a re-discovery of the phenomenon reported first by Yarwood (1950) and more recently by Bent (1967), Hislop (1967) and by Clifford *et al.* (1970).

Two other pyridine analogues were examined by Spurr and Chancey (1970). Diphenyl-3-pyridylmethane (LXX: $R^1 = R^2 =$ phenyl) also exhibited fumigant activity of the same order as the oxatricyclo compound, but diphenyl-3-pyridylmethanol (LXIX: $R^1 = R^2 =$ phenyl) was inactive as a fumigant.

Yet another analogue involves the pyrimidine nucleus. 2.4-Dichloro-α-pyrimidin-5-yl-benzhydrol (LXXII), common name triarimol,† is a broad-spectrum fungicide which is particularly effective for the control of scab

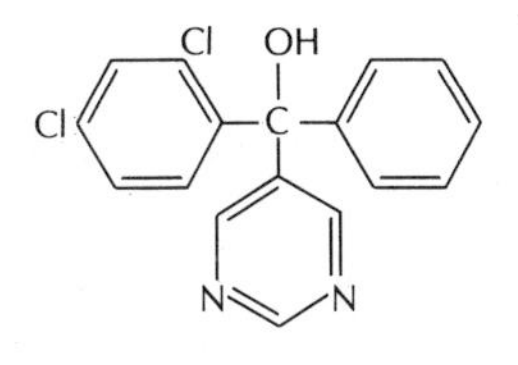

LXXII

on apple (caused by *Venturia inaequalis*) and pear (caused by *V. pirina*) at application rates of 30–40 ppm, whilst the powdery mildews of grape (*Oidium* spp.), cucurbits (*E. cichoracearum*) and rose (*S. pannosa* var. *rosae*) have been controlled at 15–20 ppm (Gramlich *et al.*, 1969). Although it does not prevent spore germination, it stops penetration and arrests any further fungus development within the leaf (Brown, 1970). This curative effect, made possible by its local systemic activity, is a valuable attribute for disease control in the field. In contrast to apple leaves, where movement is limited to the translaminar type, movement in cucumber leaves has been reported as acropetal. Uptake of triarimol from the soil is a slow process and there has been no mention of fumigant activity despite its structural similarity to compounds which have been reported as fumigant in action.

Triforine

Yet another heterocycle has been involved recently in systemic fungicide development – piperazine. N,N′-Bis-(1-formamido-2,2,2-trichloroethyl)-piperazine (LXXIII) has recently been introduced as a foliage fungicide, and it is presumably the best of some 63 compounds listed in the original patent (Ost *et al.*, 1969), and prepared from piperazine by condensation with substituted acid amides. It is particularly active against powdery mildews on cereals, apples, cucurbits and ornamentals and against apple scab (Schicke and Veen, 1969). Despite its low aqueous solubility (4 ppm) it was found to control *E. graminis* on rye seedlings by root uptake, and it

† Please see Note on page ix

also acts systemically when used as a seed dressing for cereals. Investigations of its absorption and translocation in barley plants using ^{3}H ring-labelled material have revealed not only the normal upward movement via the xylem, following root absorption, but also a downward movement

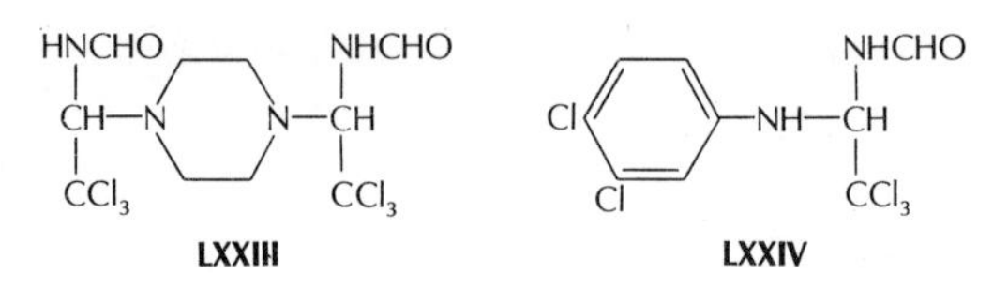

in the stem to the roots. Less than 1 ppm distributed throughout the plant was found to be sufficient to prevent *E. graminis* (Ost *et al.*, 1972). 1-(3,4-Dichloroanilino)-1-formylamino-2,2,2-trichloroethane (LXXIV) – proposed common name chloroaniformethan – also contains a formylamino group (Malz *et al.*, 1968). It is a specific for certain powdery mildews, being particularly effective in the control of mildew on spring barley (Davis *et al.*, 1971) and has also been used against mildews on wheat and oats. Glasshouse studies showed that the chemical possessed both curative and protective properties in addition to being taken up by roots and showing distal movement within leaves.

N-substituted tetrahydro-1,4-oxazines

The increase in winter barley acreage in N. Germany, with the consequent escalation of overwintering potential inoculum of *E. graminis*, provided the impetus for the development of this new class of fungicidal compounds based on morpholine, which are both systemic and eradicant in action (Kradel and Pommer, 1967; Pommer *et al.*, 1969). The effects of variations in

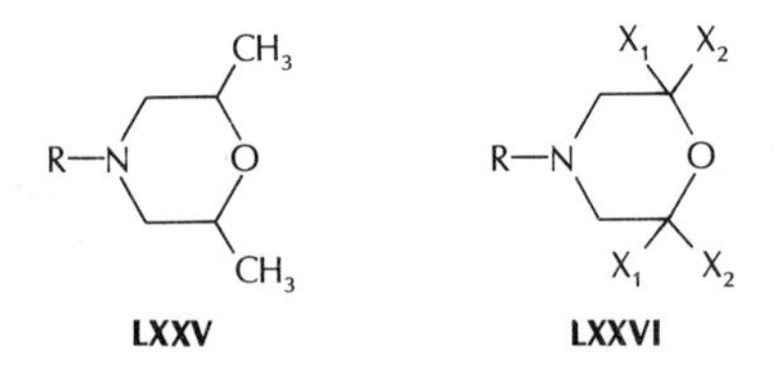

the nature and size of the N-substituent, and of substituents in the heterocyclic ring, on fungicidal activity were examined in several hundred compounds; other structural variations included N-oxide and salt formation, and complexing with heavy metals (König *et al.*, 1965; Pommer and Kradel, 1967). Table 3.21 shows the influence of the N-(*n*-alkyl)-substituents with 9–18 carbon atoms on the fungicidal activity of 2,6-dimethylmorpholine against barley mildew. The effectiveness improves with increasing chain length, reaching a maximum for C_{13} (LXXV: R = $C_{13}H_{27}$) (tridemorph) but unfortunately there is a tendency for phytotoxicity to increase at the same time. Higher members still show some phytotoxicity but fungicidal effectiveness is also reduced. With cycloalkyl N-substituted 2,6-dimethylmorpholines, there is a marked fungicidal effect when the ring substituent

TABLE 3.21 **Influence of the N-alkyl substituent on the fungicidal activity of 2,6-dimethylmorpholines (LXXV) against barley mildew (Pommer and Kradel, 1967)**

Substituent R	% *Active ingredient applied*				
	0·2	0·1	0·025	0·006	0·0015
	Leaf infection†				
i-C_9H_{19}	§	3	5	9	9
i-$C_{10}H_{21}$	1§	2§	9	9	9
n-$C_{12}H_{25}$	§§	§§	2§	3	7
i-$C_{13}H_{27}$	§§	§§	0	0	1
n-$C_{18}H_{37}$	0‡	0	0	2	3
Cyclohexyl	2	3	4	5	9
Cycloheptyl	2	3	3	3	6
Cyclooctyl	2	2	3	3	4
Cyclododecyl	§	0	0	0	1

† 0 to 9 No leaf infection to total infection
‡ Slight leaf damage,
§ greater damage; §§ Total damage

contains 7–8 carbon atoms and the distinguishing feature of the N-cycloalkyl analogues is their low phytotoxicity. N-Cyclododecyl-2,6-dimethylmorpholine (LXXV: R = cyclododecyl) (dodemorph) was highly effective with only a hint of phytotoxicity at the highest test concentration and its acetate was found to be even more fungicidal than the parent compound.

The effect of substitution in the heterocyclic ring is shown with N-cyclooctylmorpholine (LXXVI: R = cyclooctyl; $X_1 = X_2 = X_3 = X_4 = H$) (Table 3.22). Fungicidal properties change only slightly with increasing alkyl substitution, while aryl substituents result in a marked reduction in activity. The optimum combination of methyl groups (2,6-dimethyl) has been incorporated in both tridemorph and dodemorph.

Since these compounds show fungicidal activity in plate tests *in vitro*, it seems likely that they are absorbed by the barley plant, either by root or foliage, and are translocated *in toto* throughout the plant, giving protection for a period of 3–4 weeks. Penetration of the leaf tissue by tridemorph is very much slower than root uptake, but after penetration a relatively rapid distribution takes place. Although tridemorph has been

TABLE 3.22 **Dependence of fungicidal activity on the substituents on the C atoms of the morpholine ring (Pommer and Kradel, 1967)**

Substituents (LXXVI: R = cyclooctyl)				*% Active ingredient applied*				
X_1	X_2	X_3	X_4	0·2	0·1	0·025	0·006	0·0015
H	H	H	H	0	1	3	3	5
CH_3	H	H	H	0†	0	2	5	6
CH_3	H	CH_3	H	2	2	3	3	4
CH_3	CH_3	H	H	0	1	2	4	7
CH_3	CH_3	CH_3	H	0	3	3	7	9
CH_3	CH_3	CH_3	CH_3	1	2	3	5	9
CH_3	H	C_2H_5	H	0‡	0†	0	3	9
C_6H_5	H	H	H	0†	2	4	5	7
C_6H_5	H	C_6H_5	H	3	5	9	9	9

† 0 to 9 No leaf infection to total infection
‡ Slight leaf damage to total damage

used safely on both spring and winter barley there are indications of some phytotoxicity on certain winter wheat varieties. The lower activity of tridemorph against powdery mildew of wheat is probably due to reduced movement within the leaf, and it is also interesting that whereas it is taken up by both surfaces of the barley leaf equally, uptake in wheat is greater on the adaxial surface.

6-Azauracil

6-Azauracil (3,5-dioxo-2,3,4,5-tetrahydro-1,2,4-triazine) (LXXVII) which possesses both cytostatic and antibacterial activity has also been shown to be systematically active, either by foliar or root application, against powdery mildew diseases (Dekker, 1962; Dekker and Oort, 1964). Of 82 purines and pyrimidines tested against *Erysiphe cichoracearum* in a floating leaf disc test, only 6-azauracil (LXXVII) (at 0·6 ppm) and 2,6-diaminopurine (LXXVIII) as sulphate (at 20 ppm) completely prevented development of mildew; 8-azaadenine (LXXIX), 8-azaguanine (LXXX) and dithiouracil (LXXXI) at 20 ppm also showed a substantial disease reduction. All other purine and pyrimidine derivatives tested were inactive at this

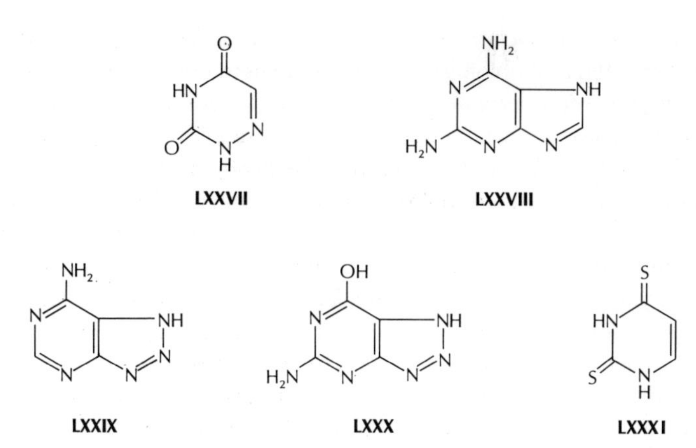

concentration and a comparison of these results with the occurrence of purines and pyrimidines in ribonucleic acid indicated that only the aza-derivatives of those bases which occur in RNA are active (Table 3.23).

Of the five active compounds, only 6-azauracil was active in the root absorption test, complete prevention of powdery mildew development on the leaves being obtained with 6 ppm. This compound also proved to

TABLE 3.23 **Activity of aza-substituted purines and pyrimidines on cucumber mildew (Dekker, 1962)**

Compound	*Activity*	*Bases occurring in RNA*
8-Azaadenine	+	adenine
8-Azaguanine	+	guanine
8-Azaxanthine	–	–
8-Azahypoxanthine	–	–
8-Aza-1,3-dimethylxanthine	–	–
8-Aza-2,6-diaminopurine	–	–
6-Azauracil	+	uracil
6-Azathymine	–	–

be active against all other powdery mildews tested by the leaf disc method, including *E. graminis* on wheat, barley and rye and *E. polygoni* DC on beet, peas and lupins. Wheat powdery mildew appeared to be most sensitive and was completely controlled by only 0·3 ppm.

In studies with *E. cichoracearum* on cucumber and *E. graminis* on wheat 6-azauracil did not inhibit germination or penetration of the fungus into the host, but interfered with the growth of the fungus during formation of the first haustorium, which was sufficient to inhibit the pathogen completely. 6-Azauridine and 6-azauridine-5′-phosphate are also active in this way, but their effects, unlike that of 6-azauracil, are not reversed by uracil. In leaf disc tests the uracil precursors, orotic acid, *L*-dihydro-orotic acid and *L*-ureidosuccinic acid, also reversed inhibition of cucumber powdery mildew by 6-azauracil. Similar observations were made in leaf disc tests with bean rust (*Uromyces appendiculatus*), cucumber anthracnose (*Colletotrichum lagenarium*) and tobacoo mosaic virus.

By contrast, inhibition of bacterial and fungal growth *in vitro* was reversed by uracil, but not by orotic acid, which thus seems to require the presence of the host. These data are consistent with the hypothesis that 6-azauracil is converted to a ribose-containing metabolite which then interferes with pyrimidine biosynthesis by inhibiting orotidylic decarboxylase, and indeed 6-azauracil has been shown to form 6-azauridine in both the cucumber plant and in the pathogen *C. lagenarium*.

Azepines

The azepines (LXXXII: R^1 and R^2 = alkyl) are rather broad-spectrum fungicides which control some powdery mildews and rusts as well as leaf spot pathogens. Two compounds, 7-dodecylamino-3,4,5,6-tetrahydro-2H-azepine and 4-t-butyl-3,4,5,6-tetrahydro-2H azepine hydrochlorides have been formulated for greenhouse and field trials (Schwinn, 1969). They also

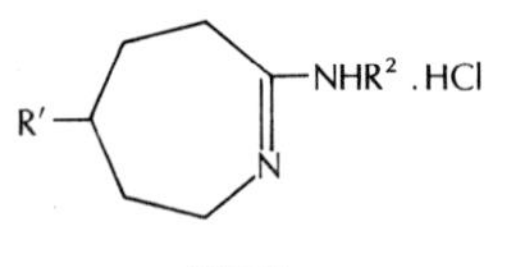

LXXXII

display some eradicant activity against *Alternaria solani* (treatments up to 24 h after infection), *Uromyces phaseoli* (up to 48 h after infection) and *Erysiphe cichoracearum* (up to more than 72 h). The systemic effect shown by 4-t-butyl-7-decylamino-3,4,5,6-tetrahydro-2H-azepine hydrochloride (LXXXII: R^1 = t-butyl, R^2 = *n*-decyl) against cucumber powdery mildew is limited to applications in a nutrient solution and is not observed with soil treatments. The azepines are in general well tolerated by fruit trees but fruit russetting may occur on certain varieties under adverse climatic conditions after application during the period of fruit formation.

AROMATIC COMPOUNDS

Although groups of purely aromatic compounds such as quinones and phenols have furnished effective protectant fungicides without being associated with an organophosphorus residue, a heterocyclic ring system or sulphur atom or atoms, the same hardly applies in the case of systemic fungicides. In the following section only two cases of systemic activity, without associated phosphorus or sulphur atoms, or heterocyclic nuclei are cited. In the first of these (chloroneb) the systemic effect is extremely specific, whilst in the other only trans-laminar or limited systemic activity is shown by the phenols concerned.

Thiourea-based fungicides

A new class of fungicides based on thiourea and containing an aromatic nucleus may however depend on conversion to the benzimidazole ring for their activity. 1,2-Bis-(3-alkoxycarbonyl-1-thioureido) benzenes are weakly acidic compounds synthesised by the reaction of o-phenylene-diamine and isothiocyanoformic esters. The conversion of 1,2-bis-(3-ethoxycarbonyl-2-thioureido) benzene (LXXXIII: $R = C_2H_5$) (thiophanate) and the corresponding methoxy compound, thiophanate methyl (LXXXIII: $R = CH_3$) to ethyl and methyl benzimidazol-2yl-carbamates respectively in aqueous solution, suggested that the fungicidal activity of

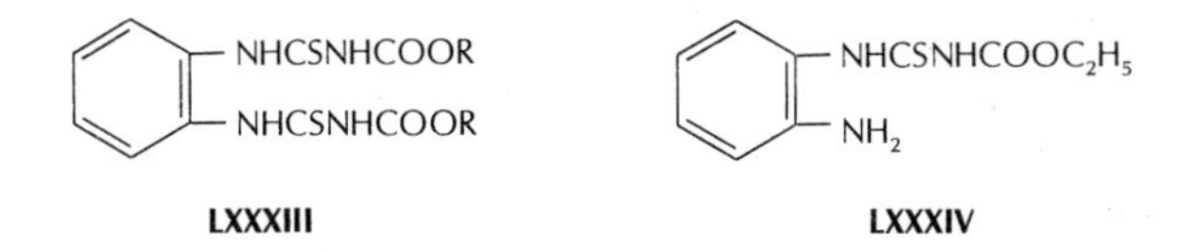

the thiophanates may well be attributed to these heterocyclic transformation products (Selling *et al.*, 1970) and the inactivity of 1,3-and 1,4-bis-(3-ethoxycarbonyl-2-thioureido) benzenes, which cannot cyclise in this way, supports this suggestion. Further evidence is provided by the behaviour of the thiophanates in plants and animals. Using both ^{14}C- and ^{35}S-labelled materials, Noguchi and Ohkuma (1971a, b) found that cleavage of the C—S bonds takes place readily and the major resultant metabolites in both plants and animals have been identified as 2-alkoxy-carbonylamino benzimidazoles by co-tlc-autoradiography. Additionally, hydroxylation of the aromatic ring of benzimidazole followed by conjugation and subsequent elimination, took place in mice.

Evaluation of the thiophanates in the UK has shown a high degree of systemic activity by root uptake against barley and cucumber mildews, together with a persistence of action, not only for thiophanate, but particularly for the analogous thiophanate methyl and compound NF48 (LXXXIV). There is little systemic activity against apple mildew, however, and the trans-laminar activity and localised movement towards the leaf apex, shown in barley and cucumber, was also much less marked in apple. These compounds were originally evaluated against tobacco mosaic virus with a view to screening for possible interference with *de novo* synthesis of polynucleotides, and it may be significant that *mono*-alkoxycarbonyl-thioureido benzenes were found to be more active against this organism.

It is interesting to recall here the systemic activity of phenylthiourea (LXXXV) which can prevent disease symptoms caused by *Cladosporium cucumerinum* by root treatment of cucumber seedlings with a solution containing 50 ppm (Kaars Sijpesteijn and Pluijgers, 1962). Despite a certain

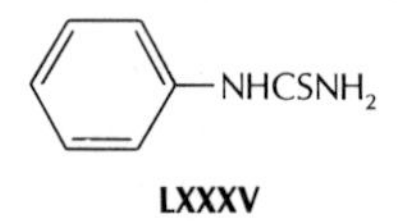

LXXXV

formal similarity with the thiophanates, any further analogy of fungicidal action seems unlikely since phenylthiourea performs very poorly against the pathogen *in vitro*, where a concentration of some 500 ppm is required to inhibit *C. cucumerinum* in the roll culture test. Since the press sap contained less than 20 ppm of phenylthiourea and did not appear to contain any other antifungal derivative thereof, it seems likely that the chemical interferes directly with plant–fungus biochemical relationship or influences plant metabolism in some way (Kaars Sijpesteijn and Sisler, 1968). Chemotherapeutic activity showed a remarkable correlation with polyphenol oxidase – inhibitory capacity and peroxides are also higher in treated than in untreated plants. Increased lignification in treated plants, especially around the penetration points of *C. cucumerinum* may well be interrelated. There are also indications of increased indoleacetic acid oxidase activity in treated plants and it has been suggested that this leads to increased resistance to cucumber scab (van Andel, 1966a).

Substituted thiosemicarbazides

In contrast to phenylthiourea, the closely related 1-phenylthiosemicarbazide (LXXXVI) was originally thought to be active *per se*, both *in vitro* and *in vivo* against cucumber mildew, apple scab, potato late blight and *Botrytis* and *Cladosporium* on tomatoes (Pluijgers and Kaars Sijpesteijn, 1966). It has recently been shown, in laboratory trials using *Aspergillus*

TABLE 3.24 **Antifungal activity of thiosemicarbazides; effect of 4-, 2- and 1-substitution (Pluijgers and Kaars Sijpesteijn, 1966)**

Substituent	*Minimum concentration causing complete inhibition of growth (ppm)*				*Systemic activity against cucumber scab after root application*
	Botrytis allii	*Penicillium italicum*	*Aspergillus niger*	*Cladosporium cucumerinum*	
—	200	100	200	100	+
4-Methyl	5	5	500	200	+
4-Allyl	5	10	500	200	+
4-*n*-Butyl	100	50	200	50	+
4-*n*-Octyl	100	50	100	20	−
4,4-Dimethyl	200	100	200	100	−
2-Methyl	500	500	> 500	50	−
1-Phenyl	5	1	2	1	+ +
1-Benzyl	500	500	500	200	+ +
1-Acetyl	500	500	500	500	−
1-Benzoyl	500	500	500	200	−

TABLE 3.25 **Antifungal activity of thiosemicarbazides: effect of aryl substitution (Pluijgers and Kaars Sijpesteijn, 1966)**

Substituent	*Minimum concentration causing complete inhibition of growth (ppm)*				*Systemic activity against cucumber scab after root application*
	Botrytis allii	*Penicillium italicum*	*Aspergillus niger*	*Cladosporium cucumerinum*	
1-Phenyl	5	1	2	1	+ +
2-Phenyl	500	500	200	50	+ +
4-Phenyl	200	200	500	100	+
1,1-Diphenyl	500	500	500	500	−
1,4-Diphenyl	500	100	500	20	−
2,4-Diphenyl	500	200	500	50	−
1-(α-Napthyl)	5	1	20	5	+ +
1-Phenylsemicarbazide (for comparison)	500	500	500	500	−

TABLE 3.26 **Antifungal activity of thiosemicarbazides: effect of substitution in 1-phenylthiosemicarbazide (Pluijgers and Kaars Sijpesteijn, 1966)**

Substituent	*Minimum concentration causing complete inhibition of growth (ppm)*				*Systemic activity against cucumber scab after root application*
	Botrytis allii	*Penicillium italicum*	*Aspergillus niger*	*Cladosporium cucumerinum*	
1-Phenyl-4-methyl	20	50	100	10	+ +
1-Phenyl-4,4-dimethyl	0·5	50	100	5	+ +
1-Phenyl-4-acetyl	500	500	500	500	−
1-Phenyl-4-benzoyl	500	500	500	500	−
1-(4′-Chlorophenyl)	0·5	0·2	0·5	0·5	+ +
1-(2′,6′-Dichlorophenyl)	50	2	5	1	−
1-(2′,4′,5′-Trichlorophenyl)	20	2	10	1	−
1-(4′-Nitrophenyl)	5	5	50	20	+
1-(4′-Methylphenyl)	2	2	5	1	+ +
1-(4′-Methoxyphenyl)	5	5	20	5	+ +

TABLE 3.27 **Structure-activity relationship of thiosemicarbazides (Kaars Sijpesteijn *et al.*, 1968)**

Compound	*Minimum concentration (ppm) causing complete inhibition of growth; glucose agar, pH 6·5*				*Coloration of medium*
	Botrytis allii	*Penicillium italicum*	*Aspergillus niger*	*Cladosporium cucumerinum*	
1. $C_6H_5.NH.NH.CS.NH_2$	5	1	2	1	+
2. $H_2N.N(C_6H_5).CS.NH_2$	500	500	200	50	−
3. $H_2N.NH.CS.NH.C_6H_5$	200	200	500	100	−
4. $(C_6H_5)_2N.NH.CS.NH_2$	500	500	500	500	−
5. $C_6H_5.CH_2NH.NH.CS.NH_2$	500	500	500	200	−
6. $4\text{-}Cl.C_6H_4.NH.NH.CS.NH_2$	0·5	0·2	0·5	0·5	+
7. $4\text{-}CH_3.C_6H_4.NH.NH.CS.NH_2$	2	2	5	1	+
8. $\alpha\text{-}C_{10}H_7.NH.NH.CS.NH_2$	5	1	20	5	+

niger, to be converted to 1-phenylazothioformamide (LXXXVII) *in vivo* and it is now clear that this is the active agent (Kaars Sijpesteijn *et al.*, 1968). In the light of this discovery the high structural specificity of substitution in

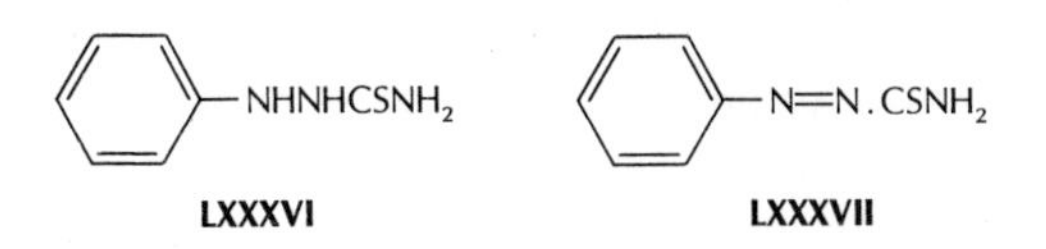

thiosemicarbazide (Tables 3.24, 3.25 and 3.26) is very interesting in retrospect. Thus only in those cases where a stable azothioformamide can be expected does the medium become orange-coloured and strong antifungal effects observed (Table 3.27) and it is clear that this requires the presence of a single suitable stabilising aromatic substituent in position 1 (Table 3.26). The presence of the thione sulphur atom would also appear to be essential since the corresponding 1-phenylsemicarbazide, possessing only an oxygen atom, is inactive (Table 3.26).

Phenols and phenol derivatives

1,4-Dichloro-2,5-dimethoxybenzene (chloroneb: LXXVIII) was introduced as an experimental fungicide in 1967 by DuPont. Being taken up by root absorption, and concentrated in both roots and lower stem, it is useful as a supplemental seed treatment for sugar beet, and as a soil treatment at planting time to control some seedling diseases of cotton and beans. It is highly fungistatic to *Rhizoctonia* spp., moderately so to *Pythium* spp., but gives only poor control of *Fusarium* spp., and is inactive against *Trichoderma* spp.

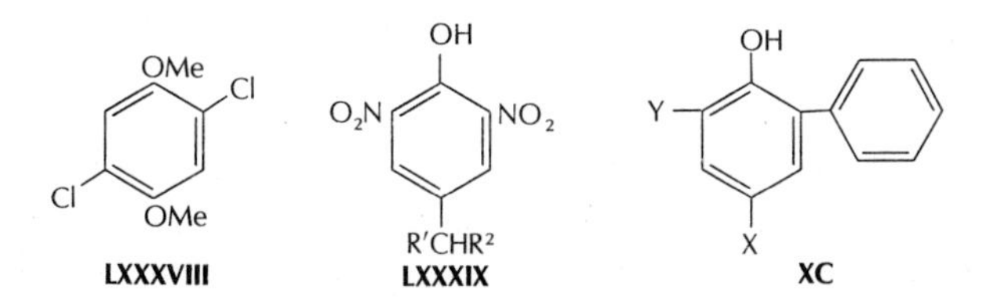

The 4-(1-substituted alkyl)-2,6-dinitrophenols (LXXXIX) show an interesting pattern of activity against cucumber and apple powdery mildews in that, in general, their partition and hydrogen bonding characteristics tend to facilitate vapour phase redistribution and also zonal activity, that is protection of large areas of the leaf surface by lateral movement within the leaf surface waxes (Clifford *et al.*, 1971). However, the values of these parameters are such that entry into the leaf, and hence trans-laminar activity and translocation are precluded. On the other hand, the 4-(1-cyclopropylalkyl)-2,6-dinitrophenols (LXXXIX: R^1 = cyclopropyl, R^2 = alkyl) differ in that their partition and hydrogen bonding characteristics are such that some homologues when applied to leaves are extensively translocated resulting in high protectant activities (Table 3.29). Translaminar activity observed with 2-bromo-6-cyclohexyl-4-nitrophenol (XC: X = NO_2, Y = Br) was also attributed to favourable combinations of these parameters, the isomeric 4-bromo-2-cyclohexyl-6-nitrophenol (XC: X = Br, Y = NO_2) showing negligible activity (Table 3.28). Since it is generally

TABLE 3.28 **Activities of 2-cyclohexyl-halogeno-nitrophenols (XC) against cucumber powdery mildew (Clifford *et al.*, 1971)**

Substituents 4	6	π	A^x	A^y	*Protectant activity* $10^5 ED_{50}$(M)	*Zonal activity*
Cl	NO_2	0·15	2·8	0·07	>40	nil
Br	NO_2	0·09	4·4	0·11	>40	nil
NO_2	Cl	0·66	2·8	0·04	7·8	nil
NO_2	Br	0·78	2·7	negligible	6·2	translocated

accepted that nitrophenols act as inhibitors of oxidative phosphorylation, those analogues which are able to penetrate will interfere with energy production in the plant with disastrous consequences, which thus preclude their exploitation commercially. These findings are in agreement with the suggestion of Kaars Sijpesteijn (1969) that all successful systemic fungicides are those which act by inhibition of biosynthetic pathways rather than by inhibition of energy production.

It is clear from a review such as this, that a wide range of (often unlikely) chemicals are effective as systemic fungicides. It is also abundantly obvious that while some compounds are freely translocated throughout the plant, others described as systemic, are really only protectant fungicides which are also capable of limited foliar movement. However, such activity can be a valuable adjunct in a protectant fungicide and can play a critical role in protection against certain pathogens. Because of the inevitable intimacy of a systemic chemical with the cells of the host plant, one of the prime requisites of a true systemic fungicide, or indeed of any compound acting systemically, must be a high specificity of action. This is obviously possible with several basic ring systems the structural and stereochemical variations possible with substituents being critical not only for initial

TABLE 3.29 **Activities of 4-(1-cyclopropylalkyl)-2,6-dinitrophenols (LXXXIX) against cucumber powdery mildew (Clifford *et al.*, 1971)**

Alkyl group	π	A^x	A^y	*Protectant activity* $10^5 ED_{50}$(M)	*Zonal activity*
Methyl	0·74	4·6	0·26	17	minimal
Ethyl	0·70	4·4	0·35	22	not tested
Propyl	0·76	4·6	0·47	3	widespread delocalised control and some chlorosis
Butyl	0·79	4·5	0·41	1·5	
Pentyl	0·90	4·1	0·37	1·0	
Hexyl	0·99	3·9	0·33	1·25	
Heptyl	1·07	4·3	0·42	6	

uptake and translocation, but ultimately for interference with biochemical processes at enzyme level. While protectant fungicides seem to act by interfering with ATP production by the fungus, the ability to inhibit biosynthesis seems to be common to many systemic fungicides.

Although conventional structure-activity studies based on screening programmes will no doubt continue to supply new fungicides, real hope for the future will surely come from the increased tempo of biochemical research into modes of action.

4

Toxicological considerations
by D. Woodcock

The prime requirement for the successful utilisation of any biologically active compound in agriculture is that it should be inherently safe both in production and in its application. This safety should extend not only to man, but also to domestic, farm and wild animals, useful plants, beneficial insects and micro-organisms. The metabolism and mode of action of the compound are no less important, and toxicological data for metabolites formed in plants, in animals and in the soil, are becoming increasingly essential.

The toxicity of chemicals can be evaluated in a number of ways using a variety of test animals, birds, fish and insects. Any classification based on acute toxicity to the various experimental subjects is very arbitrary and much depends on other factors. The mean dose in mg/kg of live weight that causes 50 per cent mortality (LD_{50}) can vary enormously between species and the trend with increasing body weight is not always extrapolatable. It is important too, that note should be taken of possible penetration routes into the test organism, and compounds of low LD_{50}'s, which are capable of being dermally absorbed or taken in via the respiratory passages are very dangerous indeed.

In the German Democratic and Federal Republics compounds are classified with respect to acute mammalian toxicity into three groups according to their LD_{50}'s: less than 100 mg/kg, 100–300 mg/kg and >300 mg/kg, and compounds of the first group are issued to users only by special permit. In the Soviet Union compounds with LD_{50}'s <50 mg/kg are regarded as 'powerful', compounds with LD_{50}'s from 50–200 mg/kg as 'highly toxic', while LD_{50}'s of 200–1000 and >1000 mg/kg are rated as of medium and low toxicities respectively.

In the UK at present there is no classification system, each new active ingredient being dealt with on its merits. However, the Council of Europe is about to publish a classification of formulated products for the user, which is likely to be the starting point for the proposed new UK legislation (Bates, JAR, private communication). The Council of Europe limits are as follows:

Class 1: 0–200 mg/kg
Class 2: 200–2000 mg/kg
Class 3: 2000–5000 mg/kg
Class 4: >5000 mg/kg

Fungicides, in general, do not pose any great mammalian hazards and few, apart from organophosphorus-based compounds, will be rated as Class 1. There are often wide variations within a single class of compound, however, and occasionally some real surprises.

ORGANOPHOSPHORUS COMPOUNDS

The action of organophosphorus compounds in phosphorylating vitally important enzyme systems, and thus inhibiting normal functions, makes them highly active as pesticides. It means, however, that toxicity to vertebrates may also be high and careful structural manipulations are required to produce compounds with suitably moderated mammalian toxicities. Although organophosphorus compounds provided the first systemic insecticides, the logical exploration of this class for antifungal activity has provided relatively few acceptable systemic fungicides. One of these is triamiphos, which was claimed as the first systemic fungicide and which, although possessing an acute oral LD_{50} to male rats of 20 mg/kg, is relatively harmless as a skin application to rabbits (LD_{50} 1500–3000 mg/kg) and mice (LD_{50} > 1000 mg/kg). Oral and intraperitoneal LD_{50} values for triamiphos and some of its formulations are shown in Table 4.1.

The phosphoric esters of pyrazolopyrimidines recently introduced by Hoechst, which are also effective against powdery mildews, appear to be rather less toxic to mammals than triamiphos. Thus the acute oral toxicities (LD_{50}'s) of O,O-diethyl-O-(6-ethoxycarbonyl-5-methylpyrazolo-[2,3-a] pyrimidin-2-yl) phosphorothioate (HOE 2873), as pure compound and commercial product, are 140 and 370 mg/kg respectively.

The rice blast fungicides, kitazin and kitazin P, likewise show only moderate mammalian toxicites with LD_{50}'s of 238 mg/kg (rats) and 660 mg/kg (mice) respectively, being quoted by Melnikov (1971).

TABLE 4.1 **Acute toxicity of triamiphos†**

Formulation	*Albino mice*	*Application*	*LD_{50}‡ mg/kg*	*(with 95% confidence interval)*
Technical	female	intraperitoneal	10·8	(7·4–15·7)
	male	intraperitoneal	10·8	(7·4–15·7)
	female	oral	27·1	(20·0–36·9)
	male	oral	23·7	(14·2–39·5)
Pure	female	intraperitoneal	7·9	(5·8–10·8)
	female	oral	20·0	(13·7–29·1)
25% w.p.	female	intraperitoneal	61·9	(48·7–78·7)
	female	oral	90·9	(54·6–151)
10% (w/w) liquid	female	intraperitoneal	92·6	(63·6–135)
	female	oral	406·0	(301·5–546·5)

† From Pesticide Information (N.V. Philips-Duphar) 25.8.1966
‡ Mortality 48 h after application

ANTIBIOTICS

Probably the safest systemic fungicide of all is streptomycin and with its widespread clinical use, it is no surprise to find an LD_{50} (acute oral for mice) of 9000 mg/kg. Griseofulvin is also one of the safest antibiotics, for Wistar rats of either sex can tolerate a daily dose of 2000 mg/kg given intraperitoneally, though there is some damage to the seminal and intestinal epithelia. Intravenous injection causes a transient depression of mitosis, most noticeable in the bone marrow, but the rats appear little affected in general condition and recovery proceeds rapidly after 24 h; the LD_{50} is 400 mg/kg. Cycloheximide is considerably more toxic than streptomycin and griseofulvin, acute oral toxicity tests giving the following LD_{50}'s (mg/kg): 133 (mice), 65 (guinea pigs), 2·5 (rats) and 60 (monkeys).

The toxicity of blasticidin is also rather high, its oral LD_{50} for rats being 39 mg/kg. It has also been reported to cause conjunctivitis in humans. By contrast, kasugamycin, the other antibiotic useful in the control of rice blast, has a very low mammalian toxicity. The LD_{50} (oral) for mice is >2000 mg/kg and TLM for carp is >1 mg/ml.

The polyoxins are also non-toxic to mammals and fish, the Japanese killifish being unaffected by a 72 h exposure to a concentration of 100 ppm. Oral administration to mice shows an LD_{50} >1500 mg/kg, and there appears to be no effect on the human eyes or mucous membranes. The simple molecule of cellocidin is also usually fairly harmless but exhibits a wide range of toxicity to mice, depending on how the antibiotic is applied.

Thus the LD_{50}'s for intravenous, oral and cutaneous administration are 11, 89–125 and 667 mg/kg respectively. It has also been reported as being less toxic to fish than DDT.

CARBOXYLIC ACID ANILIDES

Members of this class of compound appear to be very safe both dermally and orally. Thus the acute LD_{50} (rats) for carboxin is as low as 3 200 mg/kg and albino rats fed 200 ppm daily for 90 days suffered no detectable symptoms; the acute dermal LD_{50} (rabbits) is 8 000 mg/kg. The corresponding values for oxycarboxin are 2000 and >16 000 mg/kg respectively. The non-heterocyclic mebenil likewise possesses very low mammalian toxicity, the acute oral LD_{50} (rats) being 6 000 mg/kg, with no evidence of skin irritation in laboratory tests. Pyran and furan analogues also provide some very safe compounds. Thus 2,3-dihydro-6-methyl-5-phenylcarbamoyl-4-pyran (HOE 2989) has an acute LD_{50} (rats) of >15 000 mg/kg and in sub-chronic tests, the no-effect level was 800 ppm. For 2,5-dimethyl-3-phenylcarbamoylfurane (BAS 3191) the acute oral LD_{50} (rats) is >6400 mg/kg, and this does not appear to change by substitution of a cyclohexyl nucleus in place of the phenyl group (BAS 3270). Of two thiazoles used in seed treatment, whereas 2,4-dimethyl-5-phenylcarbamoyl thiazole (G 696) has very low mammalian toxicity (LD_{50} 5 600 mg/kg), replacement of one methyl group by an amino group to give 2-amino-4-methyl-5-phenylcarbamoyl thiazole results in a dramatic increase in toxicity, the LD_{50} falling to 141 mg/kg.

HETEROCYCLIC COMPOUNDS

The substituted benzimidazoles benomyl, fuberidazole and thiabendazole are all apparently very safe compounds. The acute oral toxicity of benomyl against male rats is 9 590 mg/kg and there appears to be no evidence of chronic toxicity in 90-day feeding tests at a dietary level of 2 500 ppm. There is also no sign of skin irritation and no hint of dermatitis. Substitution of heterocyclic nuclei in position 2 results in slightly increased toxicity but both thiabendazole and fuberidazole containing thiazolyl and furyl substituents respectively are still virtually innocuous, which is not surprising in the case of the former since it was initially developed as an anthelminthic. LD_{50}'s (acute oral) for thiabendazole are 3 320 (rats), 3 810 (mice) and 3 850 mg/kg (rabbits) and daily doses of 100 mg/kg given orally over a two-year period showed no toxic symptoms. Fuberidazole had the following LD_{50}'s (mg/kg) for male rats: acute oral, 1100; acute dermal (seven day exposure), 1000; acute peritoneal ~100. Five doses per week for eight weeks at 120 mg/kg caused an increase in liver weight but did not affect other organs.

The breakdown product of benomyl, methyl benzimidazol-2-yl carbamate, which has recently been introduced by BASF as a systemic fungicide (BAS 346F) in its own right, also seems to be a reasonably safe compound. Approximate median lethal doses (ALD_{50}'s) are >6400, >8000 and >8000 mg/kg for rats, rabbits and dogs respectively, and the 50 per cent formulation examined dermally has ALD_{50}'s of > 2500 and >4000 mg/kg for rats and rabbits.

The hydroxypyrimidines also appear to be very safe compounds. The acute oral LD_{50} (rats) for dimethirimol is more than 4000 mg/kg and three-day dermal applications of 400 mg/kg produced neither skin damage nor toxic symptoms. Ninety-day feeding tests at dietary levels of 1000 ppm also produced no adverse effects in rats or dogs. Ethirimol has the same acute oral LD_{50} for rats and was also harmless in ninety-day feeding trials at 200 mg/kg. Other LD_{50}'s (acute oral) are 1000–2000 mg/kg (male rabbits); >1000 mg/kg (female cats); 500–1000 mg/kg (female guinea pigs); 4000 mg/kg (hens).

Triarimol was only slightly less safe, the acute oral LD_{50} for rats being 600 mg/kg. There was a slight positive reaction at 100 mg/kg when applied as a 10 per cent suspension to the skin of rabbits but this had cleared in 72 h. Three-month feeding tests showed that the safe level for rats and dogs is 400 and 800 ppm respectively.

The piperazine derivative, triforine, is also extremely safe, acute oral LD_{50}'s for mice and rats being >6000 mg/kg. It is also reported to be harmless to fish *(Lebistes reticulatus)* at 50 ppm in water, and to bees at 1000 ppm in feeding experiments. Another compound, of proposed common name, chloraniformethan, which also contains the formyl-amino group, also appears reasonably safe after extensive testing at the Institüt für Toxicologie, Elberfeld. Oral toxicity tests showed LD_{50}'s of >2500, >1000, > 500 and 250–500 mg/kg for male rats, female mice, cats and guinea pigs respectively. In feeding tests, up to 100 ppm in the daily diet for a period of 14 weeks caused no observable changes in rats, and dermal toxicity appeared to be slight (LD_{50} >500 mg/kg). A similar dose applied daily to the clipped dorsal skin of male and female rabbits over a period of 14 days also caused no change in general condition, body weight, blood picture, liver and kidney functions, and application to the ear and conjunctival sac also appeared to have no ill-effects.

Dodemorph must also be regarded as safe, LD_{50}'s having been reported as 4800 mg/kg (acute oral) for rats and as 600 mg/kg (intra-peritoneal) for mice. Values of 4500 mm^3/kg for the LD_{50} (acute oral) of the

40 per cent formulation against rats, and of 40 ppm for the LC_{50} against guppies *(Lebistes reticulatus)* have also been given. It has also been reported to cause severe skin and eye irritation to rabbits, but it does not appear to induce skin sensitivity. Tridemorph is marginally more toxic, the acute oral LD_{50}'s for the 75 per cent (w/v) formulation being 1270 and 750 mm^3/kg for rats and rabbits respectively. Relative dermal activity is somewhat difficult to assess: the stated LD_{50} (acute dermal) for rabbits is >1000 mm^3/kg. Both these 1,4-oxazines are harmless to bees.

The azepines provide yet another systemic fungicide which has a very low mammalian toxicity. 4-t-Butyl-7-decylamino-3,4,5,6-tetrahydro-2H-azepine hydrochloride has an acute oral LD_{50} of 1100 mg/kg and one-month feeding studies have shown that a daily intake of 150 mg/kg has no effect on rats nor has 15 mg/kg on dogs. This compound is also non-toxic to honey bees in concentrations up to 0·5 mg.

The experimental soil fungicide chloroneb possesses very low mammalian toxicity, the acute oral LD_{50} for rats being 11000 mg/kg. The dermal lethal LD_{50} for rabbits is > 5000 mg/kg and a 50 per cent aqueous suspension of the 65 per cent wettable powder caused no irritation to guinea pigs and repeated applications failed to result in skin sensitisation.

From the foregoing brief survey it is clear that systemic fungicides rank among the safest pesticides in current use. Although their metabolites, both enzymic and non-enzymic, and possible conjugates have not been mentioned here, there is no doubt that their toxicology is of prime importance and will undoubtedly continue to be given exhaustive scrutiny by pesticide manufacturers.

5

Translocation

by S. H. Crowdy

There are a number of general descriptions of translocation in plants (Richardson, 1968; Crafts, 1961) and also more recent reviews of more specialised aspects (Milthorpe and Moorby, 1969; Esrich, 1970). The movement of chemicals in plants can conveniently be considered in three stages:

(*a*) Entry into the free space within the tissues.

(*b*) Movement in the apoplast. This occurs in the non-living parts of the cell, outside the protoplasm, and within the lumen of the dead vessels and tracheids of the xylem, where long distance transport takes place. Apoplastic movement is 'passive' and does not require the expenditure of metabolic energy.

(*c*) Movement in the symplast, which is within the living parts of the cell. This is 'active' and requires the expenditure of metabolic energy. Long distance transport in the symplast system takes place in the phloem, probably in specialised cells, the sieve tubes.

The 'plasmalemma', the membrane bounding the protoplast, separates the symplast from the apoplast. The tissues involved in translocation are described in general textbooks of Botany; in addition there are more detailed monographs on plant anatomy such as Esau (1965).

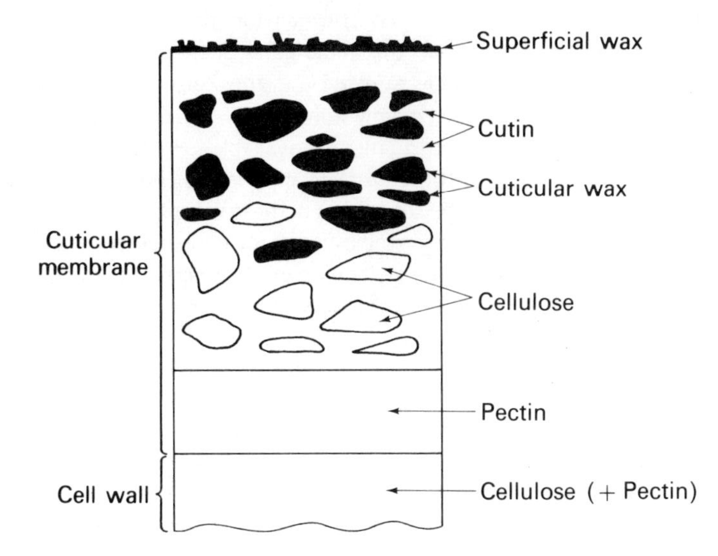

Fig. 5.1. Diagrammatic representation of the plant cuticle. Holloway (1971)

ENTRY INTO THE FREE SPACE WITHIN THE TISSUES

Entry

The initial barrier to the penetration of materials into the leaves and aerial parts of the plant is the cuticle and its associated structures. These have been described in a number of reviews (Martin and Juniper, 1970; Hallam and Juniper, 1971; Holloway, 1971; Baker, 1971). The diagram in Fig. 5.1 gives a general idea of the structure of the cuticle; the whole structure is dead and is built on the outer walls of the epidermal cells, which are themselves in contact with the living protoplasm. The layers grade into each other; the cellulose of the cell wall gradually grades into the pectin layer, which may be continuous with the pectin of the middle lamellae of the anticlinal walls of the epidermal cells. Above the pectin layer is a matrix of cutin, a polymer consisting of fatty and hydroxy-fatty acids. At the base of the cutin matrix there seems to be a network of cellulose fibres which fade out in the upper layers. Nearer the surface the cutin contains platelets of wax, which also disappear in the topmost layers of the cuticular membrane. Above the cuticular membrane is commonly a layer of cuticular wax; this may be strikingly sculptured, exhibiting patterns which are frequently characteristic of different plant species or of different parts of the same species. The main function of the cuticle is to control the loss of water from the aerial parts of the plant; the cutin matrix seems to be permeable to water and the water-proofing is provided by the wax component, mainly the wax platelets embedded in the upper layers of the cutin matrix. The area available for water loss within the cuticle and, naturally, the area available for the entry of aqueous solutions is determined by the separation of these wax platelets, which in its turn depends on the degree of imbibition of the cutin matrix. When the cutin matrix is fully imbibed the separation will be maximal; if the loss of water from the surface exceeds the supply from the roots, the cutin matrix will lose water and contract, bringing the wax platelets closer together and reducing the area available for the passage of water (van Overbeek, 1956). The extent to which this type of regulating system controls water flow has been estimated by Moreshet (1970) who showed that the resistance of sunflower leaves to water loss was inversely proportional to the relative humidity of the surrounding air, rising from an average value of 9·6 mm at a relative humidity of 70 per cent to about 20 mm at relative humidity of 30 per cent. Moreshet also varied the water tension within the plant by varying the water tension in the soil: with a high relative humidity (70 per cent) and a low water tension in the soil (0·2 atm), the leaf resistance was only 1·6 mm while when the relative humidity was low (30 per cent) and the soil water tension was high (9·9 atm), the resistance rose to 18·7 mm. These data illustrate the very considerable control which the plant exercises over the loss of water from the leaves. Moreshet also calculated the resistance offered by the cuticle in the varied conditions and suggests that cuticular resistance alone rises from about 60 mm at a relative humidity of 70 per cent to about 300 mm at a relative humidity of about 30 per cent.

The freedom with which water and substances dissolved in it will traverse the cuticle is shown by the number of substances which will enter or can be leached from leaves (Franke, 1967; Tukey, 1970); these include a variety of foreign organic chemicals used as insecticides, herbicides and fungicides and natural products such as sugars, sugar alcohols and numerous acids including amino acids. Water solubility seems to be the only absolute requirement for this trans-cuticular movement. It is doubtful whether the channels by which water and solutes enter and leave the leaves can be distinguished by present techniques; however, structures in the cuticle, referred to as ectodesmata, have been described and the recent work related to them has been reviewed (Franke, 1967). It is suggested that the ectodesmata provide an almost direct connection between the leaf surface and the protoplast of the epidermal cells. In section they appear to be outgrowths of the epidermal cell wall reaching into, but never through, the cuticle. In surface view they can be seen as dots which are clustered over the anticlinal walls of the epidermal cells particularly around the stomata and the stomatal accessory cells. These structures are more obvious when solutes are diffusing into the leaf and their distribution appears to be associated particularly with the areas of entry. It can also be shown with tritiated water that the ectodesmata are commonly associated with water loss. Ectodesmata are only revealed when the material is treated with Gilson fixative, which contains mercurous chloride and it has been suggested that the structures have no real existence and are artifacts resulting from the method of treatment, which simply demonstrate the penetration patterns of mercurous chloride (Schönherr and Bukovac, 1970a, b). The balance of evidence seems to support this view. Crowdy and Tanton (1970) and Tanton and Crowdy (1972b) have reported very similar patterns of lead accumulation in leaves treated with the chelate of lead and ethylenediaminetetra-acetic acid, but these have never been associated with cell wall structures which could be mistaken for ectodesmata. However, it should be noted that the technique used would not have revealed structures associated with direct entry from the leaf surface into the protoplasts of the epidermal cells.

Although solutes will enter and leave through the leaves in appropriate conditions, the common pathway by which nutrients and other solutes enter the plants is through the roots. The absorbing surfaces of roots are also covered with a protective layer analogous to the cuticle, but which may consist of precursors of suberin rather than cutin (Martin and Juniper, 1970). The surfaces of older roots are suberised and impervious to water. The function of young roots is to absorb water and solutes and the covering to the absorbing zones offers little resistance to water flow to the xylem (Anderson and Reilley, 1968). The main absorbing regions are relatively restricted and may extend only a few centimetres from the root tip. However, the branching of the root system ensures a multiplicity of absorbing points. In the absorbing zone the ratio of surface to volume may be greatly extended by the presence of root hairs, which grow out of the epidermal cells.

The older parts of plants, particularly perennials, are normally covered with bark. This is a heavily suberised multicellular layer, which provides an impervious covering except where the surface is damaged or is penetrated by specialised structures, the lenticels. Aqueous solutions can enter uninjured bark to a small extent and the penetration can be enhanced considerably by scraping or by maintaining a supply of solute on the surface in a gauze pad or by combining both these treatments (Ticknor and Tukey, 1957). There is some evidence that insecticides are systemically distributed after application to the bark (Mitchell *et al.*, 1960).

However, it seems unlikely that the bark will in most cases provide satisfactory entry for adequate quantities of systemic fungicides, unlike the roots and the leaves.

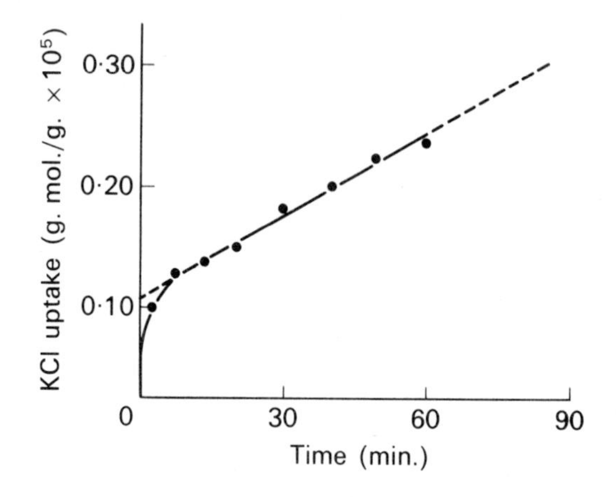

Fig. 5.2. Relation between KCl uptake and time with extrapolation to give free space uptake. Briggs and Robertson (1957)

Free space

Water and solutes entering either through the roots or the aerial parts of the plant first pass into free space. This can be defined as that part of the plant tissue which communicates freely, by diffusion, with the external environment. The term free space was introduced by Hope and Stevens (1952); the space has also been referred to as the outer space (Epstein, 1955) and the diffusion free space (Johnson and Bonner, 1956). The literature relating to this volume has been reviewed by Kramer (1957) and Briggs and Robertson (1957). Entry into this volume is by diffusion which, in the case of the roots of a transpiring plant, may be assisted by mass flow. It is difficult, if not impossible, to define precisely the free space as an anatomical entity, which precludes direct measurements of the volume. Instead, a number of indirect methods of estimation have been used. One commonly employed is based on ion uptake experiments and is illustrated in Fig. 5.2, reproduced from Briggs and Robertson (1957). These estimates assume a two-phase uptake; a rapid diffusion into the free space, which is complete in a few minutes, followed by a slow, prolonged accumulation, the amount accumulated being directly proportional to the time of exposure. This accumulation probably reflects the active transfer of the ions from the free space to the cell vacuole. The estimate of the free space volume was obtained by plotting uptake against time and extrapolating the line representing the prolonged steady uptake back to zero time; this provided an estimate U of the amount of salt not already present and diffusing rapidly into the free space. If the concentration of the external solution is C, the estimate of the free space is U/C. The free space estimated in this way includes more than the actual volume of water available for diffusion and has been termed the Apparent Free Space (Briggs and Robertson, 1957). The apparent free space thus includes, in addition to the free space, any systems associated with the free space which interact with the solute such as a Donnan system, which has been considered by Briggs and Robertson (1957) or partition into a lipoid phase (Crowdy *et al.*, 1958). The volume of the apparent free space is markedly influenced by the material under study and the conditions in which the experiment was carried out. Some estimates of apparent free space are included in Table 5.1.

TABLE 5.1 **Estimates of the volume of the apparent free space in various plant tissues (% of tissue volume)**

Authority	*Tissue*	*Apparent free space, %*
Hope and Stevens (1952)	Broad bean roots	13
Butler (1953)	Wheat roots	24·5–33
Epstein (1955)	Barley roots	23
Johnson and Bonner (1956)	Oat coleoptiles	19–22
Kylin and Hylmö (1957)	Wheat roots	27·5
Crowdy *et al.*, (1958)	Wheat leaves	33–65†
Crowdy and Tanton (1970)	Wheat leaves	3–5
Tanton and Crowdy (1972a)	Carrot root discs	4
	Wheat roots	5

† Referred to as Effective Volume

The estimates of free space in roots incorporate the uncertainties associated with the concept of apparent free space and, in addition, experimental errors associated with the handling of material. These include failure to make due allowance for the surface film of solute (Levitt, 1957) and adsorption into dead material in the older parts of the roots. A number of these sources of error have been examined systematically by Tanton and Crowdy (1972a) whose conclusions are summarised in Table 5.2. In this table the uncorrected estimate of apparent free space corresponds to many of the estimates commonly published and agrees reasonably well with the figures published in Table 5.1. A further correction can be applied to the final value in Table 5.2 since the tissue concerned was mainly the cortex and occupied only 75 per cent of the root volume. If allowance is made for this, the apparent free space in the cortex occupies about 5 per cent of the tissue.

TABLE 5.2 **Sources of error in estimates of the apparent free space in roots (% of tissue volume)**

Component	*Estimate as % of issue volume*
Apparent free space uncorrected	18·7
Error due to surface film of solute	7·4
Error due to dead cortical cells	7·5
Apparent free space, corrected	3·8

The concept of apparent free space is not readily interpreted in terms of plant anatomy and it is probably more useful to consider uptake in three stages; the first being diffusion into free space water, the second removal from the aqueous phase of the free space by physical processes and the third metabolically active removal from the first two phases. Phases 1 and 2 comprise the apparent free space. The anatomical entity is the aqueous phase of stage 1 which comprises the free space and seems to be the water of hydration of the cell wall, especially the pectins in the middle lamella. The movement of foreign materials in this volume has been demonstrated with sols of noble metals (Strugger, 1949; Strugger and Peveling, 1961) and with the chelate of lead with ethylenediaminetetra-acetic acid (Crowdy and Tanton, 1970; Tanton and Crowdy, 1972a, b) all of which give deposits which can be examined in the electron microscope. Since anatomical studies show that the lead chelate is confined to the middle lamella, the estimate of apparent free space of 3–5 per cent of the tissue probably approaches the volume of the free space closely. The sols of noble metals are of special interest since their particles seem to be near the maximum size which will move in the free space; they will move from the shoot to the leaves of *Helxine soleirolii*, but not in *Beta vulgaris* or *Eucalyptus globulus*. Even in *H. soleirolii* the movement of particles with diameters much larger than 5 nm is restricted (Gaff, Chambers and Markus, 1964). The second stage of uptake involves a physical removal of material from the aqueous phase in the free space, possibly by immobile ions or lipoid phases in the cell walls or the protoplasm. Material held in this way will be in equilibrium with the aqueous phase of the free space and, in suitable conditions, can return to it. The prolonged uptake into the cell vacuole is under metabolic control and is in most cases irreversible so that material taken up in the third stage may not be available for translocation.

The distribution of chemical in the free space of the leaves appears to be much the same whether entry has been via the cuticle, or from the roots in the transpiration stream (Crowdy and Tanton, 1970). The leaching of material from plant leaves has already been noted; this loss is adequately described by a diffusion model (Crowdy and Tanton, 1970) and introduces interesting possibilities for re-cycling and re-distributing chemicals applied to the roots.

Since the initial phases of entry are by diffusion, Fick's law will define some of the conditions which will affect entry. This applies particularly to leaves, which will be considered first. The rate of entry is directly proportional to a permeability constant, the area available for diffusion and the concentration gradient; within the limits set by these, the amount of a substance entering a leaf will increase with time until the external and internal concentrations are the same. Since we are considering diffusion from water on the leaf surface into water in the free space, diffusion will be limited to the period while there is liquid spray on the leaves. The permeability of the cuticle does, in fact, vary with external conditions, but this is a factor which is not readily controlled except, possibly, by selecting the time for spraying; the area available for diffusion will be influenced by the extent to which the leaf is wetted and can be maximised by the inclusion of appropriate surface-active agents. The concentration gradient can also be regulated to a large extent by formulation and choice of chemical, since it depends partly on the aqueous solubility of the chemical used and also on the extent to which the solute is removed from the free space in phase 2 of uptake. If removal is rapid and the potential reservoir large, relatively insoluble compounds may enter to quite a large extent. However, retention in phase 2 may impede translocation. The time of exposure can also be increased by formulation, possibly by adding humectants to delay drying. Humectants do not seem to have been used commercially, but glycerol and polyethyleneglycols are reported to have increased the uptake of streptomycin (Gray, 1956; Goodman and Dowler, 1958), various growth regulators (Rice, 1948; Holly, 1956) and possibly zineb (Rich, 1956). Humectants will only be effective when drying is a limiting factor and even then only if the humectant does not adversely affect the phase distribution of the applied chemical.

The same general considerations apply to root uptake, but normally their operation is obscured by the mass flow of the transpiration stream. Most solutions entering the roots are dilute and are not concentrated by evaporation; the transpiration stream prevents accumulation of solute in the free space though material may be removed from the transpiration stream either passively or actively and accumulated in the roots: this has been demonstrated for a number of antibiotics and sulphonamides and examples have been reviewed (Crowdy, 1959). This accumulation can occur in dead root tissue as well as living (Tanton and Crowdy, 1972a) and the material so accumulated may be released when water is substituted for the treating solution, suggesting that in appropriate circumstances the roots could provide storage for a useful chemical (Crowdy and Tanton, 1970).

APOPLASTIC MOVEMENT

The plant cell walls provide a system of water pathways outside the living protoplasm which communicates with the environment by free diffusion. Within this sytem the bulk of the movement is a mass flow from the roots to the leaves in the transpiration stream, via the xylem vessels. This is a passive movement and depends for its motive force on the difference in water potential between the root environment and the leaf environment. Since no living processes are involved, this system will transport poisons (Kurtzmann, 1966), though naturally these can be fatal to the living tissue adjoining the free space of the roots and the leaves. The actual course of the apoplastic pathway starts in the cell walls of the root hairs and is confined to the cell walls and flooded intercellular spaces in the root cortex. The cell wall pathway is blocked at the endodermal cells by an impervious suberised band, the Casparian strip, in the radial walls, which effectively prevents further movement. At this point the solutions moving in the cell walls are diverted into the endodermal cell itself and must pass through the protoplast, which is firmly attached to the Casparian strip. After traversing the protoplast, the solution returns to the cell walls in the vascular cylinder and passes into the xylem vessels through unlignified parts of their walls. The vessels also communicate with each other through the bordered pits (Tanton and Crowdy, 1972a). The protoplast of the endodermis seems to be more permeable to water than to the lead chelate in the solution which was used in these investigations; it was estimated that only about 10 per cent of the lead chelate in the solution presented to the roots crossed the endodermis. There seems little doubt that the protoplast of the endodermis constitutes a major barrier to the movement of chemicals in the transpiration stream and would reward a more detailed study.

Some information on the extent to which the movement of chemicals is impeded in the roots can be derived from experiments in which accumulation in the shoots is related to water uptake. There is usually a linear relationship between these two and the slope of the linear regression is the average concentration of the transpiration stream. The movement to the shoots can be quantified by expressing the concentration of the transpiration stream as a percentage of the concentration of the solution supplied to the plants. Some data are presented in this form in Table 5.3.

TABLE 5.3 **Concentration of chemicals in the transpiration stream in shoots as a percentage of the concentration supplied**

Plant	*Chemical*	*Transpiration conc. as % of treating sol*	*Ref.*
Vicia faba	4,4′-Diaminodiphenyl sulphone	38·3	1
	Sulphanilamide	22·7	1
	Sulphadiazine	14·8	1
	Sulphacetamide	10·0	1
	Sulphathiazole	1·6	1
	Griseofulvin	9·1	2
Triticum aestivum	Lead-EDTA chelate	1·2	3
		1·5	3

Refs. 1 Crowdy and Rudd Jones (1956); 2 Crowdy *et al.* (1956); 3 Crowdy and Tanton (1970)

In addition to the filtering effect of the endodermis, there is evidence that retention in the root system is also correlated with an ability to partition into non-polar solvents (Crowdy, Grove and McClosky, 1959).

Once within the xylem vessels there seems to be little check to the flow of neutral and acidic substances. In contrast basic substances are adsorbed to negative charges on the xylem walls. This effect has been noted in basic dyes (Charles, 1953), basic antibiotics (Crowdy and Pramer, 1955), quaternary ammonium compounds (Edgington and Dimond, 1964; Salerno and Edgington, 1963) and a variety of amino acids (Hill-Cottingham and Lloyd-Jones, 1968). Some of this adsorption is passive and can be reversed by supplying competing cations such as calcium, but there are indications of an active adsorption also (Hill-Cottingham and Lloyd-Jones, 1968).

The xylem vessels carry the transpiration stream to the leaves; within cereal leaves the water in the xylem is retained in the main vascular bundles by the mestome sheath (O'Brien and Carr, 1970), and released through the walls of the secondary vascular bundles into the free space of the leaves, which again is located in the cell walls (Crowdy and Tanton, 1970). Within the cells walls the transpiration stream moves to the cuticle and solutes are deposited at the air–liquid interface, where evaporation occurs. The pattern of this deposition is described by Crowdy and Tanton (1970) and Tanton and Crowdy (1972b). If root treatment is continued for a long time, or if treatment is followed by treatment with water, solutes are washed to the margins of net-veined leaves or to the tips of leaves with parallel veins, where even safe chemicals may attain high enough concentrations to cause a scorch. This pattern of accumulation has been illustrated for benomyl and it can be simulated in filter paper cut to leaf-shaped patterns (Peterson and Edgington, 1970; 1971). The extent to which a fungicide distributed in the apoplast will accumulate in a plant organ will be related to the loss of transpiration water from that organ, which, in turn, is usually related to the frequency of stomata. Peterson and Edgington (1971) have demonstrated this with benomyl, which does not accumulate, or only accumulates to a very small extent, in organs such as tomato fruit, geranium petals or poinsettia bracts, in which stomata are absent or non-functional. This pattern of movement implies that continuous root treatment with a systemic fungicide is needed if a leaf is to remain well protected unless the fungicide is bound in the leaf tissue. It should also be noted that fungicides which enter following leaf application will also travel to the tips and edges in this way.

SYMPLASTIC MOVEMENT

Chemicals which have reached the leaves in the apoplast cannot be redistributed in the plant in the same system except in conditions which occur rarely outside experiments. Movement from the leaves into the main body of the plant occurs in the symplast and requires metabolic energy both to transfer material from the free space into the living parts of the cell and to move material along the symplastic pathways. Movement from the apoplast to the symplast seems in many cases to be associated with specialised transfer cells (Gunning and Pate, 1969). These are located where trans-membrane fluxes might be restricted by adverse surface-to-volume ratios. In the transfer cells the ratios are improved by ingrowths of the cell wall which cause a considerable increase of the surface of the protoplasm which lines it and the interface between the symplast and the apoplast. The living protoplasm forms effectively a continuous network throughout the tissues of the plant, hence the use of the term symplast; protoplasmic connections between adjacent cells are provided through specialised pores, the plasmodesmata. The fine structure of these pores in willow has been described by Robards (1968). The symplastic pathway is continuous with the phloem, the vascular tissue in which long-distance transport occurs. The actual cells involved are the sieve tubes, which are long trains of cells communicating with each other through perforated plates which look like and are called sieve plates. Phloem structure and function have been discussed recently by Crafts and Crisp (1971), who have provided an excellent bibliography of literature relating to this tissue. The sieve tubes look initially like normal parenchyma cells but divide early to form a characteristic pair of cells, the sieve tube and the companion cell, which develop together. As the sieve tube develops many of the structures, including the nucleus, degenerate and may disappear, but despite these deficiencies, the sieve tube remains a living cell as long as it is functional. The pores in the sieve plates are derived by enlargement of the pores originally occupied by plasmodesmata (Wark and Chambers, 1965). In contrast to the sieve tubes, the companion cells have normal intracellular structure and dense protoplasm: they are the only cells connected to the sieve tubes with plasmodesmata and these connections are well developed. This close association suggests that the companion cells may have taken over many of the functions of the degenerate protoplast of the sieve tubes.

The sieve tubes are delicate structures which are easily damaged by manipulation and their fine structure is still a matter for debate, since it is always difficult to decide whether the structures seen in the optical or electron microscopes reflect faithfully the conditions *in vivo* or are artifacts. Competing theories of phloem movement require different organisation of the sieve tube trains with the result that proponents of different models find it difficult to agree as to the true status of the artifacts. It seems likely that the full interpretation of the structure of the sieve elements must wait on a better understanding of the mechanisms of phloem movement.

The main function of the phloem is to carry food materials, mainly carbohydrates, from organs in which they are plentiful, 'sources', to organs which require nutrients, 'sinks'. This has been demonstrated by ringing experiments, in which sections of the phloem were cut out or killed by steam and the effect on the distribution of nutrients noted. In its simplest form this implies conveying carbohydrates from the leaves in which they are photosynthesised, to the growing zones in which they are used for tissue building and to storage organs. This is an oversimplification since growing leaves may be net importers of nutrient and may even import and export simultaneously. Equally, carbohydrates may be stored temporarily in various tissues along the route, such as the parenchyma of the phloem. These may act as sinks or as sources depending on the conditions in the tissue and act as a buffer protecting the tissue from extreme fluctuations. In addition to long-distance transport, the phloem must also provide nutrient to the living tissues of the mature plant body, which also implies a capacity for free exchange of nutrients with the surrounding tissues. The concept of sources and sink is useful for describing movement in the phloem provided that the actual tissues involved are defined when necessary.

Material in the symplast is actively accumulated in the sieve tubes against a concentration gradient (Mason and Phillis, 1933) and there are some indications that light is necessary to this loading process. Movement within the sieve tubes is also dependent on metabolic energy; this has been demonstrated by treating zones of the phloem with respiratory inhibitors or by denying them oxygen. General observations on movement have been supported by micro-autoradiography, which makes it possible to locate suitable isotopes in specific cells. The concentration of solutes in the sieve tubes is usually high, much higher than the concentration in the surrounding tissue, so that the contents of the sieve tubes are under pressure and will exude if they are punctured. Phloem exudation can be collected from the cut ends of stems, which provides a general sample of phloem contents, or from aphid stylets, which are commonly located in a single sieve tube and provide a sample from a restricted zone of phloem, possibly from a single train of cells. These methods have provided various data on the contents of sieve tubes and their behaviour. In general the content and concentration of the phloem sap varies with the type of plant used and the environmental conditions, high concentrations being particularly dependent on active photosynthesis. Crafts and Crisp (1971) have summarised data on the composition of phloem exudate. The exudate contains 10–25 per cent dry matter, of which about 90 per cent is sugar. In some species this is entirely sucrose, but other sugars and sugar alcohols may occur. About 2 per cent of the dry matter may consist of mineral nutrients such as potassium, which is present in relatively large amounts, magnesium, calcium and phosphorus. Amino acids may contribute about 0·5 per cent of the dry matter (Tammes and Die, 1966). Hormones and a variety of foreign chemicals including herbicides and insecticides, may also be transported in the phloem. Crafts and Crisp (1971) list 122 chemicals which have been recorded in the phloem of various plants. Data derived from a study of exudates must be treated with some reservations, since the phloem is concerned with distribution as well as transport and this requires a capacity to exchange materials with neighbouring cells. The leakiness of the system, which this implies, confuses the interpretation of translocation patterns.

The actual distribution of the assimilated carbohydrate has commonly been studied by feeding ^{14}C-labelled carbon dioxide or sucrose to leaves and following its distribution by counting or autoradiography of the whole plant, or of tissues, if the aim is location at the cell level. Other phloem-mobile tracers such as ^{35}P, tritium or fluorescein have been used less frequently, but are valuable in special experiments such as those demonstrating bidirectional movement. This work has been summarised by Wardlaw (1968) and Crafts and Crisp (1971). In general young, growing leaves are net importers of carbohydrate and newly matured leaves are net exporters; the changeover takes place when the leaf is about half its final area. Leaves tend to export to the sinks which are nearest to them, subject always to limitations imposed by the anatomy of the vascular system and the rather restricted movement of the photosynthate around the stem. Thus the upper leaves tend to supply the growing point and the associated young leaves; the lower leaves the roots, while those which are intermediate may export in either direction. This pattern is not absolute and can be varied experimentally by pruning or removing leaves. In some cases photosynthate may be translocated downwards to the roots and then redistributed in the xylem and apoplast to the top of the plant; this can complicate the general picture of distribution. Developing fruits may provide very strong sinks and grapes may accumulate photosynthate from leaves up to 4 m from the cluster (Meynhardt and Malan, 1963). In wheat and rye the source of supply is much more restricted, 80 per cent of the material for grain growth being derived from the flag leaf, stem and ear; in rice the source is rather less restricted and lower leaves may be involved.

Work on patterns of translocation, speeds of movement and possible mechanisms has been reviewed recently by Canny (1971) and Crafts and Crisp (1971). The movement of materials is rapid, probably in the range 0·39–1·09 m h^{-1}, which is too fast for simple diffusion, and can occur simultaneously in both directions. Two general types of mechanism have been suggested to account for this movement, those based on mass flow, which operate best with empty tubes and unimpeded pores, and those which require specialised sieve tube structure or properties. The mass flow hypotheses are historically the oldest and were first clearly formulated by Münch (1927). Basically these envisage a high concentration of soluble carbohydrate in the sources, such as the leaves, and low concentrations in the sinks, where active metabolism occurs, or in storage organs, where soluble carbohydrate is removed from the system either by metabolism or by conversion into insoluble forms. This concentration gradient tends to produce a mass flow of solution from sources to sinks, which would incidentally carry with it other solutes which were present in the phloem. There would only be flow in one direction in a single train of phloem cells, but flow in adjacent trains could be in different directions.

There is some evidence for greater turgor pressures at the top of the phloem than at the bottom and it is claimed that the pressure differences are sufficient to maintain the required flow in an unimpeded system. In its simplest form, this model required high pressure differences to account for the speeds and the distances involved and to overcome the resistance of the sieve plates. This model did not call for living processes to maintain the movement between the sources and the sinks although evidence from metabolic inhibitor experiments indicated that active processes were involved in movement throughout the length of the phloem system. Some of these objections have been met by proposing electro-kinetic or electro-osmotic systems, maintained by metabolic energy, to boost the mass flow in the sieve tubes. These essentially involve 'sap-carrying' ions circulating either over an appreciable length of phloem (Fensom, 1957, 1959) or circulating around the sieve plates via either the companion cells (Spanner, 1958) or via adjacent sieve tubes (Spanner, 1970). Potassium, which is in fairly high concentrations in the sieve tubes has been suggested as the ion involved.

In spite of these modifications, there are a number of observations which throw doubt on the validity of the various mass flow hypotheses. There is evidence from aphid stylet experiments, which indicate bi-directional flow in a single train of sieve tubes in *Vicia faba*. Eschrich (1967) collected honeydew from an aphid feeding on the stem above a leaf treated with fluorescein and below one treated with ^{14}C. Both tracers appeared in the honeydew, presumably derived from a single train of sieve tubes. This experiment did not exclude the possibility of some mixing of materials travelling in adjacent paths and supports rather than establishes bidirectional movement in a single sieve tube. There is also some evidence that the water in sieve tubes is relatively immobile (Gage and Aronoff, 1960; Peel, 1970). In addition Tyree and Fensom (1970) and Canny (1971) find the mass flow models supported by electro-osmotic systems theoretically unsatisfactory. Canny (1971) considers the kinetics of a number of alternative model systems and suggests that the most satisfactory is one indicated by Mason and Maskell (1928a, b) in their study of carbohydrate movement in cotton. This suggested that the direction of the transfer of dry material was dictated by the gradient of sucrose concentration in the phloem and that the rate of transfer was directly proportional to this gradient over a wide range of values. This leads to a model which is formally identical to the steady state form of Fick's law of diffusion but with an apparent diffusion coefficient 3×10^4 times greater than the coefficient for the diffusion of sucrose in water. They referred to this process as accelerated diffusion and models of this type are referred to as diffusion analogue models. Diffusion analogue models require specialised structures within the sieve tube to convey the solute and exchange between these structures and a stationary water phase. Suitable structures would be the trans-cellular strands reported by Thaine (1962) in the phloem of *Cucurbita* and *Primula obconica*, which he suggested later were tubules which might carry sucrose solutions by peristaltic contractions (Thaine, 1969). The subsequent lack of agreement as to whether these strands are genuine structures or artifacts as suggested by Esau and her associates (1963) illustrates the difficulties inherent in studying the fine structure of phloem. Jarvis and Thaine (1971) have recently reported strands in quick-frozen material of *C. pepo* confirming the original observations. Milthorpe and Moorby (1969) have reviewed some of the evidence relating to these strands and the studies by Fensom and his associates (1968) on the living phloem of *Heracleum mantegazzianum* indicating a network of micro fibrils, which seem to be in active motion and which carry particles which stain very rapidly when the vital stain, Janus Green, is applied to a cut strand, should also be considered in this connection. Aikman and Anderson (1971) published a theoretical analysis of Thaine's model and concluded that it could function.

The source–sink relationships feature in all the models of phloem transport and there is evidence that the sinks are to some extent under hormonal control. The apical bud, which is a typical sink, normally maintains adequate concentrations of indole-3-acetic acid; if the apical bud is removed, the cut end of the stem will only continue to act as a sink when exogenous indole-3-acetic acid is supplied (Booth *et al.*, 1962). In some plants indole-3-acetic acid can be replaced by foreign growth regulators such as naphthylacetic acid 2,4-D or 2,4,5-T (Bowen and Wareing, 1971). Gibberellic acid and cytokinins have no influence on the direction of transport in the absence of indole-3-acetic acid, but when this hormone is supplied, both enhance its effect (Seth and Wareing, 1967; Morris and Thomas, 1968). It is not possible to say how far this effect is directly attributable to the hormone or is indirect and follows hormonal stimulation of growth.

Symplastic and apoplastic transport are physiologically and anatomically distinct within the plant. In most cases movement in these two pathways can be distinguished readily in experiments, since symplastic movement is away from sources towards sinks and can be prevented by treatments which inhibit metabolic activity at the source, or immobilise the phloem, while apoplastic movement takes place passively on the water stream and is directed to zones where water is lost by evaporation, or where it is required to maintain growth or extension (Potter and Milburn, 1970). However, there are circumstances in which solutes will move from a leaf into the body of the plant in the apoplast; then their distribution will have a superficial similarity to symplastic movement; these must be recognised when designing and interpreting experiments. Centripetal movement in the apoplast occurs when the apoplastic pathway is exposed to atmospheric pressure by cutting, abrasion or chemical damage; the tension is released and the solution in the apoplast system sucks back, drawing with it air, or solution, applied to the break. This effect has been used to assist the deep penetration of herbicides into *Convolvulus arvensis* (Crafts, 1933) and in certain conditions may lead to the movement of desiccants into the tubers of potatoes, causing a stem end necrosis (Headford and Douglas, 1967). It is also possible to reverse apoplastic movement by immersing a leaf in water or a solution. This effectively stops evaporation at the immersed surface, which can then supply solution to the main transpiration stream.

Although symplastic movement is mainly concerned with distributing physiologically important chemicals, it will also transport those which are foreign to the plant. The range of chemicals transported is more restricted than in apoplastic movement, since there is a need for active transfer from the apoplast to the symplast; this may involve the ability to use specialised carrier mechanisms. In spite of this restriction, symplastic movement has been demonstrated for a variety of chemicals; these have been listed by Crafts and Crisp (1971). Research has tended to concentrate on chemicals important in agriculture such as herbicides, insecticides and nutrient salts. Movement from the leaves is active, at least at some stage, and does not take place in leaves which have been starved during a period in the dark. This has been demonstrated in the case of 2,4-D, by Hay and Thimann (1956), who also found that exogenous sucrose could be substituted for light while mannitol, arabinose and urea could not. Further, added sucrose had no effect on transport in the light. Ring barking inhibited 2,4-D movement, indicating a symplastic pathway. The sugar is probably needed to provide energy for the transport system; it may also be needed to establish nutrient gradients. Hay and Thimann also showed that excessive 2,4-D inhibited translocation, suggesting that too high a concentration may poison the metabolic processes involved. When foreign chemicals are moving in the symplast, their pattern of distribution indicates that they are moving with the nutrient supply though there are considerable differences in the distribution and in the extent of movement. In general 2,4-D tends to be absorbed during its passage through the plant, while other chemicals such as dalapon (2,2-dichloropropionic acid), amitrole (3-amino-s-triazole) and maleic hydrazide move more freely. Dalapon and maleic hydrazide transfer from the phloem to the xylem, which allows circulation in the symplast–apoplast systems, and are very generally distributed within the plant (Crafts and Crisp, 1971). Foy (1961) even claimed that the effect of dalapon treatment could be carried into a second generation through the seed of wheat. Chemicals, such as dalapon, which are carried in the symplast to the roots may also be released from the roots into the medium surrounding them. This also occurs to a limited extent with 2,4-D, but the effect is more striking with α-methoxyphenylacetic acid (Preston *et al.*, 1954) and 2,3,6-trichloro- and 2,3,4,6-tetrachlorobenzoic acids (Linder *et al.*, 1958), who showed that enough growth regulator was released into the root medium to produce malformations in adjacent plants. Six other assorted di-, tri-, tetra- and pentachlorobenzoic acids and some phenoxy and naphthyloxy acids did not migrate in this way.

The pattern of movement of these foreign chemicals is also markedly influenced by the growth stage and vigour of the plant. Yamaguchi and Crafts (1957) noted that the characteristic retention of 2,4-D in the translocation pathway in *Zebrina pendula* as compared with amitrole and maleic hydrazide was marked in slow-growing plants, but was virtually obscured if growth was rapid. Van der Zweep (1961) was also able to demonstrate an effect of age on the direction in which 2,4-D was translocated in barley, *Hordeum vulgare*, plants. At the two- and three-leaf stages there was good movement to the roots while at the four- and five-leaf stages movement to the roots was much less marked.

TRANSLOCATION OF SYSTEMIC FUNGICIDES

The evidence which has accumulated relating to the translocation of systemic fungicides is much more sparse than that related to the movement of herbicides and insecticides and, in most cases, the reader must infer the pathways involved from data derived from disease control trials. Frequently also disease control or a non-specific bioassay, are the only evidence that the chemical is translocated; this inevitably raises doubts as to the identity of the actual chemical moving. The translocated chemical has only seldom been extracted from the treated plant and characterised, though this has been done for the antibiotics griseofulvin and chloramphenicol (Crowdy *et al.*, 1955). The pure antibiotics, recovered from *Vicia faba* after root treatment, were identified by mixed melting point determinations and comparison of their infra-red spectra. More commonly the identity of the translocated compound is based on the similarity of its behaviour on chromatograms to standard pure chemicals. In some cases, such as benomyl and thiophanate, disease control is claimed for the compound applied when it seems likely that the compound actually translocated is a fungicidal metabolite. In most cases the data presented below are supported at least by chromatographic evidence as to the actual chemical translocated or, in the case of some of the antibiotics by bioassays with bacteria specifically requiring the antibiotic for growth. This area has recently been surveyed by Wain and Carter (1967).

Most of the older chemicals, which might be described as near, or not so near, misses showed typical apoplastic movement. These included a variety of antibiotics, of which the most important for the present discussion were probably griseofulvin and streptomycin, whose behaviour has been surveyed briefly by Crowdy and Pramer (1955). Apoplastic movement has also been shown for cycloheximide (Wallen and Millar, 1957) and for certain of its derivatives (Wallen, 1958). A very varied selection of synthetic chemicals have also shown an apoplastic pattern of movement, though in most cases the actual chemical translocated has not been identified. For various reasons only a few of these have even emerged from greenhouse tests before being abandoned. Fungicidal chemicals have only rarely given evidence of symplastic movement, though it is difficult to judge at this stage how far it must be attributed to dilution of not very active chemicals and lack of attention to the patterns of symplastic movement. Crowdy and Wain (1951) provided some evidence of the control of *Botrytis fabae* on *Vicia faba* associated with symplastic movement of 2,4,6-trichlorophenoxyacetic acid. However the effect was too small to be of commercial interest and the translocated chemical was not identified. Gray (1958) found that the antibiotics streptothricin and pleocidin were translocated from the sprayed leaves of *Phaseolus vulgaris* and would protect younger leaves from infection with *Xanthomonas phaseoli*; he found no such movement with streptomycin. In contrast, Napier *et al.* (1956) reported that a prophylactic spray of streptomycin on the primary leaves of *Phaseolus vulgaris* would prevent infection due to *Pseudomonas medicaginis* as high as the 4th trifoliate leaf for periods of 7–11 days.

The first requirement for symplastic movement is that the chemical should cross the plasmalemma and enter the protoplasm; little effort has been devoted to examining this aspect of translocation with potential fungicides. Pramer (1955, 1956) studied the entry of penicillin, chloramphenicol and streptomycin into the cells of the alga *Nitella clavata*. Penicillin did not enter the cells in detectable quantities, chloramphenicol entered slowly, apparently by diffusion, and attained a concentration of less than half the external solution in 24 h. Streptomycin entered rapidly and accumulated to a concentration of seven times that of the external solution in 18·5 h, which indicated that the uptake was active, a conclusion which was supported by further experiments which examined the effects of competitive ions, respiratory inhibitors and glucose.

The first generation of commercial systemic fungicides also appears to be transported in the apoplast; again, in most cases, this must be inferred from the results of trials designed to study the control of diseases. Peterson and Edgington (1970, 1971) have published a classic series of studies with benomyl and its fungicidally active metabolic produce *MBC*, the methyl ester of 2-benzimidazolecarbamic acid (Clemons and Sisler, 1969), which clearly demonstrate that these are mainly transported in the apoplast of *Phaseolus vulgaris*. Benomyl is also translocated in sugar beet, *Beta vulgaris* (Soel, 1970), tomato, *Lycopersicon esculentum*, cucumber, *Cucumis sativus* and watermelon, *Citrullus laratus* (Thanassoulopoulos *et al.*, 1970; Ebben and Last, 1969), seedlings of apple, *Malus pumila* and cherry, *Prunus cerasus* (Gilpatrick, 1969) and cotton *Gossypium* sp. (Erwin, *et al.*, 1968a). Thiophanate methyl also appears to translocate in the apoplast in sugar beet (Soel, 1970); since it also has MBC as a fungicidal metabolite (Selling *et al.*, 1970), it behaviour is likely to be similar to benomyl in many respects. Staron and his associates (1966) demonstrated systemic activity following apoplastic movement of thiabenzadole. This compound is also mobile in cotton (Erwin, Sims and Partride, 1968b), sugar beet (Soel, 1970) and soybean, *Glycine max* (Gray and Sinclair, 1971); in soybean and cotton it appears to translocate unchanged and in the former it tends to remain at the base of the plant (Gray and Sinclair, 1971). The two pyrimidine fungicides, ethirimol and dimethirimol, appear to move unchanged in the symplast to the leaves, where they are subject to degradation. Ethirimol, following root treatment, is generally distributed in cereal leaves, but is retained in the veins of apple leaves: this probably accounts for the failure of the chemical to control mildew, *Podosphaera leucotricha* (Cavell *et al.*, 1971). Triarimol appears to have been used only as a spray, but the eradicant activity of this compound indicates some systemic activity (Gilpatrick and Szkolnik, 1970). The oxathiin derivatives, carboxin and oxycarboxin, have also shown typical apoplastic translocation in cereals (Schmeling and Kulka, 1966) and in other grasses (Hardison, 1967), in cotton (Borum and Sinclair, 1967), tea *Camellia sinensis* (Venkata Ram, 1969) and *Phaseolus vulgaris* (Snel and Edgington, 1970). Comparable evidence has also been presented for the apoplastic movement of tridemorph in barley (Pommer, Otto and Kradel, 1969), triforine in rye, *Secale cereale* (Schicke and Veen, 1969) and three organophosphorus fungicides kitazin and kitazin P in rice, *Oryza sativa* (Yoshinaga, 1969) and HOE 2873 against apple mildew (Hay, 1971). Apoplastic transport seldom seems to be specific to a particular plant and, in general, success in treating one herbaceous plant indicates that success can be expected in treating others; this observation may well cover quantitative differences which the present investigations have failed to reveal. It would be interesting to know if the behaviour of ethirimol in apple is an isolated anomaly, or represents a genuine difference between woody and herbaceous plants.

Most of the experiments noted above provide no evidence that the active chemical moves back from a treated leaf into the rest of the plant. However, it appears that this may occur in some cases, though symplastic movement has not been demonstrated clearly in any of them. This type of movement has been reported for the benzimidazole derivatives thiabendazole and benomyl, or, more probably, its metabolite MBC. Both these fungicides markedly reduced infection on untreated leaves of sugar beet inoculated with *Cercospora* three days after the plant had been sprayed (Soel, 1970). Even more striking results were obtained when sprays of these chemicals, acidified with hydrochloric acid, were applied to cotton leaves. Both sprays were effective against a virulent strain of cotton wilt, *Verticilium albo-atrum*, inoculated into the stem either before or after spraying and fungitoxic materials were detected by bioassay both in leaves above the point of application and in xylem tissue. Material was extracted from the xylem in the stems of plants sprayed three times with these acid formulations and was examined by ultra-violet light spectrophotometry. Thiabendazole was indentified in the extract of the xylem of the plant sprayed with this compound, while the fungicidal metabolic product, MBC, was identified in the xylem of plants sprayed with benomyl. The solutions sprayed were very acid, pH 1·5–1·7 for benomyl and 2·7–3·0 for thiabendazole so one must bear in mind the possibility of damage to the apoplast system and suck-back in the xylem although the plants were not visibly damaged (Buchenauer and Erwin, 1971). A rather similar doubt surrounds the translocation of certain nitrophenols (El-Zayat *et al.*, 1968). When these are applied to the leaves, they will protect leaves which are formed after treatment. No visible damage appeared to be associated with the treatment, but the chemicals are too phytotoxic to be applied to the roots and the possibility of local damage and xylem suck-back must again be considered. The ability to translocate from treated leaves to younger growth has also been shown by 4-*n*-butyl-1,2,4-triazole, which is effective against *Puccinia recondita* on wheat (Meyer *et al.*, 1970); this chemical can also be used for root treatment.

On general principles, one would expect to find systemic fungicides which are transported in the symplast and it is possible that the present lack of information in part reflects the difficulty of testing for this ability. Before symplastic movement is assumed, other possibilities must be rigorously excluded and it is also necessary to determine whether the chemical is actively accumulated in the symplast or enters the cells by diffusion, as was suggested for the entry of chloramphenicol into *Nitella clavata* by Pramer (1955), since this could well influence the extent and pattern of translocation.

CONCLUSION

Most protectant fungicides are general cell poisons and are selective because they remain on the surface of the plant; the fungicidal chemicals are selected and formulated with this in view. When these chemicals penetrate the cuticle, they cause damage since there may be little margin of safety between phytotoxicity and fungitoxicity. A systemic fungicide must co-exist with the cells in the living tissue and this requires an entirely different type of selectivity, which discriminates between the living tissue of the host and of the pathogen. Chemicals showing systemic action must also be chosen on their ability to enter the plant and must be formulated to exploit this ability to the full. These requirements are a specification for an eradicant fungicide and the systemic fungicides now being launched show a remarkable ability to eradicate established infections; with most the recommended uses exploit this feature alone and, with one or two exceptions, the emphasis is on spraying for the control of leaf diseases. Since chemicals which are mobile in the apoplast migrate to the leaf edges and tip, the main leaf surface may only be protected for quite a short time unless a continuous supply of chemical is maintained. When the material is applied to the leaf surface, it may well be necessary to balance entry with retention on the surface and aim at a protectant spray with exceptional eradicant properties. This may prove an unsatisfactory compromise, since the properties required for retention may well militate against entry. Disease eradication implies movement between the host and the pathogen, which will only occur in the absence of an impervious wound barrier isolating the lesion; a better knowledge of this aspect of host–parasite relations would provide a base for planning disease control experiments.

Residue problems with systemic chemicals used in this way may be different to those arising from the use of conventional protectant sprays; they need not necessarily be more intractable. Material on the surface of the leaf would weather off in the usual way and material within the tissue would migrate to the leaf margin and, in due course, leach out of the tissue. A material which has penetrated the cuticle of a fruit might present a more difficult problem than one which is retained on and might weather off the surface. The entry of individual systemic chemicals used as sprays and their leaching from fruit surfaces requires a more detailed study both from the point of view of residues and from possible effects in reducing storage rots.

Treatment through the roots presents a rather different pattern of problems. Assuming that the chemical is stable in the soil, the problem reduces itself to providing sufficient chemical and distributing it so that it is accessible to the roots at the relevant stages of growth. The plant appears to have little capacity to regulate intake through its roots; there appears to be a barrier to movement at the endodermis, but there is no evidence that this responds to conditions in the top of the plant where the translocated material is accumulating. In these circumstances, the proper dose can only be supplied by ensuring a slow and steady release of the chemical into the soil water supplying the transpiration stream. This could be achieved, either by ensuring the correct degree of insolubility or by using a chemical which is adsorbed to the soil and released slowly, such as ethirimol or dimethirimol (Graham-Bryce and Coutts, 1971). The second course has certain possible advantages since desorption from the base exchange system may, to some extent, be related to root activity. A limited mobility in the soil would also be an advantage. An understanding of this system must be based on a detailed knowledge of the interactions between the plant, the soil and the fungicide. The necrotic tissues, which surround living roots, are also capable of accumulating and, later, releasing some chemicals into the transpiration stream (Tanton and Crowdy, 1972a): these might provide a useful store and a buffer against violent fluctuations in the concentration in the soil solution passing through the roots.

Chemicals translocated in the apoplast are not redistributed within the plant; however, they may be leached from the leaves, taken up by the roots and recirculated in the apoplast. This effect has been reviewed by Tukey (1970); the recirculation of nutrients may be of some importance in crops growing in marginal conditions and may reduce toxic accumulations in the leaves. The transfer, in nature, of physiologically active chemicals is illustrated by the interaction between *Camelina alyssum* which produces a toxin which is washed from the leaves and will harm a plant such as flax following root uptake (Grümmer and Beyer, 1959). This process would provide a continuing supply of material to transpiring leaves. Continuing root treatment would be associated with residues in the transpiring parts of the plant; water loss from fruit is probably not great and residues in fruit might not be high. This has been shown for tomato fruit by Peterson and Edgington (1971).

The introduction of chemicals with low inherent phytotoxicity introduces the possibility, which is now being realised, of treating pathogens in the soil and of treating established internal infections such as vascular wilts. If the fungicides are moved in the apoplast, this involves soil treatment, but chemicals distributed in the symplast could be used to control vascular and root diseases, if applied to the leaves. A chemical with the mobility of α-methoxyphenylacetic acid could be sprayed on the leaves, translocated to the roots and released into the soil surrounding them, which could well be the ideal way of treating root diseases. Chemicals distributed in the symplast would open up an entirely different pattern of disease control since they would be transferred from the treated leaves to young growing tissue, possibly reducing the need for repeated spraying. One would expect these chemicals to be distributed in a pattern similar to carbohydrates and this pattern can be profoundly influenced by infection with plant disease; the subject has been reviewed recently by Smith *et al.* (1966). There seems to be a clearly established pattern of accumulation at the lesions of a number of obligate parasites, which occurs both in the host tissue and in the fungal mycelium and may involve retention of carbohydrate in the infected leaf and short-range transport within the leaf to the pathogen. In most cases also there is a general disturbance of the translocation patterns in the plant and active infections may attract photosynthate from adjacent leaves and possibly from a greater distance. In addition to the work reviewed by Smith *et al.* (1966), this phenomenon has been studied further in wheat infected with *Puccinia striiformis* (Siddiqui and Manners, 1971) and with *Ustilago nuda* (Gaunt and Manners, 1971) and in apple scab (Hignett and Kirkham, 1967), who implicated a melanoprotein in polarising the nutrient flow. The mechanism of this disturbance of the translocation pattern has not been elucidated. It is sometimes possible to produce very similar effects by applying growth-regulating chemicals and there is ample evidence that growth-regulating chemicals are produced by fungi and may occur in high concentrations in fungal lesions in higher plants (Wood, 1967): these may well be implicated. Fungicides transported in the symplast with carbohydrates could well accumulate high residues in fruit and storage organs and further work will be aimed at finding out how far the transport of these two groups of chemicals is necessarily related.

6

Effects on physiology of the host and on host/pathogen interactions

by A. E. Dimond

INTRODUCTION

In broadest terms, systemic compounds act to control disease in one of three ways: they may kill or inactivate the pathogen in the host, they may increase host resistance to infection, or they may interfere with pathogenic processes and, in this manner, block the development of symptoms in the invaded plant. These several mechanisms have been discussed from differing points of view on earlier occasions (Dimond, 1965; Dimond and Horsfall, 1959; Howard and Horsfall, 1959; Grossmann, 1968b; Oort and van Andel, 1960; Woodcock, 1971).

As systemic compounds enter the plant, they encounter a variety of active biochemical systems, and they are often altered in this encounter. Some compounds are converted into fungitoxic molecules after relatively minor modification. Others enter as raw materials for biosynthesis of fungitoxic structures. Yet others alter the physiology of the plant or features of its anatomy. Some treatments mimic the mechanisms of natural resistance to disease.

One cannot discuss treatments that affect physiology of the host without also dealing with those that alter pathogenesis. The two kinds of treatment involve different facets of the same relationship; frequently the two are complementary. For example, one treatment may modify host tissue so that it is less readily attacked by a fungal enzyme. Another treatment may inhibit this enzyme, an enzyme that is important in pathogenesis, but not essential for growth or survival of the pathogen. Again, the spread of vascular wilt pathogens through a large host often depends upon migration of propagules with the transpiration stream. The spread of spores through the vascular system can be prevented by use of treatments that induce tylose and gel formation in conductive vessels of the host. Alternatively, use of antisporulants or of compounds that inhibit the budding of fungal cells may equally well stop reproduction of the pathogen and spread of its spores through xylem vessels. The one treatment acts upon the host; the other affects pathogenic processes. Either method has the same objective. For such reasons, the discussion to follow may not always be confined to cases where physiology of the host is altered. Rather, discussion will be presented around principles underlying control of disease with systemic compounds.

Although a variety of modes of action for chemotherapy are recognised, the greatest successes to date have been with compounds that are true systemic fungicides. Some of these compounds, in addition to being fungitoxic, also affect the physiology of the host.

EFFECTS OF SYSTEMIC FUNGICIDES ON THE HOST

Benomyl and other systemic fungicides based upon benzimidazole may effect changes in and be changed by the host plant. They are readily absorbed by roots from treated soil and, to a slightly lesser extent, through leaves.

Benomyl is converted by hydrolysis to the fungitoxic compound methyl 2-benzimidazol-2-yl-carbamate (Clemons and Sisler, 1969; Sims *et al.*, 1969). The latter compound may be formed in treated soil or in the plant and persists in either for a considerable period of time without further modification (Biehn and Dimond, 1969; Hine *et al.*, 1969; Peterson and Edgington, 1969).

Plants treated with benomyl are altered in turn. Taylor (1970) has noted that benomyl-treated tobacco plants are protected against ozone injury. Other systemic fungicides based on benzimidazole also alter ozone susceptibility of plants. In fact, Pellissier *et al.* (1971) have shown that protection of plants against ozone damage follows treatment with these systemic fungicides or with benzimidazole itself. Neither benomyl nor benzimidazole reacts with ozone directly (Rich and Tomlinson, 1970). Therefore, the protective action results from an effect upon the physiology of the plant.

Benomyl probably affects nucleotides in fungi (Kaars Sijpesteijn, 1970) and interferes with DNA synthesis (Clemons and Sisler, 1971). Benomyl and related benzimidazoles probably also affect these same compounds in higher plants but to a lesser degree.

PRODUCTION OF FUNGITOXICANTS IN PLANTS

Natural disease resistance in plants has sometimes been correlated with the presence of fungitoxic chemical barriers. More often it is believed to be associated with the formation of fungitoxic products in host tissue as the tissue is invaded.

Some systemic compounds are converted into fungitoxic molecules by host tissue. An example is the fate of 6-azauracil in cucumber leaves. Dekker (1962) showed that when this compound is introduced into leaves, powdery mildew infections fail, even though 6-azauracil is not fungitoxic. Spores germinate and enter leaves normally, but growth of the fungus ceases upon development of the first haustorium.

Uracil reverses the chemotherapeutic action of 6-azauracil. In the plant, 6-azauracil is converted to 6-azauridine and then to the monophosphate (Dekker and Oort, 1964). Applied to plants, the latter two compounds also inhibit haustorial development, but their action is not reversed by uracil. The 6-azauridine monophosphate blocks nucleic acid biosynthesis. These reactions occur in some fungi and in higher plant tissue as well. However, failure of infection from the time when the first haustorium invades a host cell suggests that the host cells have produced the 6-azauridine monophosphate, which then kills the developing haustorium.

6-Azauracil is also effective in preventing development of lesions of cucumber scab, caused by *Cladosporium cucumerinum*, and strains of this fungus are resistant to the chemotherapeutant (Dekhuijzen and Dekker, 1971). In the resistant strain only one-third as much 6-azauracil is converted via 6-azauridine to the monophosphate as is done in the susceptible strain of the fungus. This is because the resistant strain incorporates orotic acid into RNA regularly but the process is partially blocked in the susceptible strain.

At least superficially, the action of procaine and kinetin on a variety of powdery mildews is similar (Dekker, 1961a, b, 1963b; Dekker and van der Hoek-Scheuer, 1964). Haustorial establishment fails when host cells are invaded. The parallel behaviour suggests that action in these cases also is based upon production of a fungitoxic molecule by host cells.

Phytoalexins are formed in plants under influence of the invading pathogen. They are also formed as a result of certain types of injury or chemical treatment. Cruickshank (1966) has suggested that in the future, chemical control of plant diseases may be based upon systemic treatments that stimulate plants to produce phytoalexins.

The induction of resistance in apple leaves to scab, caused by *Venturia inaequalis,* following injection of DL-phenyl alanine may be an example.

Kuć *et al.* reported in 1957 that injection of phenyl alanine into petioles of apple leaves converted their reaction to scab from susceptible to resistant. Subsequently, Holowczak *et al.* (1962) found that phenyl alanine is converted, at least in part, to phloridzin, phloretin and phloretic acid. α-Amino isobutyric acid, which induces a resistant reaction to *Venturia inaequalis* in susceptible apple leaves, is also converted to phloridzin and phloretin (McLennan *et al.*, 1963). Resistance of apple leaf tissue to scab has been associated with the phenolic aglycone, phloretin. This undergoes oxidation and polymerisation as the pathogen enters the leaf and activates the polyphenol oxidase system (Noveroske *et al.*, 1964a, b; Kuć, 1968; Raa, 1968). The intermediate and reactive products of oxidation and polymerisation react non-specifically with enzymes and other proteins and denature them. These reactions are associated with the pin-point lesions associated with a resistant reaction (Kuć, 1968; Raa and Kaars Sijpesteijn, 1968).

The reactions of plant tissue to invasion by a pathogen are similar for a variety of diseases. Phenolic compounds appear and phenol oxidases become activated. The reactions of resistant varieties to invasion are fast and acute, whereas those in a susceptible variety are slow and prolonged, but in a number of respects the reactions occurring in resistant and susceptible varieties are similar (Kosuge, 1969).

The series of responses of tomatoes to invasion by *Fusarium oxysporum* f. sp. *lycopersici* includes the appearance of indole acetic acid, a variety of phenolic compounds, and a very considerable activation of the phenol oxidase system. In resistant varieties, these changes are intense and occur rapidly (Matta *et al.*, 1967). Naphthalene acetic acid, a chemotherapeutant for *Fusarium* wilt, caused a marked activation of the polyphenol–polyphenol oxidase system in susceptible varieties but caused little change in resistant varieties (Matta, 1963). These events suggest how chemical treatment may elicit a non-specific resistance reaction that is closely akin to the hypersensitive reaction.

STOMATAL CLOSURE AS A PHYSICAL BARRIER TO HOST ENTRY

By their action upon stomata, a variety of compounds may create a physical barrier to entry by pathogens that enter through stomates, either exclusively or preferentially. Among compounds that close stomata on leaves are 8-hydroxyquinoline sulphate (Stoddard and Miller, 1962), phenyl mercury acetate (Zelitch, 1961), and the alkenyl succinates (Zelitch, 1963). Some of these compounds are themselves directly fungitoxic, whereas others are not. The incidence of infections on bean leaves by

Uromyces phaseoli was materially reduced following application to foliage of sprays containing non-fungitoxic α-hydroxy succinates that close stomata (Dimond, 1965). The extent to which this principle is useful in disease control in the field needs to be explored systematically.

SYSTEMIC COMPOUNDS THAT ALTER GROWTH OF THE HOST

The ability of growth regulants to reduce severity of vascular wilt diseases has been known for some time (Davis and Dimond, 1953). Many of these compounds, synthetic as well as natural, have chemotherapeutic activity against a range of pathogens in their hosts. More than one mode of action is now recognised for members of this group.

Alterations in vascular anatomy

No gymnosperm suffers from a vascular wilt disease. In fact, among trees, it is the diffuse porous species that are primarily affected by this group of diseases (Dimond, 1970).

Among species of elms and within varieties, susceptibility to *Ceratocystis ulmi* is correlated with what McNabb *et al.* (1970) have called the vessel group size. Vessel group size is the product of the average vessel diameter and the average number of contiguous vessels in an annual ring.

According to studies of Pomerleau (1970), *C. ulmi* crosses from vessel to vessel only through pit pairs between vessels, when vessels are contiguous. At least during pathogenesis, mycelium of *C. ulmi* is not found in xylem parenchyma or fibres, according to Pomerleau (1970). His findings give the rationale for the concept of vessel group size. This function clearly is related to the ease of invasion by mycelium of the vessels in an annual ring. If this thesis is correct, then any factor that changes the vessel group size in a tree should modify its resistance to vascular wilt diseases such as Dutch elm disease and oak wilt.

Perennial hosts occasionally recover naturally from vascular wilt diseases. When elms recover from Dutch elm disease, for example, the pathogen continues to grow in the originally invaded vessels in an annual ring but does not cross into vascular tissues of subsequently developed rings (Banfield, 1968). Natural recovery of American elms from infections by *C. ulmi* occurs frequently in young trees and rarely in large ones. The reason for this relationship is believed to be associated with the thickness of the annual ring, the number of vessels that must be crossed in a year to keep pace with the growth of the tree. It is also related to the size and distribution of vessels, both across the annual ring and longitudinally.

The action of certain growth regulants as chemotherapeutants in Dutch elm disease and oak wilt is related, probably, to their ability to modify vessel group size. Amino trichlorophenylacetic acid retards development of Dutch elm disease, according to Edgington (1963). When applied early, before the onset of the main flush of cambial growth, this treatment induced rapid cell division, and caused formation of a layer of dense, small, starch-filled cells between earlier and later developing vessels, an effect that significantly reduced vessel group size and, at the same time, reduced severity of disease in inoculated trees.

Trichlorophenylacetic acid applied to oaks affected the characteristics of sapwood and altered susceptibility of trees to oak wilt, caused by *Ceratocystis fagacearum*, according to Venn *et al.* (1968). Young vessels that were formed prior to the time of treatment rapidly filled with tyloses. Wood that formed subsequently consisted of fibres, mainly, and fibre tracheids of small diameter, together with xylem parenchyma. Symptoms rarely progressed following treatment. This treatment also altered vessel group size.

Alteration of the lumina of vessels

The fungi causing vascular wilt diseases spread within their hosts by producing spores or bud cells, which are transported through xylem vessels. On lodging, these spores germinate and establish new centres of infection. In this way, a large host plant can become extensively invaded more rapidly than it could through mycelial growth alone. This process occurs in a variety of vascular wilt diseases in both annual and perennial hosts (see Beckman, 1964; Dimond, 1970). For example, Pomerleau and Mehran (1966) have isolated spores of *C. ulmi* from shoots and leaves of American elm, harbouring Dutch elm disease, while Talboys (1962) had earlier called attention to the same process in the case of *Verticillium* wilt of hop.

When wilt-inducing pathogens invade through roots, the process of spore production and migration may pass through several cycles before the plant as a whole becomes invaded. Beckman *et al.* (1962) have observed how spores of *Fusarium oxysporum* f. sp. *cubense* in the plate region of banana are carried to a cross wall, become lodged, germinate and grow through pores, then produce more spores, which are dislodged and carried to the next obstruction. Beckman *et al.* (1961) have correlated resistance to Panama disease with the rapid development in xylem vessels of tyloses and gels in response to the presence of the pathogen. When these obstructions are formed quickly and are both stable and persistent, spores cannot migrate and the host cannot be invaded by the pathogen (Beckman, 1964).

In Dutch elm disease, a similar situation prevails. In addition to the correlation of susceptibility in elms with vessel group size, Elgersma (1970) has correlated susceptibility with conductivity of vessels in two-year-old shoots and related conductivity to length and diameter of vessels. Conversely, resistance is related to lower conductivity, and Elgersma (1967) has noted how spore transport is prevented in resistant elms by the rapid deposition of gum in vessels.

The vascular gels contain both pectic substances and hemicelluloses (Beckman, 1969a, b; Beckman and Zaroogian, 1967). Tyloses are formed simultaneously and provide a stable network for holding gels.

Hyperauxiny is characteristic of wilt-diseased plants as is true of some other diseases (Beckman, 1964); and the tyloses that abound in vessels of oaks suffering from oak wilt have been attributed to the effects of auxins (Beckman *et al.*, 1953). In Panama-diseased banana, Mace and Solit (1966) have isolated indole acetic acid and have experimentally demonstrated its ability to induce the formation of tyloses in vessels.

The foregoing provides a background for explaining a second type of action of growth-regulating compounds as chemotherapeutants in the vascular wilt diseases. They may inhibit growth of sapwood, as Beckman has shown (1958) when auxin supplies are normal, and by doing so, change the vessel group number. But in addition, they induce tylose formation and reduce the conductivity of vessels before infection takes place. Thus, in Dutch elm disease, Smalley (1962) reported the occlusion of large vessels by tyloses following treatment of trees with 2,3,6-trichlorophenylacetic acid and suggested that the protection of trees given by this treatment is at least partially attributable to its preventing movement of spores through vessels. After inoculation, pre-treated elms developed tyloses in from two to four days, whereas tyloses in untreated but inoculated trees took from seven to nine days, according to Brener and Beckman (1968). This treatment delayed formation of spring wood also. Similarly, Venn *et al.* (1968) reported that oak treated with this same compound developed heavy tylosing in vessels existing prior to treatment and a reduction in diameter and numbers of vessels formed after treatment. Severity of disease was reduced.

Halogenated benzoic acids have also been employed to reduce Dutch elm disease and oak wilt on an experimental basis. Geary and Kuntz (1962) noted a delay in formation of spring vessels, along with reduction in severity of oak wilt. However, the chemotherapeutic effects and phytotoxicity have been erratic with this treatment.

In herbaceous plants, growth-regulating compounds that induce tylose formation have likewise reduced the severity of vascular wilt diseases. 2-Chloroethyltrimethylammonium chloride (cycocel) had little effect upon growth of *Verticillium albo-atrum in vitro* but induced tylose formation in tomato plants and reduced severity of disease in treated plants (Sinha and Wood, 1964). 3-Indoleacetic acid, and α-naphthalene acetamide had similar effects and reduced invasion of stems by the pathogen (Sinha and Wood, 1967).

Modification of cell walls

Among plant diseases, those causing rotting of tissues involve the degradation of cell walls by hydrolytic enzymes that are produced by pathogens. In addition to the rots, a wide variety of other diseases causes decomposition of cell walls. The involvement of pectic and cellulolytic enzymes in pathogenesis is now well established (Bateman and Millar, 1966). A variety of hydrolytic enzymes is produced during pathogenesis that reflects the wide variety of chemical bonds and the variety of carbohydrate monomers that are represented in cell walls (Albersheim *et al.*, 1969).

As the composition of host cell walls varies with stage of host maturity and with species, so also must the hydrolytic enzymes produced by different pathogens vary. The nature of some of these enzymes has been reviewed by Bateman and Millar (1966) and by Wood (1967). Two types of enzymes effect major changes in the molecular weight of pectic substances: the polygalacturonases and the pectate lyases. The former tend to be inhibited when their substrate is largely in the form of calcium pectate or when Ca^{++} is present in quantity. The pectate lyases require calcium (Bateman and Millar, 1966). Thus, a pathogen that degrades pectin primarily by means of polygalacturonase should damage host tissues according to the level of calcium present.

This has proven to be so. *Fusarium oxysporum* f. sp. *lycopersici* produces polygalacturonase both *in vitro* (Waggoner and Dimond, 1955) and in the infected tomato plant (Patil and Dimond, 1968; Mussell and Green, 1968). Tomato plants that are calcium-deficient are severely attacked by *Fusarium*, whereas plants grown on a normal calcium supply suffer less (Edgington and Walker, 1958).

How growth regulators act as chemotherapeutants in *Fusarium* wilt began to be apparent when Corden and Dimond (1959) correlated the chemotherapeutic effectiveness of a series of naphthalene-substituted aliphatic acids with their activity as growth regulants by each of several criteria. Chemotherapeutic activity was highly correlated with the ability of compounds to inhibit root elongation, but with no other index of growth hormone activity. This result suggested that the compounds acted upon pectins in cell walls by causing increased calcium bonding, which in turn decreased root elongation. Corden and Edgington (1960) then showed that the chemotherapeutic effect of naphthalene acetic acid was calcium-dependent. Calcium-deficient plants were just as susceptible to wilt whether or not they had been treated with naphthalene acetic acid!

The effects of calcium nutrition and treatment with naphthalene acetic acid upon the pectins in cell walls were then examined (Edgington *et al.*, 1961). Although total pectins were not altered, calcium-deficient plants contained more water-soluble pectins than plants grown on a normal calcium supply. Plants treated with naphthalene acetic acid contained less water-soluble pectin than normal plants. The pectic compounds of calcium-deficient plants were readily attacked by pectolytic enzymes from the tomato wilt *Fusarium*, but those from naphthalene acetic acid-treated plants on a normal calcium supply were resistant to enzymatic hydrolysis. Thus, susceptibility of plants to *Fusarium* wilt and susceptibility of their pectic components to enzymatic attack were well correlated. Evidently, when applied in growth-inhibiting concentrations, growth regulants reduce methoxylation of pectic substances and, in the presence of a normal supply of calcium, components in cell walls are highly cross-bonded with calcium bridges which are polygalacturonase-resistant. In calcium-deficient plants, this cannot occur, so the pectic components of cell walls are readily hydrolysed.

This interpretation is consistent with what is known of the role of auxins in maintaining plasticity of cell walls in healthy plants, a property long associated with effects on pectins. Bonner (1961) has offered evidence that its site of action is the same as for calcium-dependent rigidity.

Ethionine has shown chemotherapeutic activity on several occasions. Zentmyer *et al.* (1962) showed its action against root rot of avocado by *Phytophthora cinnamomi* and Moje *et al.* (1963) demonstrated that methionine antagonised this effect. Action by ethionine and its reversal by methionine (or the converse in appropriate host–pathogen combinations) can be interpreted as an effect upon cell walls, much as auxins can modify the pectins in cell walls. Methionine in the presence of auxins donates methyl groups to pectins in wall formation (Cleland, 1963a) and ethionine prevents this donation of methyl groups (Cleland, 1963b). Thus, for diseases in which cell wall attack is by comparable mechanisms, ethionine or auxin could produce similar effects. Jones and Woltz (1969) have provided the data to confirm this by showing that ethionine reduces the severity of *Fusarium* wilt of tomato. Edgington *et al.* (1961) showed that naphthalene acetic acid reduced *Fusarium* wilt of tomato through its action upon pectins of the cell wall.

Other growth effects

The correlation between growth-regulant activity and chemotherapeutic activity has been used as a basis for seeking new compounds that may be useful in control of plant diseases. A number of compounds have been found in this way, but as yet their mode of action is not known. The action of salts of 2-benzothiazolyl thioglycolate against *Fusarium* wilt and Dutch elm disease is a case in point (Dimond and Davis, 1953). These compounds produce formative effects in the host. Whether they stimulate tylose formation or act in some other way is not known. The same is true of the morphactins (methyl, *n*-butyl, and *n*-heptyl esters of 9-hydroxy-fluorene-9-carboxylic acid and the 2-chloro-substituted analogues). Buchenauer and Grossmann (1969) have reported moderate reductions of *Fusarium* and *Verticillium* wilts and of late blight of tomato as a result of treatment, but the mode of action is not known.

In three instances, at least, the use of growth retardants has reduced disease severity. The amount of brown spot, caused by *Alternaria longipes*, was reduced in tobacco to which maleic hydrazide had been applied for sucker control (Ramm *et al.*, 1962). Van Andel (1966a, b, 1968) has reported a reduction in cucumber scab, caused by *Cladosporium cucumerinum*, as a result of applying compounds that reduced auxin through increasing IAA oxidase activity. Among them were L-threo-β-phenyl serine and 2-chloroethyltrimethylammonium chloride. In addition, several unnatural D-amino acids retarded root extension in assays and reduced cucumber scab (van Andel, 1962).

This evidence suggests that growth regulants and retardants affect cucumber scab and *Fusarium* wilt of tomato oppositely. Growth-promoting substances tend to reduce wilt and growth retardants, such as ionising radiation or maleic hydrazide greatly increase severity of disease (Waggoner and Dimond, 1957).

The severity of powdery mildews is reduced by treatment, not only with ethionine, but with methionine also. In fact, L-methionine is the only natural amino acid having a chemotherapeutic effect (van Andel, 1966a). Dekker (1969a) reported poor development of powdery mildew (caused by *Sphaerotheca fuliginea*) on cucumber leaves, an effect confirmed by Covery (1971) with apple powdery mildew, caused by *Podosphaera leucotricha*.

In similar fashion L-methionine prevents root rot of peas, caused by *Aphanomyces euteiches* and ethionine is also active (Papavizas and Davey, 1963). The basis of this action remains unexplained. L-Methionine also reduces root rot of sugarbeet, caused by *Aphanomyces cochlioides* (Winner, 1966).

The apparently diverse behaviour of amino acids and their analogues in reducing severity of disease or in negating effects induced by an antagonist surely arises because of the many ways in which these compounds can act in biological systems. In each disease, two organisms are involved, a host and a pathogen, and observed response can be from either organism. Depending upon the mechanism of pathogenesis, response to a given amino acid will vary. Impose, in addition, response to natural (L) and unnatural (D) amino acids, and the multiplicity of possible responses increases yet more.

The amino acids can be grouped in families that share a common pathway of biosynthesis. A more orderly picture of amino acid interrelations, for bacteria, at least, has been presented by Dawes and Large (1968). Sands and Zucker (1971) have compared phytopathogenic with saprophytic pseudomonads in respect to their amino acid metabolism. Pathogenic species proved frequently to have impaired biosynthetic pathways, whereas saprophytic species rarely do. Toxic effects of an administered amino acid may be related to a faulty step in the biosynthetic pathway. A host plant can compensate in its own amino acid supply for the faulty pathway in a pathogen. If the compensation by the host can be blocked, then therapy is possible.

TREATMENTS THAT BLOCK PATHOGENIC PROCESSES

Antisporulants

Blocking the spread of propagules of vascular wilt pathogens through xylem vessels has already been mentioned as a potential method of chemotherapy. Two approaches are possible. One can induce tyloses and gums in vessels to prevent ready movement of spores through vessels. Alternatively, one can inhibit the development of propagules by fungi through use of antisporulants. Horsfall and Lukens (1968) have explored the possibility of disease control through using antisporulants to reduce the aerial spread of plant pathogens in the field and have sought compounds that reduce spore formation *in vitro*. Howard and MacHardy (1969) have discussed this principle in reference to vascular wilt organisms. Biehn (1972) has examined the extent to which the antisporulant *p*-fluorophenylalanine prevents bud cell formation by *Ceratocystis ulmi* with therapy in mind. Keen *et al.* (1971) have examined the effectiveness of various antisporulants, including 5-fluorodeoxyuridine, on cultures of *Verticillium albo-atrum*.

The idea of controlling disease through controlling the amount of available inoculum is appealing, and with the perfection of systemic antisporulants, a new method of control through chemotherapy will be available.

Inactivation of pectolytic enzymes

Just as maceration of tissues can be reduced by means of treatments that make cell walls more resistant to enzymatic hydrolysis, so can maceration be reduced through compounds that inactivate these wall-splitting enzymes. Grossmann (1958b) sought to inactivate the pectolytic enzymes present in culture filtrates of *F. oxysporum* f. sp. *lycopersici*. He found a number of phenolic compounds, among them rufianic acid (1,4-dihydroxyanthraquinone-2-sulphonic acid), that would do this *in vitro* (Grossmann, 1962a). Rufianic acid is also effective *in vivo* as a chemotherapeutant against *Fusarium* wilt of tomato (Grossmann, 1962b). This work demonstrated the feasibility of chemotherapy by enzyme inhibition and offered a means of disease control on a clearly demonstrable scale *in vivo*. It was also a necessary part of the demonstration of the role of pectolytic enzymes in pathogenesis of the wilt diseases.

In principle, the inhibition of macerating enzymes is also a means for controlling soft rots and such leaf-spotting diseases as depend upon hydrolysis of cell wall polysaccharides. Grossmann (1968a) found that rufianic acid reduced maceration of plant tissue by *Rhizoctonia solani* and that non-fungicidal inhibitors of pectolytic enzymes also prevented infection of tomato by *Alternaria solani*.

The phenolic compounds in plant extracts often auto-oxidise and polymerise to form highly reactive products. These react non-specifically with proteins generally, a process that leads to inactivation of pectolytic and other enzymes. Byrde (1956, 1957) had called attention to this as a natural process, accounting for resistance to rotting in some cider apples and Cole (1958) in this way accounted for lack of detectable pectolytic activity when some cultivars of apples are attacked by rotting fungi. Grossmann (1962c), in his assays, noted loss of pectolytic activity when plant extracts containing phenolic compounds had auto-oxidised. Byrde (1963) has traced the course of this idea and the role of natural phenolic compounds in enzyme inactivation. Products of chlorogenic acid or caffeic acid, arising from the action of potato phenolase, were shown to inactivate polygalacturonase of *Verticillium albo-atrum* (Patil and Dimond, 1967a).

In many diseases, plant tissues exhibit an activation of phenol oxidases and phenols become oxidised and polymerised. Such reactions are now believed to be involved in disease resistance, the hypersensitive reaction, and, to some degree, in phytoalexin formation. These reactions occur both in susceptible and resistant tissues, but are far more rapid in resistant tissue and generate fungitoxic materials. The response can be evoked by pathogens, non-pathogens and certain chemical and physical treatments. Hopefully, such treatments will one day be used in disease control to convert susceptible into hypersensitive plants.

Regulation of enzyme production by chemotherapy

The role of hydrolytic enzymes in breakdown of host cell walls is now recognised to be an important aspect of pathogenesis. Historically, attention was first focused on the pectolytic enzymes and their role in maceration of tissues.

As attention was directed experimentally toward these enzymes, the conditions favouring induction and repression began to be stressed. Deese and Stahmann (1960) reported the higher polygalacturonase activity when various *formae* of *F. oxysporum* were grown on tissues from susceptible hosts than when grown on tissues from plants that are resistant to these *formae*. Deverall and Wood (1961) compared pathogenesis of *Botrytis cinerea* and *B. fabae* in respect to polygalacturonase production under comparable circumstances.

Repression of polygalacturonase synthesis was explored in *F. oxysporum* f. sp. *lycopersici* in the presence of glucose or galacturonic acid (Patil and Dimond, 1967c). In contrast, polygalacturonase production by *V. albo-atrum* showed end-product repression but the enzyme was produced in considerable amounts when glucose or galactose were present (Patil and Dimond, 1967b; Mussell and Green, 1970).

Patil and Dimond (1968) then examined the effect of applying regulators of enzyme production to infected plants. Glucose, 2-deoxy-D-glucose, and monogalacturonic acid all interfered with production of polygalacturonase by *F. oxysporum* f. sp. *lycopersici*. Each was introduced into the vascular system of tomato plants previously inoculated with *Fusarium*. For treated plants and inoculated controls, the amount of polygalacturonase activity and the rate of advance of disease symptoms were compared. In plants treated with a repressor, less polygalacturonase was detected and symptoms increased at a slower rate than in controls. Plants treated with glucose showed fewer disease symptoms than plants treated with water, despite the fact that the former contained more culturable mycelium.

Such studies are being extended to other types of enzymes that are involved in degradation of cell walls. Thus, Keen *et al.* (1970) have examined the conditions affecting induction and repression of β-galactosidase synthesis in *V. albo-atrum*. As such studies are extended to a wider range of enzymes involved in the attack of cell walls, greater variety of treatments will be available for interfering with this process of pathogenesis.

The diversity of wall-splitting enzymes produced by a given pathogen is greater than was formerly realised, thanks to the work of Albersheim *et al.* (1969). He has shown how enzymes are induced in a pathogen in a prescribed sequence, according to the sequence of carbohydrates in wall components of the host. As tissues in a growing host plant develop, the components of its cell walls change and the sequence of enzyme induction must also change. This relationship is thought to account for the short-lived susceptibility of juvenile tissues to certain pathogens (Nevins *et al.*, 1968). If the wall-digesting enzymes are not induced in the correct sequence, pathogenesis is impaired.

These ideas suggest that attack of cell walls could be inhibited were it possible to interfere with processes of enzyme induction and especially to interfere with the sequence.

Synthesis of a variety of wall-hydrolysing enzymes can be regulated. Both cellulase and polygalacturonase are inhibited by the presence of glucose in *F. oxypsorum* f. sp. *lycopersici*. Polygalacturonase synthesis is affected oppositely by glucose in *F. oxysporum* f. sp. *lycopersici* and *V. albo-atrum* (Patil and Dimond, 1967b, c; Mussell and Green, 1968). These differences aid in accounting for susceptibility of plants to infection by organisms, according to the amount of sugar in their tissues. Horsfall and Dimond (1957) have discriminated between 'high sugar' and 'low sugar' diseases, according to how sugar levels in tissue affect susceptibility to infection by different pathogens. In general, the rusts and powdery mildews are best able to infect tissues that are high in sugar, whereas low-sugar diseases are caused by *Helminthosporium* and *Altenaria* spp. These differences are consistent over a variety of circumstances. When pathogenesis involves pectolytic or cellulolytic enzymes to a considerable extent and when these enzymes are consistently repressed in synthesis by the presence of glucose or sucrose or other sugar, the regulation of enzyme synthesis dictates a low-sugar disease. Moreover, the end product of hydrolysis of cellulose will repress synthesis of polygalacturonase or of any other enzyme repressed by glucose. When cell-wall destruction is unimportant and when pathogenesis depends upon the synthesis of other enzymes that are not affected by sugars, high sugar levels do not interfere with pathogenesis.

End-product repression of synthesis of wall-attacking enzymes and its absence in the regulation of synthesis of enzymes in other species of pathogens accounts for some, but not all, high-sugar diseases. It does account for how *Botrytis cinerea* and *V. albo-atrum* (Patil and Dimond, 1968) can attack tissues containing more sugar than those attacked by

B. fabae (Deverall and Wood, 1961) or *F. oxysporum* f. sp. *lycopersici* (Patil and Dimond, 1967c). But it fails to account for why diseases caused by the obligate parasites, the rusts and powdery mildews, are high-sugar diseases as a group.

Two low-sugar diseases have been controlled by chemotherapy by taking advantage of the hypothesis. Lukens (1970) noted that the melting-out disease of turf, like all diseases caused by species of *Helminthosporium*, behaves as a low-sugar disease. He has used this information to alleviate *Helminthosporium vagans* on Kentucky bluegrass turf. High cutting, elimination of shading or use of glucose sprays on turf increased reducing sugar levels in leaves and reduced severity of melting-out disease. Kaars Sijpesteijn (1961a) has offered evidence in a second case. Cucumber scab, in her studies, behaved as a low-sugar disease, just as *Alternaria* does. The chemotherapeutant carboxymethyldimethyl dithiocarbamate in treated seedlings reduced sugar levels and increased resistance to infection by *Cladosporium cucumerinum*. When seedlings were treated with both sugar and the same chemotherapeutant, they remained susceptible to infection.

Counteracting toxins in plant disease

To an increasing degree, specific symptoms associated with plant disease are being ascribed to the action of toxins originating in the pathogen. Toxins have in a number of cases been isolated from cultures of the pathogen and from diseased plants, and, in a few cases, they have been identified.

The idea that toxins can be neutralised or antidoted and so bring about chemotherapy is an old one (Howard and Horsfall, 1959). For example, in infections by *Pseudomonas tabaci* on tobacco leaves, chlorotic haloes surround the lesion and the chlorosis has been attributed to a toxin on numerous occasions. Lovrekovich and Farkas (1963) called attention to the similarity between development of chlorosis and of altered protein metabolism in leaf tissue and processes of senescence, and they demonstrated that these changes in infected leaves were reversible by administration of kinetin. Here is a potentially useful illustration of therapy. But there are few cases where therapy, based on toxin inactivation, has been achieved in the field. This possibility remains to be considered, however, especially when a specific toxin is known to play a critical role in the induction of disease.

An example where a toxin plays a critical role in disease involves pyricularin, the toxin of *Pyricularia oryzae*, cause of rice blast. Tamari *et al.* (1966) found that ferulic acid detoxifies pyricularin and increases resistance of the rice plant to blast. Extracts of treated plants also inactivated pyricularin. Pyricularin appears to act by suppressing the critical hypersensitive reaction of host tissues, and ferulic acid in the rice plant causes respiratory changes that are not related to the hypersensitive reaction. Under such circumstances, the neutralisation of a toxin can make a critical difference in the success of an infection.

The host-specific toxins illustrate a second case where toxins play a critical role in disease induction – to the extent that Scheffer and Pringle (1967) refer to them as primary determinants of disease. As yet, among the various diseases where host-specific toxins are known to participate, no compound has been reported that will neutralise the effect of these toxins or suppress their synthesis by the pathogen. These possibilities could prove useful in combating such important diseases as southern corn blight, caused by *Helminthosporium maydis* or *Helminthosporium* blight of oats, caused by *H. victoriae*.

FUTURE PROSPECTS

The foregoing discussion has dealt with many instances where disease can be alleviated through treatments that are not in themselves fungitoxic. Primarily, these treatments interfere with one or more phases of the process of pathogenesis. These treatments generally delay the development of disease, but in only a few instances has treatment effected a cure.

Postponing an epidemic sometimes makes the difference between success and failure in harvesting an annual crop. Among perennial plants, such as shade trees, cure of disease is more pertinent than alleviation of symptoms.

A systemic fungicide can eliminate infection in a plant and can effect a cure. A systemic chemotherapeutant can only slow the progress of the pathogen or make the host resistant to its invader. As yet, systemic fungicides would seem to hold the key to control of disease by chemotherapy.

7

Effects on fungal pathogens

by A. Kaars Sijpesteijn

INTRODUCTION

Knowledge of the mechanism of action of fungicides is not only important from a purely scientific point of view but it seems also a necessary prerequisite for a proper evaluation of the general toxicology of these compounds. In addition, knowledge of the mode of action may in many cases help to elucidate the mechanism by which resistant mutants differ from the normal sensitive strain. For obvious reasons the fundamental study of the mode of action of antifungal agents used in plant protection has always lagged far behind that of antibacterial agents applied in medicinal chemotherapy. For this reason much can be learned from the achievements of this related field of study (cf. Franklin and Snow, 1971).

This chapter attempts to deal with a variety of compounds in a comparative way.

The description of the effect of a fungicide on a fungus requires morphological, physiological as well as biochemical observations. Generally we may distinguish at least three different methods by which fungicides can exert their action.

(a) *Inhibition of energy production,* or in other words, of ATP production; for instance by inhibition of respiration or by uncoupling of oxidative phosphorylation. The energy-producing processes are located in part in the cytoplasm and in part in the mitochondria. Strong inhibition of such processes will eventually have a fungicidal effect.

(b) *Interference with biosynthesis,* i.e. with processes leading to the production of new cell material required for growth and maintenance of the organism. Biosynthesis of low molecular weight compounds such as amino acids, purines, pyrimidines and vitamins takes place in the cytoplasm, whereas protein synthesis proceeds mainly on the ribosomes, and DNA- and part of RNA-synthesis in the nucleus. Inhibition of biosynthesis sometimes has only a fungistatic effect.

(c) *Disruption of cell structure,* for instance by affecting permeability of the cell membrane resulting in leakage of the cell contents. This effect is fungicidal.

One has of course to distinguish between primary and secondary effects of a toxicant. Thus, rupture of cell structure will lead inevitably to inhibition of respiration; alternatively, inhibition of energy production will result in a decrease of biosynthesis for lack of ATP and often also for lack of necessary intermediates normally produced during respiration.

At present far more is known of the antifungal action of protectant fungicides than of systemic fungicides and one must admit that the exact biochemical mode of action is not yet fully understood for any systemic fungicide. For several compounds, however, the site of action can be indicated and the results obtained up to now have already revealed that systemic fungicides, unlike protectant fungicides, act predominantly on biosynthetic processes.

When a new fungicide becomes available for study a variety of experimental results will be required to build up a picture of how the compound acts. Primary questions are the level of activity, the antifungal spectrum, the influence of pH and composition of the medium on activity. It must be determined whether respiration is inhibited and it is useful to investigate whether certain compounds, as for instance growth factors, amino acids, –SH compounds, can act as antagonists of the fungitoxic agent. One needs to know whether the agent exerts a fungistatic or a fungicidal effect. Also the structure of the antifungal agent can sometimes give a clue as to its mode of action, and there is the question whether the fungicide applied is itself the actual toxic agent or whether the true toxicant is only formed in the medium or by metabolic activity of the fungus. Only from a careful analysis of such data can a hypothesis on the biochemical mode of action of a compound be built up and checked.

Since at higher concentrations further processes may become inhibited, it is advisable to use low growth-inhibiting concentrations when studying the primary mode of action. In such studies whole cells should be used as much as possible, one of the reasons being that in cell-free preparations the concentration of the toxicant at the site of action may deviate considerably from that in whole organisms.

Obviously all these studies can often be carried out more efficiently with saprophytes than with parasites. For systemic compounds that act exclusively on obligate parasites like mildews, rusts and smuts the *in vitro* study of the action of the fungicide can only be of limited scope.

The following sections deal with the *in vitro* effects on the fungi of a variety of systemic fungicides used in practice as well as of some other systemic fungicides. The information available from the literature is still very limited and it is noticeable that some manufacturers appear quite willing to publish results on their compounds whereas others are very reluctant to do so and leave publication entirely to independent workers. Unfortunately the scarcity of data on the action of systemic compounds does not permit us to group these according to their mode of action as was done for fungicides in general in an earlier publication (Kaars Sijpesteijn, 1969). The various compounds have therefore been grouped according to structure, the antibiotics being treated separately.

For comparison, the minimum concentrations inducing complete inhibition of visible growth on glucose mineral salts agar pH 6·8 obtained in our standard roll-culture test (Pluijgers and Kaars Sijpesteijn, 1966) are given in Table 7.1 for a variety of compounds.

TABLE 7.1 **Minimum inhibitory conc.† (ppm)**

	Botrytis cinerea	*Penicillium italicum*	*Aspergillus niger*	*Cladosporium cucumerinum*
Benomyl	0·02	0·05	0·5	0·2
MBC	0·05	0·05	0·5	0·2
Thiophanate-methyl	0·2	0·5	20	2
Thiabendazole	0·2	0·1	2	0·5
Fuberidazole	0·5	0·5	10	2
Carboxin	20	100	100	10
F427‡	20	5	10	1
Dimethirimol	>500	>500	>500	>500
Ethirimol	>500	>500	>500	>500
Triarimol	2	2	20	0·2
Tridemorph	5	200	>500	10
Triforine	>500	>500	>500	100
Chloroneb	>500	>500	>500	>500
Pyrazophos	>500	>500	>500	>500
Kitazin	100	500	500	200
Cycloheximide	20	20	100	10

† On glucose mineral salts agar pH 6·8, three days after inoculation with conidial suspension
‡ See p. 139

EFFECT OF INDIVIDUAL COMPOUNDS ON THE FUNGUS

Benzimidazole and thiophanate fungicides

This group of systemic fungicides comprising benomyl, thiabendazole, fuberidazole and thiophanate-methyl arouses most interest at present. Table 7.1 illustrates their high activity *in vitro*. The compounds lack antibacterial activity but are active towards a great many fungi including the powdery mildews.

Benzimidazoles Investigations by Bollen and Fuchs (1970) on the antifungal spectrum of benomyl reveal a striking selectivity in action within the group of the Ascomycetes: whereas many species studied are highly sensitive, those belonging to the Porosporae or the Annelsporae proved to be quite insensitive. The Oomycetes and Phycomycetes equally were insensitive.

Edgington *et al.* (1971) obtained similar results and reported, moreover, a great similarity in the antifungal spectrum of benomyl, thiabendazole and fuberidazole. No experimental data are given, however, for the latter two compounds.

Clemons and Sisler (1969) have shown that in water benomyl decomposes very rapidly to methyl benzimidazol-2-yl carbamate (MBC). Since moreover, the antifungal effect of benomyl and MBC is similar, benomyl is assumed to act only after conversion into this latter compound, MBC.

Similarity in structure and antifungal spectrum of MBC and thiabendazole has led many investigators to suggest a common biochemical mode of action, the benzimidazole moiety being the active part of the molecule. The same may be expected for fuberidazole, but data on its antifungal spectrum are very limited so far.

The finding of Bollen and Scholten (1971) that a benomyl-resistant strain of *Botrytis cinerea* isolated from benomyl-treated cyclamen was resistant to thiabendazole and fuberidazole as well supports the idea of the common mode of action. Similarly, benomyl-resistant isolates of *Penicillium brevicompactum* and *P. corymbiferum* were observed by Bollen (1971) to be resistant to thiabendazole and fuberidazole. On the other hand Bartels-Schooley and MacNeill (1971) report on a fuberidazole-resistant mutant of *Fusarium oxysporum* obtained by ultra-violet irradiation, which was normally sensitive to benomyl and thiabendazole. No quantitative data are given on the degree of resistance. The exceptional position of fuberidazole would be explicable if this resistance depended on breakdown of the toxicant or on another process unrelated to the inherent action of the compound.

Thiophanates In their performance against plant diseases thiophanate-methyl and thiophanate show a striking similarity to benomyl (Aelbers, 1971; Matta and Gentile, 1971). Moreover, the antifungal spectrum of the thiophanates closely parallels that of benomyl (Bollen, 1972). We suggested that thiophanate-methyl might be transformed into MBC, the toxic principle of benomyl (see below) and indeed in tap water, buffer or

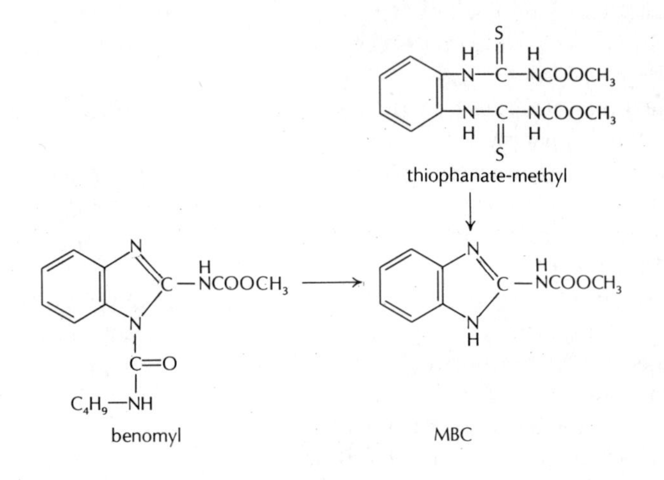

sterile nutrient medium thiophanate-methyl slowly formed MBC (Selling *et al.*, 1970; Vonk and Kaars Sijpesteijn, 1971) (as illustrated by Fig. 7.1). Antifungal activity as well as amount of MBC formed increased with the age of

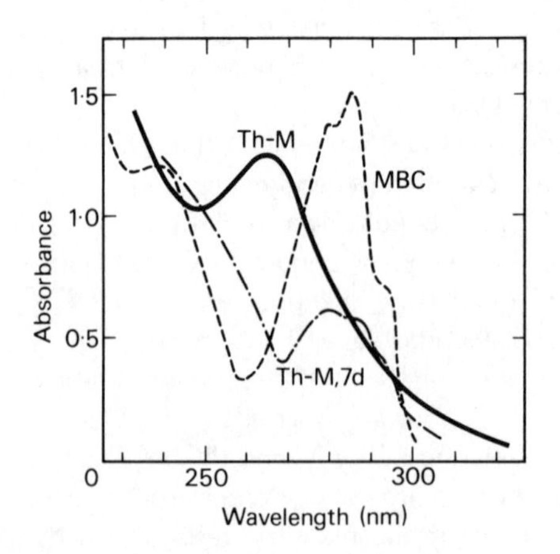

Fig. 7.1 UV-spectra of thiophanate-methyl (Th-M), MBC and of thiophanate-methyl shaken for 7 days in sterile glucose-mineral salt solution (Th-M, 7 d.)

a solution of thiophanate-methyl; moreover, fungitoxicity as well as the rate of MBC formation (see Fig. 7.2) strongly decreased with pH. From this correlation we concluded that the antifungal effect of thiophanate-methyl depends entirely upon the formation of MBC. Similarly, the antifungal effect of thiophanate could be assigned to formation of ethyl benzimidazol-2-yl carbamate (EBC) which is slightly less fungitoxic than MBC.

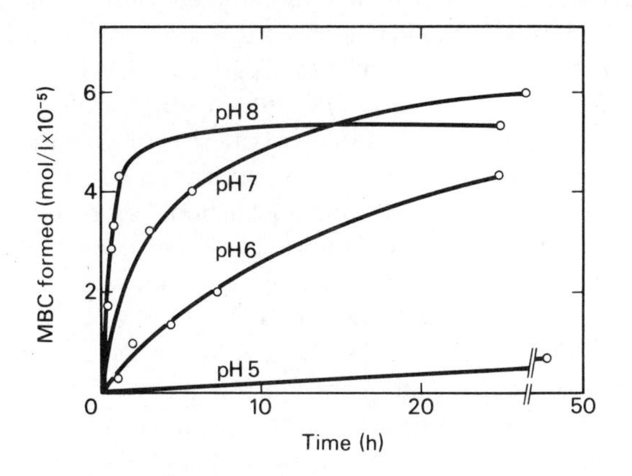

Fig. 7.2 Influence of pH on the formation of MBC from thiophanate-methyl in phosphate buffers at 70°C

Because fungitoxicity of both benomyl and thiophanate-methyl apparently depends on formation of the very same compound, MBC, it is obvious that benomyl and thiophanate-methyl have in fact the same mode of action. The observation that the transformation into MBC of benomyl proceeds far more rapidly than that of thiophanate-methyl, even at pH 7, explains the stronger fungitoxicity of the former compound. In view of this common fungitoxic principle benomyl-resistant fungal strains should of course also be resistant to thiophanate-methyl. This was in fact observed by Bollen and Scholten (1971) for *Botrytis cinerea* and by Bollen (1971) for *Penicillium brevicompactum* and *P. corymbiferum*.

In contrast to the above Noguchi and Ohkuma (1971b) reported that the thiophanates, though active in combating plant disease are inactive *in vitro* against fungi. No mention is made, however, of the concentrations examined nor of the pH of the medium used. It seems remarkable that, although the authors observed transformation of the thiophanates into MBC and EBC in phosphate buffer pH 7·0 as well as in plants (Noguchi and Ohkuma, 1971a), and incubation in rice sheath sap was found to activate the former two compounds (Noguchi and Ohkuma, 1971b), MBC or EBC were not recognised by the authors as the actual fungitoxic principles.

From the above observations and those of Matta and Gentile (1971) that in contact with plant tissue or sap the thiophanates become more fungitoxic than they are in water, one must conclude that the rate of MBC or EBC formation increases under these conditions. Autoclaved sap did not influence fungitoxicity. These results point to a metabolic conversion of thiophanates which is superimposed on the conversion taking place in water. Also fungi were found to increase the rate of MBC formation from thiophanate-methyl (Vonk and Kaars Sijpesteijn, 1971).

Mode of action of benzimidazole and thiophanate fungicides

As shown above benomyl and thiophanate-methyl have the same biochemical mode of action, both being converted into MBC. For thiabendazole and fuberidazole the mode of action is expected to be quite similar to that of MBC in view of the similarity in structure and antifungal spectrum as well as of the observed cross-resistance with MBC.

A thorough study of the action of MBC was made by Clemons and Sisler (1971). They report that mycelium growth is inhibited rather than spore germination. The primary effect of low concentrations of MBC on growing cultures is an increase in dry weight which, however, is not accompanied by an increase in cell number. A study of the effect of benomyl on the fine structure of *Botrytis fabae* was made by Richmond and Pring (1971). In the presence of sublethal concentrations the conidia produced swollen and distorted germ tubes. In these the orientation of organelles towards the hyphal tips was disorganised. The endoplasmic reticulum as well as the nuclei were abnormal, whereas the mitochondria appeared unaffected. The insensitivity to benomyl of Phycomycetes is supposed to be associated with the different organisation of their endoplasmic reticulum.

Growth-inhibiting concentrations of MBC did not affect oxidation of glucose by conidia of *Neurospora crassa* or sporidia of *Ustilago maydis* in buffer solution during a 12 h period (Clemons and Sisler, 1971). The toxicant was, moreover, not lethal to these organisms when incubated in buffer solution; in media supporting growth, however, it was lethal after about 10 h, as judged by loss of ability to form colonies.

Of the several aspects of cell metabolism studied, DNA synthesis was the process most rapidly and severely inhibited by MBC as measured by the total amount of DNA formed. Clemons and Sisler conclude from their results that the mode of action of MBC is an interference with DNA synthesis or some closely related process such as nuclear or cell division. No experiments have yet been made on the rate of incorporation of nucleotide precursors of DNA in the presence of MBC.

The observation of Hastie (1970) that low concentrations of benomyl can cause haploidisation of diploids of *Aspergillus nidulans* may also point to an interference of MBC with DNA. He suggests an interference with the normal process of chromosome segregation, or induction of chromosome breaks which lead directly to genetic instability.

Sisler (1969) proposes that, by analogy with purines and pyrimidines, MBC and thiabendazole might be upgraded by fungal metabolism to nucleotide derivatives. The deoxyribose-phosphate derivatives would then be the actual toxic agents much in the same way as described for 6-azauracil (Dekker and Oort, 1964) which is metabolised to a ribose-phosphate. The lag period shown for all effects caused by MBC might be the time required for this nucleotide formation. After addition of ^{14}C-labelled MBC, Clemons and Sisler (1971) observed formation of a compound with lower R_F-value than MBC itself in cells of *Ustilago maydis*, but its identity was not established.

Experimental results on the mode of action of thiabendazole in several respects run parallel to those of benomyl or MBC. Thus we found no inhibition of glucose oxidation by washed suspensions of *Aspergillus niger* (Kaars Sijpesteijn, 1969) and *Cladosporium cucumerinum* treated for a 24 h period with 100 ppm of thiabendazole, which is 100 times the growth-inhibiting concentration. Similar observations were made by Lyda and Burnett (1970) for *Phymatotrichum omnivorum*. Gottlieb and Kumar (1970) report a greater effect on elongation of germ tubes and on mycelial growth than on spore germination of *Penicillium atrovenetum* and other fungi. Germ tubes showed malformation (Koch, 1971).

These findings are not in contradiction to a mode of action similar to that described for benomyl. Yet Allen and Gottlieb (1970) conclude from their experiments with *P. atrovenetum* that inhibition of respiration probably is the primary effect of thiabendazole. Growth of this fungus was completely inhibited by 10 ppm of this compound: for inhibition of oxygen consumption, 20 ppm was required. Thiabendazole proved a potent inhibitor of various enzymes in isolated mitochondria. Unfortunately the experimental details given on inhibition of respiration are very limited. Since no inhibition of respiration by thiabendazole has been observed in other fungi the conclusion of the authors at any rate has no general application.

As regards the mode of action of fuberidazole we found no inhibition of respiration of washed suspensions of *Aspergillus niger* in glucose phosphate buffer containing 100 ppm of this compound over a 16 h incubation period. No further data are available.

For the time being it appears most likely that MBC, thiabendazole and fuberidazole have the same or a very similar mode of action.

Oxathiins and related compounds

Like the benzimidazoles the oxathiins display a remarkable selectivity in their antifungal spectrum. This was shown for carboxin (Edgington *et al.*, 1966), as well as for its sulphone, oxycarboxin, and for various other derivatives (Snel *et al.*, 1970). Fungi belonging to the Basidiomycetes are particularly sensitive, e.g. loose smut of barley, bean rust and *Rhizoctonia solani*. Of fungi belonging to other groups only *Verticillium albo-atrum* and *Monilia cinerea* f. *americana* were highly sensitive (Edgington and Barron, 1967).

According to Mathre (1968) the selectivity in action of these compounds may be related to the degree of uptake and binding. Thus he found that the sensitive species *Rhizoctonia solani* and *Ustilago maydis* rapidly absorbed both carboxin and oxycarboxin whereas two insensitive species *Fusarium oxysporum* f. sp. *lycopersici* and *Saccharomyces cerevisiae* took up little of either fungicide from solutions. Similarly the greater effectiveness of carboxin over oxycarboxin may be due to the greater uptake of the former compound.

Edgington and Barron (1967) reported that the antifungal spectrum of the 2′-phenyl substituted analogue of carboxin, 2,3-dihydro-6-methyl-5-(2′-diphenylyl) carbamoyl-1,4-oxathiin (F 427), is much wider than that of the other oxathiins: apart from inhibiting Basidiomycetes it also is highly active on several representatives of various other groups of fungi. This observation also would support the view that the selectivity in the oxathiins is not connected with their mode of action but rather depends on differences in their availability at the site of action, caused by selective absorption or permeation. By disrupting cells of *Ustilago maydis* treated with ^{14}C-labelled carboxin Mathre (1968) showed that most of the compound was located in the ribosomal and soluble fraction; only a minor part was associated with the cell wall fraction or the mitochondrial fraction.

In phosphate buffer containing glucose the respiration of mycelium of *Rhizoctonia solani* was inhibited to about 60 per cent by a growth-inhibiting concentration (74·8 M or ca 18 ppm) of carboxin (Mathre, 1970). Similar results were obtained by Ragsdale and Sisler (1970). It is remarkable that in spite of this inhibition of respiration the effect of carboxin is not fungicidal, for Mathre (1968) found that mycelial discs of *Rhizoctonia solani* placed on potato dextrose agar containing 1000 ppm of carboxin even after seven days immediately resumed growth when placed on agar without fungicide. Inhibition of glucose oxidation of *Ustilago maydis* was less severe (Mathre, 1970) and according to Ragsdale and Sisler (1970) declined with age of the sporidia. The latter authors found respiration of conidia of *Neurospora crassa* inhibited by growth-inhibiting concentrations when acetate, but not when glucose was the substrate (Table 7.2). The present author observed 70 per cent inhibition of oxygen uptake by mycelial pellets of *Cladosporium cucumerinum* in the presence of low growth-inhibiting concentrations (10 ppm); the compound F 427 (1 ppm) induced a 55 per cent inhibition of oxidation of glucose.

TABLE 7.2 **Effect of carboxin on oxidation of glucose by sporidia of *Ustilago maydis*, mycelium of *Rhizoctonia solani*, and conidia of *Neurospora crassa* and *Verticillium albo-atrum***

Organism	*µg/ml Toxicant*	*% Inhibition*†
U. maydis (sporidia)		
12 hr old	2	38
19 hr old	2	5
R. solani (mycelia)	10	48
N. crassa (conidia)	50	0
	100	0
V. albo-atrum (conidia)	10	46

† Based on total O_2 consumed in 4 hr by *U. maydis* and *R. solani* and in 3 hr by *N. crassa* and *V. albo-atrum*

It was shown that carboxin also strongly interferes with the synthesis of protein and of RNA and DNA in rapidly metabolising cells of all organisms studied (Mathre, 1970; Ragsdale and Sisler, 1970), but this can be due to a lack of energy as a consequence of inhibition of respiration.

Further studies by Mathre (1971) and by White (1971) revealed that carboxin inhibits mitochondrial respiration of *Ustilago maydis* at or close to the site of succinate oxidation and does not greatly affect the remaining portion of the electron transport system. Georgopoulos and Sisler (1970) made the remarkable observation that certain carboxin-resistant mutants of this fungus at the same time had become quite sensitive to antimycin A; this may point to a change in the electron transport system and it would be another indication that the primary mode of action of carboxin is in this system.

The related compounds mebenil and 2,3-dihydro-6-methyl-5-phenylcarbamoyl-4H-pyran may act in a similar way. No experiments have been reported so far; these products are recommended for combat of Basidiomycetes namely *Rhizoctonia solani* and rusts (Pommer and Kradel, 1969), smuts (Jank and Grossman, 1971) respectively.

Pyrimidine derivatives *Dimethirimol and ethirimol* The pyrimidines dimethirimol (Elias *et al.*, 1968) and ethirimol (Bebbington *et al.*, 1969) have been recommended respectively for the control of cucumber powdery mildew caused by *Sphaerotheca fuliginea* and of cereal powdery mildews caused by *Erysiphe graminis*. They have low mammalian toxicity and act selectively on mildews only.

According to Bent (1970) they inhibit spore germination of mildew of wheat and barley *in vitro*. Complete suppression of spore germination was achieved by 50 ppm of either compound but 5 ppm was quite effective (cf. also Table 7.4). In contrast, a concentration of 50 ppm did not affect germination of conidia of *Venturia inaequalis* or of uredospores of *Puccinia recondita* and the germination of conidia of *Botrytis fabae* and of zoosporangia of *Phytophthora infestans* was reduced by only 7 and 26 per cent, respectively. Mildew of vine *(Uncinula necator)* was less sensitive to

TABLE 7.3 **Effects of metabolites on the systemic action and on the fungitoxicity of dimethirimol in tests on wheat powdery mildew**

Chemical treatment	*Disease on leaf segments*† *(% of disease on untreated segments)*		*Germination of spores in vitro (% of germination of untreated spores)*	
Dimethirimol alone	0	0	3	4
+ adenine	9	—	12‡	—
+ folic acid	64	31	38	34
+ guanine	0‡	—	12	—
+ riboflavin	88	97	47	8‡
+ thiamine	13	—	24	—
+ thymine	0‡	—	28	—
+ uracil	17	—	24	—
None	100	100	100	100
Dimethirimol concentration (ppm)	13	10	10	10
Metabolite concentration (ppm)	1,300	100	100	100
Disease on untreated segments (% area)	100	63	—	—
Germination of untreated spores (%)	—	—	22	31

† Floated on solutions of the chemicals
‡ No significant reversal ($P = 0{\cdot}05$); all other treatments gave a significant decrease in dimethirimol activity

the compounds than the other powdery mildews tested. Whereas cucumber mildew proved more sensitive to dimethirimol than to ethirimol, the reverse is true for wheat and barley powdery mildew.

In a study of the effect of various metabolites on fungitoxicity of the two pyrimidine fungicides Bent (1970) observed that inhibition by dimethirimol (10 ppm) of *in vitro* spore germination of the wheat mildew fungus was markedly antagonised by the incorporation of riboflavin or folic acid (100 ppm) into the solution of the fungicide (Table 7.3). Slighter reversal of activity was found by addition of guanine, thiamine, thymine and uracil. Somewhat similar effects were observed *in vivo*. With ethirimol the same observations were made. The effect of riboflavin could be interpreted as a photochemical inactivation of the fungicide taking place only in the light. It was not stated whether folic acid also reverses when used in a lower, more physiological range of concentrations.

Present evidence points to an interference with tetrahydrofolic-acid-directed C-1 transfer reactions essential for a wide range of cellular components, as for instance purines (cf. Slade *et al.*, 1972). The effect of the two pyrimidine fungicides on enzymes involved in tetrahydrofolate metabolism was examined by Sampson (1969). No inhibition was found, however, of dihydrofolic acid reductase or of N^5N^{10}-methylene tetrahydrofolate reductase in enzyme preparations of cucumber powdery mildew. Limited evidence suggests an inhibition of some pyridoxal-requiring enzymes. Clearly more work will have to be done to reveal the true nature of the action of these fungicides.

6-Azauracil In studying the action of the related compound 6-azauracil, Dekker found that it has systemic activity against powdery mildew of cucumber, as well as against *Cladosporium cucumerinum* of cucumber and several other diseases (Dekker and Oort, 1964); but it never reached the stage of practical application.

Its structure raises the question whether its mode of action is related to that of dimethirimol and ethirimol. A comparison of the properties of azauracil with those of the latter two compounds demonstrates, however, that this cannot be so. Extensive investigations (Dekker and Oort, 1964) have shown that azauracil undergoes metabolic conversion into azauridinemonophosphate. This latter compound inhibits the enzyme orotidylic decarboxylase, which catalyses the decarboxylation of orotidinemonophosphate to form the nucleotide uridinemonophosphate.

Its action is reversed by uracil, which can act as an alternative precursor for uridine monophosphate. 6-Azauracil does not inhibit spore germination, but strongly suppresses the development of haustoria in the host plant. These facts as well as the fact that dimethirimol-resistant cucumber mildew is not resistant to azauracil (personal communication of Dr J. Dekker) clearly demonstrate the different mode of action of azauracil as compared with dimethirimol and ethirimol.

Triarimol This pyrimidine derivative is active in the field against *Venturia* of apple and pear and against powdery mildew of apple, cucumber, cereals and other diseases (Gramlich *et al.*, 1969). It is not effective against Phycomycetes. The action is curative (Brown and Hall, 1971). Unpublished results by the present author revealed that the compound, even at 500 ppm, does not inhibit oxygen uptake by *Aspergillus niger*. Houseworth *et al.* (1971) found no inhibition of CO_2 production; spore germination appeared less affected than germ tube development.

Tridemorph

Tridemorph performs well against powdery mildew of barley and of roses (Pommer and Kradel, 1967). Pommer, Otto and Kradel (1969) described its antimildew action as eradicant. Schlüter and Weltzien (1971) found inhibition of haustorial formation of barley mildew on tridemorph-treated barley leaves.

Against various phytopathogenic fungi the compound had direct fungicidal activity in agar tests (Pommer, Otto and Kradel, 1969). Our results given in Table 7.1 confirm that the compound also inhibits growth of certain non-mildew fungi; moreover, growth of certain gram-positive bacteria was inhibited by 10 ppm of the compound. The long alkyl chain of tridemorph might suggest that this compound interferes with fungal cell membranes resulting in leakage of the cell contents as is known for dodine and polyene antibiotics, but experimental evidence is lacking so far. Low growth-inhibiting concentrations did not inhibit glucose oxidation by *C. cucumerinum* over a 16 h period. A concentration of 100 ppm caused slight inhibition.

Triforine

Triforine shows activity against powdery mildews on cereals, apples and cucurbits, as well as against rusts on cereals, and apple scab (Schicke and Veen, 1969). Results of Fuchs *et al.* (1971) have demonstrated that the compound acts *in vitro* on a variety of Ascomycetes and Fungi Imperfecti. It inhibits the germination of conidia of *Aspergillus niger* and *Cladosporium cucumerinum*. In the experiments the formulated compound was used because of the very low solubility of triforine. The formulating material itself also showed fungitoxicity; this somewhat confuses the interpretation of the results. Haustoria formation by barley mildew was impaired *in vivo* (Schlüter and Weltzien, 1971). Unpublished results by the author revealed that 500 ppm of triforine does not inhibit respiration of *C. cucumerinum*.

Chloroneb

Hock and Sisler (1969a) found that this compound inhibits growth of *Rhizoctonia solani* and *Sclerotinia rolfsii* at a level of 5–8 ppm, whereas it is less active on *Neurospora crassa* and *Saccharomyces pastorianus*. A further study revealed that this selectivity is not based on selectivity in uptake or on metabolic conversion of the compound (Hock and Sisler, 1969b). Both *S. pastorianus* and *Scl. rolfsii* proved unable to metabolise chloroneb, whereas *R. solani* and *N. crassa* are able to do so. Growth inhibition by chloroneb of heavy mycelial suspensions of *R. solani* could even be overcome by quick metabolic conversion of the compound, 2,5-dichloro-4-methoxyphenol being formed.

A study of the mode of action of chloroneb by the same authors (1969a) showed that at a level of 20 ppm respiration of *R. solani* is not inhibited. Incorporation of thymidine into the DNA fraction of the hyphae was 65–90 per cent inhibited by 8 ppm of the compound. Incorporation of phenylalanine into protein and of uridine into RNA were far less inhibited. Since the conversion of thymidine into its triphosphate was not affected and addition of the four deoxyribonucleosides found in DNA to cultures of the fungus did not reduce the toxic effect, chloroneb appears to act by a direct, or indirect inhibition of DNA synthesis at the nucleotide polymerisation stage. Studies on DNA synthesis in a cell-free system from *R. solani* will be necessary to clarify this problem.

Organophosphorus compounds

Wepsyn (triamiphos) This compound is active against powdery mildew of barley and of apple (Koopmans, 1960). Microscopic observations by Magendans and Dekker (1966) revealed that on wepsyn-treated barley plants the main barrier to the development of powdery mildew lies at the stage of haustorial formation. The compound only slightly reduces germination of the conidia placed on glass slides.

De Waard (1971a) developed a technique which facilitates the *in vitro* study of germination of powdery mildew conidia. It is based on his observation that a very high percentage of the conidia of cucumber and barley mildew germinates when dusted onto a cellophane membrane which had been placed on agar in a petri dish. The medium used was Czapek-Dox agar to which 3 per cent sucrose and 0·1 per cent yeast extract had been added. The effect of fungicides could be studied by incorporating them into the agar (de Waard, 1971b). As Table 7.4 shows, neither wepsyn (WP 155) nor the related compound WP 356 inhibits germination of conidia at low concentration. The same was observed by Koopmans (1960) for another derivative of wepsyn and by Dekker and Van de Hoek-Scheuer (1964) for procaine, a systemic antimildew compound not to be described here any further.

TABLE 7.4 **Percentage of germination and estimated percentage of collapse of conidia of *Sphaerotheca fuliginea* on cellophane membranes, laid on an agar medium, mixed with systemic and non-systemic fungicides. Percentage of germination based on control as 100%**

Concentration	10^{-6} M		10^{-5} M		10^{-4} M	
	germ.	*coll.*	*germ.*	*coll.*	*germ.*	*coll.*
Karathane	48	0	0	>90	0	>90
Acrex	15	0	1	0	1	0
Morestan	1	>90	1	>90	1	>90
WP 155	94	0	88	0	39	20
WP 356	92	0	69	0	21	5
PP 149[1]	2	0	2	0	0	0
PP 675[2]	11	0	0	0	0	0
Benomyl	62	0	60	0	57	30
Hoe 2873[3]	57	32	39	37	17	47

Note. The concentrations of the pure compounds in the formulations used were 22·5%, 98%, 25%, 25%, 25%, 80%, 10%, 50% and 100% respectively.
[1] ethirimol [2] dimethirimol [3] pyrazophos

The fact that these compounds are only active on mildew in a stage where they rely entirely on the plant for further development has led to the suggestion that they act by changing host metabolism, thus rendering the plant more resistant to mildew, or by disturbing the intimate relation between host and parasite which is imperative for growth of mildew.

Kitazin This synthetic phosphorus-containing compound inhibits growth of *Pyricularia oryzae* at 50 ppm (Yoshinaga, 1969; Maeda *et al.*, 1970). Mycelial respiration, protein or nucleic acid synthesis are not affected, but the incorporation of ^{14}C-glucosamine into chitin is impaired by this concentration. It was, moreover, observed that UDP-N-acetylglucosamine accumulated in the cells under these conditions. It thus appeared that the compound inhibits the biosynthesis of an important cell wall constituent chitin. Those analogues that were fungitoxic also inhibited glucosamine-^{14}C incorporation whereas the inactive analogues failed to do so.

These results might suggest an inhibition of chitin synthetase by kitazin. However, Maeda *et al.* (1970) found that not only chitin formation is impaired but also the incorporation of ^{14}C-glucose into cell wall glucans was markedly inhibited in kitazin-treated mycelium. Therefore they suggest that the primary action of kitazin is not on chitin synthetase itself. They rather believe that kitazin limits the outward permeation through the cytoplasmic membrane of the substrates for chitin synthetase and glucan synthetase. Since the formation of glycolipids in the membranes was also markedly inhibited in kitazin-treated mycelium, the authors hypothesise that this inhibition may be the primary cause of the impairment of chitin synthesis and growth inhibition, glycolipids being necessary for membrane transport of chitin precursors. Since kitazin can inhibit acetyl cholinesterase, interference with esterases connected with glycolipid metabolism may well be the actual mode of this organophosphorus fungicide.

The action of kitazin, at least in intact fungi, is highly selective as also illustrated by its very low activity towards the four fungi mentioned in Table 7.1. No information is available on activity of the compound towards fungi which lack chitin in their cell walls.

The related compound hinosan (see below) according to de Waard (1972) also affects chitin synthesis.

Pyrazophos This organophosphorus compound, like wepsyn, is highly effective against powdery mildew of cereals and cucumber (Mariouw Smit, 1969). Experiments with barley mildew showed an effect on appressorial formation and penetration into the plant (Schlüter and Weltzien, 1971). By the technique described above de Waard (1971b) observed that it affects *in vitro* in low concentrations the germination of conidia of cucumber and barley mildew, thus suggesting a different mode of action from wepsyn.

Further investigations of this author (de Waard, 1972) revealed the high sensitivity of *Pyricularia oryzae* for pyrazophos. It had no effect on chitin synthesis by this fungus. Since, moreover, a pyrazophos-resistant strain of *P. oryzae* showed no cross resistance towards kitazin or hinosan, its mode of action is regarded to be different from that of these latter compounds.

Antibiotics *Polyoxin D* The antibiotic polyoxin D is a pyrimidine derivative. It inhibits growth of *Cochliobolus miyabeanus, Alternaria kikuchiana, Pyricularia oryzae, Neurospora crassa* and other fungi. Sasaki *et al.* (1968) found that it inhibits neither respiration nor synthesis of nucleic acids or proteins of the former fungus. Incorporation of ^{14}C-glucosamine into cell walls was, however, strongly inhibited.

The biochemical mode of action was studied more closely by Endo *et al.* (1970), using *Neurospora crassa*. A concentration of polyoxin D which strongly inhibits growth (100 ppm) inhibited the incorporation of ^{14}C-glucosamine into chitin while UDP-N-acetylglucosamine accumulated. Also synthesis of chitin from ^{14}C-UDP-N-acetylglucosamine by cell-free preparations of the fungus was strongly inhibited. In such preparations the compound was even far more effective than *in vivo* possibly due to the presence of a penetration barrier in intact mycelium (Endo and Misato, 1969). Thus the antifungal action of polyoxin D appears due to inhibition of the enzyme chitin synthetase. The fact that the antibiotic is a competitive inhibitor for UDP-N-acetylglucosamine in this reaction, is explained by the structural resemblance between these two compounds.

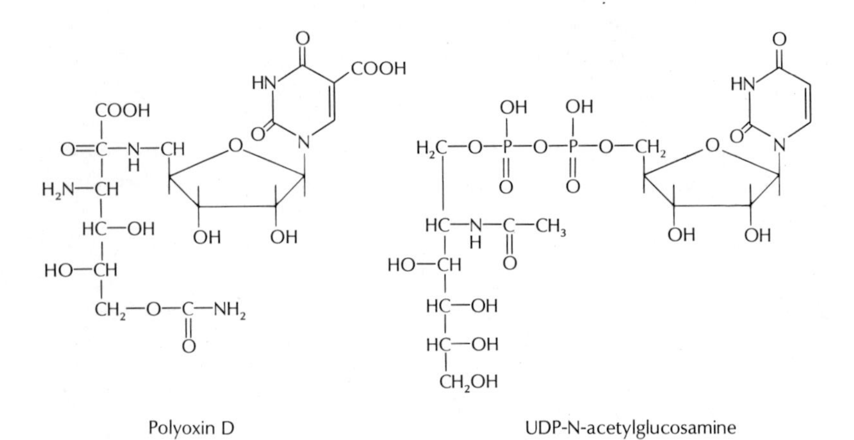

No information is available whether fungi lacking chitin in their cell walls are insensitive to polyoxin D. Endo *et al.* (1970) report that even several fungal species with a very high content of chitin are insensitive to polyoxin D, but possibly the compound does not reach the site of action in these organisms.

The antifungal mode of action of polyoxin D parallels in certain respects the antibacterial mode of action of penicillin, for this compound inhibits the final step in the biosynthesis of murein, a constituent of bacterial cell walls. A structural resemblance of penicillin with an intermediate in murein synthesis is supposed to account for this effect.

When *Cochliobolus miyabeanus* was grown in the presence of polyoxin D it formed giant, osmotically sensitive, protoplast-like structures (Endo *et al.*, 1970), a phenomenon similar to that obtained with bacteria grown in the presence of penicillin.

Cycloheximide This compound acts on a wide variety of fungi including the Phycomycete *Pythium debaryanum* (Whiffen, 1950), but it often shows a great difference in activity towards closely related organisms. Respiration is only slightly inhibited in most organisms. A thorough study of its mode of action was made by Sisler and Siegel (1967). Using *Saccharomyces pastorianus* they found that low fungitoxic concentrations (3·16 ppm) specifically inhibit protein synthesis at the stage where in a complex with the ribosomes amino acids are transferred from tRNA into polypeptides. The precise mechanism and site of action of cycloheximide is still obscure. Although DNA synthesis is also strongly inhibited, protein synthesis is regarded as the primary basis for toxicity. This conclusion finds support in the observation that cell-free protein-synthesising systems of resistant strains of *S. cerevisiae* are resistant to the toxicant whereas those from sensitive strains are sensitive to the compound.

In contrast to *S. pastorianus* and *S. cerevisiae, S. fragilis* is highly insensitive to cycloheximide. This is not due to lack of penetration of the compound or to metabolic conversion, since cell-free protein-synthesising systems from the latter species are far less sensitive to the compound than those of the former species. Cross-over experiments with the 60 *S* and 40 *S* sub-units of ribosomes from the sensitive *S. cerevisiae* and the insensitive *S. fragilis* show that sensitivity to cycloheximide is associated with the 60 *S* ribosomal unit (Rao and Grollman, 1967). These apparently differ in the sensitive and insensitive organisms.

A detailed discussion of the work on cycloheximide is given by Sisler (1969).

Blasticidin Blasticidin is a pyrimidine derivative. *Pyricularia oryzae* is particularly sensitive to the antibiotic, which, unlike most compounds dealt with, also inhibits growth of certain bacteria (Misato, 1967). Since mycelial growth is affected by lower concentrations than spore germination, blasticidin can be applied to rice plants after infection has occurred. The compound is somewhat phytotoxic. Respiration of glucose is 50–60 per cent inhibited both by growth-inhibiting concentrations (0·1 ppm) and by far higher concentrations. At the level of 1 ppm the incorporation of amino acid-^{14}C into the protein fraction of intact mycelium and of cell-free extracts of *P. oryzae* was completely inhibited; nucleic acid synthesis was undisturbed. Since amino acid activation and the transfer of activated amino acids to soluble RNA were less inhibited than overall incorporation, Misato and co-workers concluded that the probable site of action is the final step of protein synthesis taking place on the ribosomes. The rather serious inhibition of respiration will presumably also be of influence on inhibition in as far as it limits energy production.

In strains of *P. oryzae* which had become resistant to blasticidin, this resistance could be ascribed to a difference in permeability of the cells, because in cell-free extracts protein synthesis was as sensitive as in those of the normal strain.

In contrast to *P. oryzae, Pellicularia sasakii* is highly tolerant to blasticidin; protein synthesis both in whole cells and in cell-free extracts was little affected by the compound. Therefore in this case insensitivity to blasticidin is not due to impermeability of the cell membrane but probably to an inherent insensitivity of the processes leading to protein synthesis.

No experiments on cross resistance between cycloheximide and blasticidin are known so far; also it is unknown whether both compounds have the same mode of action.

Kasugamycin This compound selectively inhibits *Pyricularia oryzae* and some bacterial species. It is active only at low pH. The antibiotic inhibits protein synthesis *in vivo* as well as in cell-free systems, nucleic acid synthesis being undisturbed. The process that is inhibited appears to be the binding of the aminoacyl-tRNA complex with the messenger RNA and ribosome to a complex (Tanaka *et al.*, 1966).

Helser *et al.* (1972) related sensitivity of *E. coli* to kasugamycin to the presence of an RNA methylase. Resistant strains lacked this methylase and were unable to methylate 16*S* ribosomal RNA. This observation is in agreement with the ribosome being the site of action, but the relation between methylase and kasugamycin activity is still obscure.

FACTORS DETERMINING ACTIVITY: MODE OF ACTION; SELECTIVITY AND RESISTANCE; METABOLIC CONVERSION

In the foregoing pages a bird's-eye view has been given of present-day knowledge on the effect of systemic fungicides on their target organisms, the fungi. Surveying this picture one is impressed by the great scarcity of data and it is quite obvious that much work remains to be done.

Mode of action of systemic fungicides

Cycloheximide, blasticidin and kasugamycin apparently act by inhibition of protein synthesis, chloroneb and the benzimidazoles – including the thiophanates – by interference with DNA synthesis, azauracil by inhibition of UMP synthesis which ultimately leads to impairment of RNA synthesis, dimethirimol and ethirimol may act by inhibition of certain C-1 transfer reactions. The action of the polyoxins has been located as obstruction of chitin synthesis and the organophosphates wepsyn, pyrazophos and kitazin may perhaps interfere with essential esterase functions. Of triforine, tridemorph and triarimol we know only that they do not or only to a very slight degree inhibit respiration. Obviously several compounds act as antimetabolites. This has been established for azauracil and polyoxin D, and for the benzimidazoles it is likely that they are antimetabolites of nucleotides or their precursors.

As far as the biochemical mode of action has been more or less elucidated, it is apparent now that most systemic fungicides affect biosynthetic processes and that the oxathiins are the only compounds which act by inhibition of processes essential for energy production in the cell. Disruption of the structure of cell membranes, as occurs after treatment with polyenes or dodine, is not yet known as a mode of action of true systemic fungicides.

In contrast most protectant fungicides act by inhibition of processes involved in energy production. This at least applies to the dialkyl-dithiocarbamates, the bisdithiocarbamates, captan, dinocap, triphenyltin acetate, tetrachloro-isophthalonitrile and phenylmercuric acetate.

A possible explanation for this very low incidence of inhibitors of energy production among the systemic fungicides may be that such inhibitors, as for instance inhibitors of respiration, or uncoupling agents, if systemic, would tend to be equally toxic for plant tissues as for fungal tissues. In contrast inhibitors of biosynthesis which act primarily under conditions of growth might often be far more effective on the fast growing parasitic fungi than on the slow biosynthetic processes going on in most plant cells. Moreover, inhibitors of energy production, on the whole, appear less selective in their antimicrobial spectrum than inhibitors of biosynthesis. Strikingly also antibacterial agents used for medical purposes do not act by inhibition of energy production. The same is true for systemic insecticides used in plant protection.

Obviously the oxathiins are exceptional in that they are systemic fungicides which do inhibit respiration and yet are not too phytotoxic to be used. Mathre (1971) has, however, revealed that the mitochondrial systems of Pinto bean plants are far less sensitive to these compounds than those of fungi. This may explain the absence of strong phytotoxicity.

The study of the biochemical mode of action appears to be almost a prerequisite for the understanding of selectivity of action and of the mechanisms of the resistance which has been observed with certain fungicides, although metabolic conversion of the compounds by the fungal cell or impermeability of the cells may also be a factor of importance. Conversely observations on selectivity of action and on the mechanism of resistance may help to establish the true nature of the biochemical mode of action. The following will show how much interwoven these features are.

Selectivity of action and resistance

Whereas protectant fungicides generally show a broad antifungal spectrum, selectivity in antifungal action is almost a characteristic feature for systemic fungicides. Little is known as yet of the basis of this selectivity. In the case of the benzimidazoles, selectivity of action bears a relation to the systematic position of the fungus, as is also known for the polyene antibiotics. In the latter case the absence of an effect on the Phycomycetes has been attributed to the absence in their membranes of cholesterol, the substance on which these compounds act in sensitive organisms; in the former the difference in organisation of the endoplastic reticulum is held responsible for selectivity (Richmond and Pring, 1971). One may expect fungi lacking chitin to be insensitive to polyoxin in view of the interference of this compound with chitin biosynthesis, but no reports on the antifungal spectrum are available so far.

Lack of activity of cycloheximide and of blasticidin on certain fungi has been related to an inherent insensitivity of the system which in sensitive organisms is affected, namely protein synthesis.

The lower activity of chloroneb on certain fungi could not be ascribed to differences in the rate of metabolic conversion of this compound.

Carboxin on the whole is more active on Basidiomycetes than on Ascomycetes and Fungi imperfecti. In this case differences in uptake by the cells appeared to account for selectivity. This view is supported by the observation that by introduction of one phenyl group a compound with a far wider spectrum was obtained. It seems possible that differences in penetration and accumulation frequently account for selectivity of fungicides.

Most antimildew agents appear to act only on these obligate parasites, whereas compounds inhibiting growth of other fungi frequently are inactive towards powdery mildews. This exceptional situation of the mildew fungi may partly be due to their rather strongly lipophilic character; on the other hand several antimildew agents, e.g. wepsyn, are supposed to act primarily on the host plant, the effect on the parasite being only a consequence of this.

Several authors have stressed the almost complete absence of activity of systemic fungicides against fungi belonging to the Phycomycetes. So far no general explanation can be given for this fact which in some way is reminiscent of the exceptional position taken up by gram-negative organisms amongst the bacteria, which on the whole are far less sensitive to bactericides than the gram-positive bacteria.

The above observations enumerate various mechanisms by which selectivity in antifungal action may be brought about. These are:

absence of a vulnerable mechanism, inherent insensitivity of the system affected in sensitive organisms, metabolic conversion of toxicant to inactive product, lack of sufficient penetration and accumulation of the compound in the cells. It seems conceivable that ineffectiveness can also be achieved by inability of an organism to convert an inactive precursor into an active toxicant: but so far this mechanism has not yet been encountered as an explanation for selectivity between different species. Selectivity in action is also met when resistant mutant strains of a sensitive species emerge. The mechanisms on which selectivity between sensitive and mutant strains can depend may in principle be the same as described above for selectivity between different species; thus inherent insensitivity of protein synthesis in the case of cycloheximide, decreased permeability of the cells in the case of blasticidin and lack of metabolic conversion into the toxicant in the case of 6-azauracil. In Chapter 8 this subject will be dealt with in more detail.

It is not surprising that fungicides often appear to be far more common biocides. After all, many biochemical mechanisms in plants, animals as well as micro-organisms show basic similarities. To mention some examples: tridemorph and blasticidin have antibacterial activity, benomyl is known as an acaricide and thiabendazole as an anthelminthic; wepsyn, pyrazophos and kitazin have relatively high mammalian toxicity possibly because they inhibit cholinesterase activity. Cycloheximide is an inhibitor of protein biosynthesis in higher plants and in animals; it is a specific inhibitor of 80 *S* ribosomes and does not act on 70 *S* ribosomes such as occur in bacteria. Accordingly, there is no antibacterial activity. From a point of view of selectivity polyoxin D is an ideal fungicide in that its antifungal action depends on inhibition of chitin synthesis, a process only known to occur in certain fungi and lower animals. Its low mammalian toxicity may be due to this selective activity.

Metabolic conversion

On the whole the low acute oral mammalian toxicity of many fungicides dealt with may often be due to metabolic conversions taking place in the animal tissues, usually followed by excretion, as observed for benomyl, thiabendazole, fuberidazole, chloroneb, dimethirimol, ethirimol, carboxin.

Also the compounds can be transformed by microbial and plant metabolism, a process which often results in loss of antifungal activity. The phenomenon of conversion is so important from a toxicological, environmental and ecological point of view that it seems appropriate to mention some observations with regard to the systemic fungicides.

MBC, the fungicide formed from benomyl and thiophanate-methyl in aqueous solution, appears to be quite persistent in plants. No break-down products have been reported so far. MBC and thiabendazole have considerable persistence in soil (Hine *et al.*, 1969; Erwin *et al.*, 1971). Wang *et al.* (1971) report that in cotton plants treated with ^{14}C ring-labelled thiabendazole 60 per cent was present as the free original compound; the remainder appeared to be bound in higher molecular weight compounds which have not yet been identified. According to Frank (1971) the furan ring of fuberidazole is split open in certain animals to yield a γ-hydroxy butyric acid derivative, but it is unknown whether this reaction also takes place in plants or microbes.

The oxathiins are somewhat more liable to conversion. Several authors have observed the oxidation of carboxin in various plants, the main product being the non-fungitoxic sulphoxide (see below) (Snel and Edgington, 1970; Allam and Sinclair, 1969); small amounts of the fungitoxic sulphone

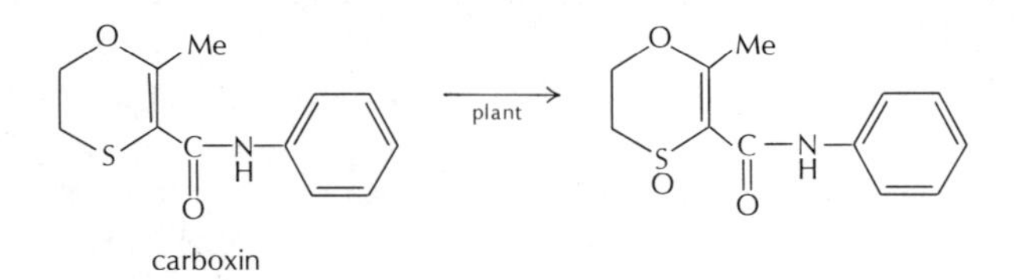

carboxin

were found as well (Chin *et al.*, 1970a). Hydrolysis of carboxin was reported for bean (Snel and Edgington, 1970) but found not to occur in wheat and barley (Chin *et al.*, 1970a). In soil the compound was eventually oxidised to the sulphoxide, but no hydrolysis took place (Chin *et al.*, 1970b). This would mean that micro-organisms were unable to break down the main structure of the compound.

Ethirimol and dimethirimol equally are quite unstable in plants. The ethyl and methyl groups are removed by metabolic activity. These compounds and the parent fungicides are subsequently converted to water-soluble conjugates, including glucosides. Some of the metabolites retain fungicidal activity (cf. Slade *et al.*, 1972). Ethirimol is known to be degraded in soil, but the end products have not been reported (Brooks, 1971a).

Metabolism of triforine in various plants was studied by Fuchs *et al.* (1972). The fungicide was converted to a series of non-fungitoxic products the terminal residue being possibly piperazine.

In cotton and bean, chloroneb is rather stable; the principal metabolite is 2,5-dichloro-4-methoxy phenol (Rhodes *et al.*, 1971). According to Hock and Sisler (1969b) the same compound is produced by incubation of *Rhizoctonia solani* and certain other fungi with chloroneb; it is not fungitoxic. This conversion by fungal metabolism may account in part for the slow breakdown in soil observed by Rhodes *et al.* (1971).

Metabolic conversion of kitazin by *Pyricularia oryzae* was demonstrated by Uesugi (1970) using the ^{32}P and ^{35}S labelled compound. The main pathway of metabolism is by hydrolysis of the S—C linkage. The metabolic fate of hinosan in the rice plant has been subject of careful studies, the results of which are given in the figure below taken from a Hinosan Information Bulletin.

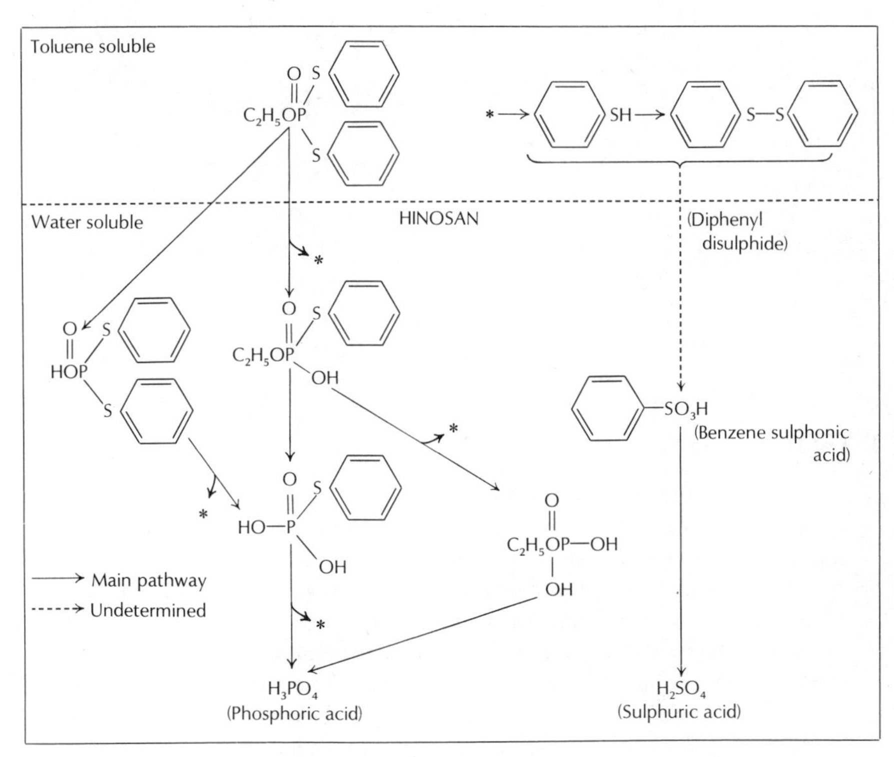

Fungal breakdown of cycloheximide has been observed for *Myrothecium verrucaria* by Walker and Smith (1952).

This survey of metabolic transformations taking place in plants and micro-organisms obviously shows great gaps in our knowledge. One may, however, assume that much knowledge gathered by industrial research has remained unpublished.

After conversion of a fungicide has led to inactive compounds the effect on the fungus has come to an end; the above survey may, however, have shown that breakdown of the compounds in most cases does not by any means go to completion. In contrast, the skeleton of the molecule often remains more or less intact and this fact may stimulate the search for compounds which after having done their job are subject to extensive decomposition.

8

Resistance

by J. Dekker

INTRODUCTION

Living organisms possess the ability to adapt to changing environmental conditions. This is illustrated in the history of life on earth. The evolution of living matter would have been impossible without this property.

In higher organisms this adaptation or evolution is a time-consuming process, but in micro-organisms with a short reproduction time and an abundant procreation, adaptive changes may be observed more readily. Well known is the adaption of bacteria to antibiotics, which became apparent shortly after the application of these biocides for control of bacterial diseases in humans. Also insects and mites may adapt rapidly to certain new organic insecticides, which has caused problems in the control of plant pests. In contrast to this, adaptation of fungi to agricultural fungicides has until recently been observed only rarely in practice. In view of this the opinion has been expressed that fungi might be less liable to adapt to biocides than bacteria or insects. From experience with new organic chemicals with selective action on fungi, it is now known that this assumption is no longer tenable. Reduced sensitivity of fungi to various of these fungicides has been encountered both in laboratory experiments and in practice.

Organisms which exhibit a reduced sensitivity or even insensitivity to a toxicant, are called 'resistant' or 'tolerant'. These terms denote the response of a fungus, of which the development is unchecked or only slightly inhibited by the fungicide concerned at concentrations which are inhibitory to other fungi or to other strains of the same fungus. Both terms are often used interchangeably, when describing the relation between fungus and fungicide, although they have a different meaning with respect to the relation between host and parasite. In the latter case the plant may resist attack or be susceptible to it, and after the attack has succeeded, may prove tolerant or sensitive. In the former case it seems of little practical relevance whether the fungal cell resists 'attack' by preventing the fungicide from entering, or is insensitive after penetration has taken place. Moreover, the mechanism of resistance or tolerance is not known for most fungicides.

Within this concept of resistance (tolerance), we may discern natural resistance, present in the whole population of a fungal species, family, class or order, and acquired resistance in strains of a normally sensitive species. Only the latter type of resistance will be discussed in this chapter.

OCCURRENCE OF ACQUIRED RESISTANCE

To conventional fungicides

The ability of fungi to adapt to fungicides may be studied in the laboratory in two ways. The first involves the transfer of fungal mycelium to agar media containing increasing concentrations of the fungicide. This so-called 'training' has been the preferred method in experiments with the conventional agricultural fungicides†. The second method consists of exposure of the fungus to mutagenic agents or rays, and subsequent selection of the survivors after seeding the spores on an agar medium containing a normally inhibitory concentration of the fungicide. Most of this work has been reviewed by Ashida (1965) for inorganic fungicides and by Georgopoulos and Zaracovitis (1967) for organic fungicides. A high degree of resistance to these fungicides was seldom observed either in the laboratory or in the field, and acquired resistance usually disappeared again after transfer of the fungus to a medium free of toxicant.

There are, however, a few exceptions to this general rule. Resistance has developed to some aromatic hydrocarbon fungicides, such as diphenyl and sodium orthophenylphenate (Eckert, 1967), chlorinated nitrobenzenes (Georgopoulos, 1963; Esuruoso and Wood, 1971; Priest and Wood, 1961), dicloran (Locke, 1969; Webster *et al.*, 1970) and hexachlorobenzene (Kuiper, 1965). Apart from these aromatic hydrocarbon fungicides, the use of which is rather limited in practice, acquired resistance has been reported to only a few other commercial fungicides. It was found that the

† The adjective 'conventional' will be used for those fungicides, which show little or no penetration into the plant tissue, such as metal toxicants, dithiocarbamates, captan, etc.

application of dodine against apple scab had become less effective in New York state (Szkolnik and Gilpatrick, 1969). Resistance to this fungicide in *Nectria haemotococca* was studied by Kappas and Georgopoulos (1970, 1971). Development of resistance in *Sclerotinia homoeocarpa* on turf grass to cadmium-containing compounds (Cole *et al.*, 1968) and to dyrene (Nicholson *et al.*, 1971) has been reported from the USA. The inheritance of cadmium tolerance in a natural population of *Cochliobolus carbonum* was studied by MacKenzie *et al.* (1971). Failure to control *Drechslera avenae (Pyrenophora avenae)* on oats with organic mercury compounds has been reported from Scotland (Noble *et al.*, 1966), Ireland (Malone, 1968), England (Dickens and Sharp, 1970), USA (Crosier *et al.*, 1970) and The Netherlands. Tolerance obtained by training of this fungus on mercury-containing media in the laboratory was, however, lost again after transfer to fungicide-free media (Greenaway and Cowan, 1970).

Considering the long period during which the conventional fungicides have been in commercial use, the number of cases in which the emergence of tolerance has become a problem for practical disease control is rather low and concerns mostly fungicides which are used on a limited scale for special purposes.

To systemic fungicides

It seems, however, that the situation with respect to acquired resistance has changed with the introduction of systemic fungicides, which are taken up and transported in the system of the plant. Shortly after these fungicides came into practical use various cases of acquired resistance were reported. A dramatic example is the rapid development of tolerance in *Sphaerotheca fuliginea* to the pyrimidine derivative dimethirimol (Milcurb). This compound was introduced in Holland in 1968 and used on a large scale for control of powdery mildew in cucumber greenhouses in the following year. A suspension of this fungicide, applied to the soil around the base of the plant, provided continuous protection of the whole plant for several weeks. The absence of residues on the leaves and the low cost of this labour-saving technique also contributed to the success of this systemic fungicide. In the autumn of 1969, and more so in the spring of 1970 the results obtained were less satisfactory in some greenhouses. Experiments with powdery mildew from sites of poor control and from sites where dimethirimol had never been used, revealed a certain degree of tolerance to this fungicide. Apparently the concentration of dimethirimol does not reach a level in the plant which is sufficiently high to inhibit the development of tolerant fungi. According to Bent *et al.* (1971) tolerant mildew was still widespread in Holland in May 1971, after a year in which very little dimethirimol had been used. They noted that resistance had been detected only in glasshouses in N.W. Europe and not yet in field-grown crops.

A second example of fungicide resistance has been encountered in the cyclamen culture in Dutch greenhouses, where *Botrytis cinerea* causes a serious heart rot. As conventional fungicides performed unsatisfactorily against this disease, the infected plant parts had to be removed by hand, a labour-intensive and therefore costly procedure. Application of benomyl provided an excellent solution to this problem as a 0·15 per cent spray with the formulated product (Benlate 50 per cent W.P.), kept the plants free from the disease for several weeks. The enthusiasm of cyclamen growers was tempered, however, when after a short period of use benomyl appeared to be less effective than originally. From diseased plants a strain of *B. cinerea* was isolated which tolerated much higher concentrations of benomyl than the wild type fungus. Even at 1000 ppm of benomyl there was not a complete inhibition of growth, while the wild type fungus was eliminated at 0·5 ppm of benomyl in the medium (Bollen and Scholten, 1971) (Fig. 8.1). The tolerant strains showed cross resistance to the benomyl-

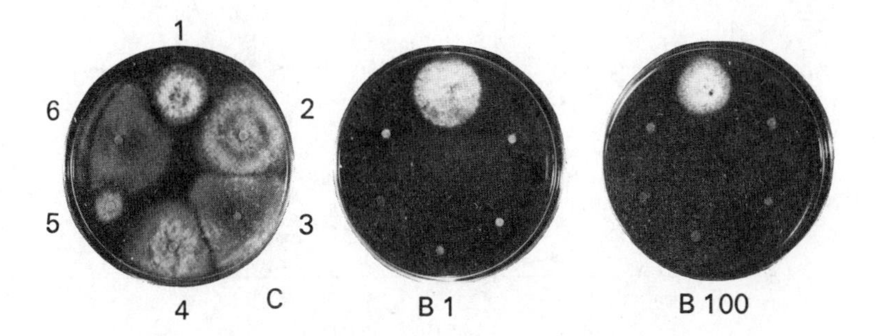

Fig. 8.1. Acquired resistance of *Botrytis cinerea* to benomyl. Growth of six isolates on PDA without fungicide (C) and with 1 (B1) and 100 (B100) µg/ml Benlate, 50 per cent W.P.; 1: resistant isolate from cyclamen. 2, 3, 4, 5, 6: sensitive isolates from cyclamen, dahlia, lettuce, sunflower and soil, respectively (from Bollen and Scholten (1971), *Neth. J. Pl. Path.* 77, 86)

related compounds thiabendazole and fuberidazole, and also to thiophanate-methyl. Cross-resistance to the latter compound, which is not a benzimidazole derivative, can be understood, since we know that both benomyl and thiophanate-methyl are converted to methyl benzimidaz-2-yl-carbamate (MBC), which is the actual fungitoxic compound (Selling *et al.*, 1970).

The remarkable observation was made that in benomyl-treated greenhouses, plants infected with the benomyl-resistant strain of *B. cinerea* were more severely diseased than plants in untreated greenhouses. This is not necessarily due to an increased virulence of the resistant strain, but might also be caused by the inhibition of benomyl-sensitive antagonistic fungi. It was noticed that after prolonged benomyl treatment disease incidence again appeared reduced in some plants. From these plants Bollen (1971) isolated two strains of *Penicillium brevicompactum*, non-pathogenic to cyclamen, which also had acquired resistance to benomyl. Both strains tolerated a concentration of 1000 ppm of benomyl in an agar medium, while growth of the wild type fungus was completely inhibited at a concentration of 0·5 ppm (Fig. 8.2). When tested on malt agar, a striking

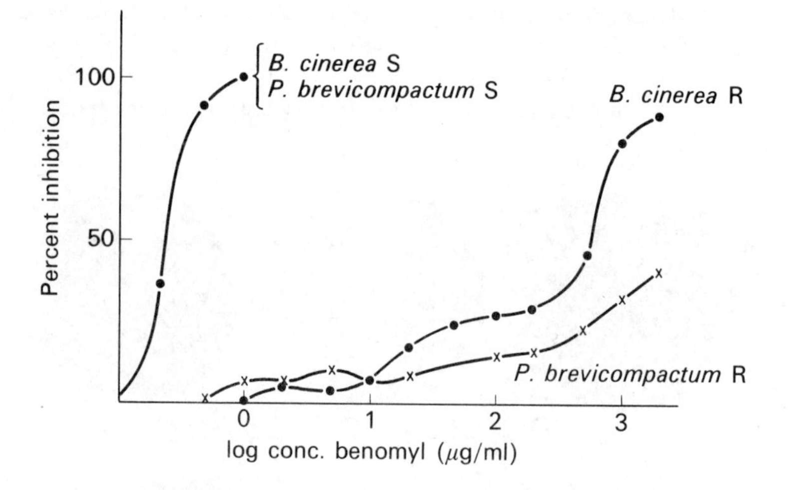

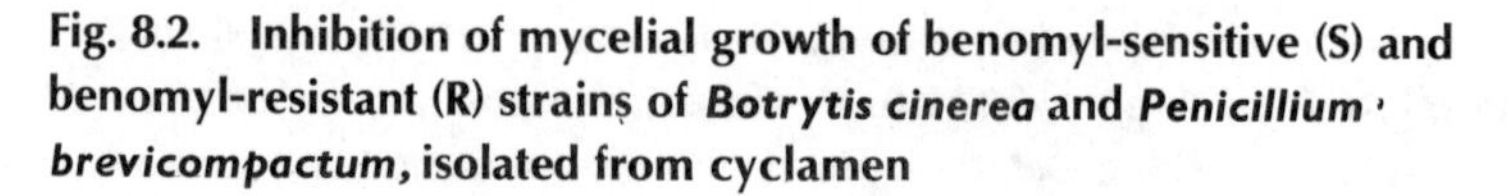

Fig. 8.2. Inhibition of mycelial growth of benomyl-sensitive (S) and benomyl-resistant (R) strains of *Botrytis cinerea* and *Penicillium brevicompactum*, isolated from cyclamen

antagonism between benomyl-resistant strains of *B. cinerea* and *P. brevicompactum* was obtained. This antagonism, which was shown by a growth-free zone between the two fungi, was stronger in the presence of high concentrations of benomyl than without this fungicide in the medium. This indicates that the use of selective fungicides may interfere with the relation between both pathogenic and non-pathogenic micro-organisms.

Other cases of acquired resistance to benomyl in practice have been reported. Schroeder and Provvidenti (1969) observed a decreasing effect of benomyl against cucumber powdery mildew, which they attributed to acquired resistance. Netzer and Dishon (1970) found resistance in fungi to benomyl in powdery mildew on muskmelon. From a cucumber greenhouse in Holland, where benomyl had been applied against powdery mildew, a strain of *Sphaerotheca fuliginea* was obtained which, in a leaf disc test, tolerated concentrations of benomyl which were a hundred to a thousand times as high as those tolerated by the sensitive fungus (Table 8.1).

TABLE 8.1 **Development of resistance to benomyl (Benlate 50 per cent W.P.) in *Sphaerotheca fuliginea*. Disease assessment after seven days, using a scale from 0 (no mildew) to 5 (covered with mildew). R = resistant, S = sensitive**

	Leaf discs floating on solution							*Roots in nutrient solution*				
	0	0·1	1	10	100	500 ppm		0	0·5	5	50	500 ppm
S	5	0	0	0	0	0	S	5	5	0	0	0
R	5	5	5	5	3	0	R	5	5	5	4	0

Before the introduction of systemic fungicides in practice, it was already known from laboratory experiments that resistance in fungi to selective fungicides, namely antibiotics and antimetabolites, could arise readily. Resistance to antimycin A in *Venturia inaequalis* was demonstrated by Leben *et al.* (1955) and different types of resistance to 6-azauracil in *Cladosporium cucumerinum* by Dekker (1967, 1968b). The number of reports concerning the development of resistance in fungi in the laboratory as well as in practice is now increasing rapidly (Table 8.2).

ORIGIN OF RESISTANCE

Resistance to fungicides may emerge as a consequence of genetic changes in the fungal cell. Although fungicide tolerance due to non-genetic factors has often been observed, the physiological and biochemical aspects of this phenomenon have been relatively little studied. Partridge and Rich (1962) accepted adaptive enzyme formation as the most probable origin of increased tolerance of *Penicillium notatum* to glyodin and captan. As this type of tolerance seldom reaches a high level, and is usually rapidly lost again after transfer to a fungicide-free medium, it seems of limited

TABLE 8.2 **Acquired resistance of fungi to systemic fungicides**

Benomyl and related compounds
- *Aspergillus nidulans* (Hastie and Georgopoulos, 1971)
- †*Botrytis cinerea* (Bollen and Scholten, 1971)
- *Cladosporium cladosporioides* (Bollen, unpublished)
- *Cladosporium cucumerinum* (Dekker, unpublished)
- †*Erysiphe cichoracearum* (Netzer and Dishon, 1970)
- *Fusarium oxysporum* f. sp. *lycopersici* (Thanassoulopoulos *et al.*, 1971)
- *Fusarium oxysporum* f. sp. *melonis* (Bartels-Schooley and McNeill, 1971)
- *Neurospora crassa* (Sisler, 1971)
- †*Penicillium brevicompactum, P. corymbiferum* (Bollen, 1971)
- †*Sphaerotheca fuliginea* (Schroeder and Provvidenti, 1969)
- *Ustilago hordei* (Ben-Yefet and Dinoor, pers. comm.)

Thiophanates
- †*Botrytis cinerea* (Bollen and Scholten, 1971)
- †*Penicillium brevicompactum, P. corymbiferum* (Bollen, 1971)

Dimethirimol
- †*Sphaerotheca fuliginea* (Bent *et al.*, 1971)

Ethirimol
- †*Erysiphe graminis* (Wolfe, 1971)

Triarimol
- *Sclerotinia sclerotiorum* (Paster and Dinoor, pers. comm.)

Oxathiins
- *Rhizoctonia* spp. (Grover and Chopra, 1970)
- *Ustilago maydis* (Georgopoulos and Sisler, 1970)
- *Ustilago hordei* (Ben-Yefet and Dinoor, pers. comm.)
- *Neurospora crassa* (Sisler, 1971)

Kasugamycin
- *Pyricularia oryzae* (Ohmori, 1967)

Blasticidin-S
- *Pyricularia oryzae* (Nakamura and Sakurai, 1962)

Chloroneb
- *Ustilago maydis* (Tillman and Sisler, 1971)

Kitazin
- *Pyricularia oryzae* (De Waard, unpublished)

Dichlozoline
- *Aspergillus nidulans* (Dekker, unpublished)

† Strains isolated from plants in the greenhouse or the field

practical importance. Therefore no special attention will be paid to it in this chapter.

Genetic changes in the fungal cell, which originate as mutations, may be distributed in the population by a variety of Mendelian and non-Mendelian processes. Fungicide resistance has been most intensively studied to some members of the aromatic hydrocarbon group. By crosses made between resistant and sensitive strains of fungi it has been shown that resistance to these fungicides develops through mutation of any one of a number of Mendelian factors. Five loci for resistance to aromatic hydrocarbons have been identified in *Nectria haematococca* (Georgopoulos and Panopoulos, 1966) and two in *Aspergillus nidulans* (Threlfall, 1968).

The level of the resistance may or may not be different for different loci, and further influences may result from the existence of more than one resistant allelomorph at each locus and also by the presence of modifying genes. The studies on resistance to aromatic hydrocarbons and to dodine in *Nectria haematococca* and to cycloheximide in *N. crassa* (Vomvoyanni *et al.*, 1971) have provided good examples of the existing possibilities in this respect. When more genes, each of which confers resistance, are brought together into the same haploid nucleus, resistance may be additive, as was found for resistance to dodine (Kappas and Georgopoulos, 1970), or non-additive, as was demonstrated for aromatic hydrocarbons, where double mutants appeared phenotypically indistinguishable from the single tolerant strain (Georgopoulos, 1963).

Resistance to benomyl, induced by ultra-violet irradiation, has been studied in nine strains of *Aspergillus nidulans* (Hastie and Georgopoulos, 1971). Five of these strains showed relatively high-level resistance, while the four others appeared less resistant. From crosses with the wild type fungus and subsequent ascospore analysis, it was concluded that benomyl tolerance in each strain was determined by a single gene mutation, ben-1 or ben-2. It was shown that ben-1 and ben-2 were not allelic and recombined freely, but that they did not have additive effects. Resistance appeared recessive in this case, but may be dominant in other cases.

Mutations for tolerance will be immediately expressed in haploid pathogens, and in diploid pathogens, when tolerance is dominant to sensitivity. When, however, tolerance is recessive in diploid pathogens, the expression of resistance may be delayed until after segregation.

The frequency with which fungicide-resistant strains will emerge in the laboratory depends on the number of mutations required for a certain level of resistance and the mutability of the genes at the loci concerned. It is possible that, when a fungicide interferes at more sites with the metabolism of the fungus, more mutations will be needed for resistance. This might be an explanation for the fact that little or no genetic resistance has been found to multi-site inhibitors, such as dithiocarbamates and metal toxicants. If, on the other hand, mutation in one gene may confer resistance, as is the case in *Nectria haematococca* to aromatic hydrocarbon fungicides, and in *Aspergillus nidulans* to benomyl, resistant strains might be obtained more readily. Hastie and Georgopoulos (1971) obtained nine benomyl-resistant mutants from a sample of 8×10^6 spores of *A. nidulans* after u.v. treatment, and Bartels-Schooley and MacNeill (1971) nine out of $4{\cdot}6 \times 10^5$ spores of *Fusarium oxysporum* f. sp. *melonis*. Considering the large numbers of spores which may be produced by fungi, the development of resistant strains in cases like this might be a common phenomenon.

The question whether a fungicide induces mutation or only acts as a selective agent cannot always be answered easily. It may be assumed that most fungicides do not possess mutagenic activity, and that mutations arise 'spontaneously', i.e. without interference of the fungicide. For benomyl, however, Hastie (1971) obtained evidence that it induces instability in *A. nidulans*, where abundant sectors were induced in colonies grown from heterozygous diploid conidia. He suggested that benomyl either interferes with the normal process of chromosome segregation or induces chromosome breaks which lead directly to genetic instability.

An example of acquired resistance due to heterokaryosis has been studied by Webster *et al.* (1970) in resistance of *Botrytis cinerea* to dicloran. On media containing this fungicide tolerant sectors appeared frequently in the colonies. No training was necessary to obtain high levels of resistance. The authors express the opinion that heterokaryosis may serve as a mechanism in maintaining low levels of tolerance to dicloran in a multinucleate fungus such as *B. cinerea*. Dicloran resistance might emerge when the number of nuclei carrying resistance increases due to selection pressure exerted by the chemical. Evidence for such a phenomenon was also presented by Meyer and Parmeter (1968), who obtained tolerance in *Thanatephorus cucumeris* to pentachloronitrobenzene after heterokaryotic association from homokaryons both of which were intolerant.

MECHANISM OF RESISTANCE

The interesting question arises as to how a genetically determined resistance becomes expressed in the metabolism of the fungus. It is obvious that an explanation of this mechanism of resistance in a fungus to a fungicide is only possible if we know the mechanism of action of the toxicant. It must be admitted that little detailed information is yet available about the mechanism of action of the systemic fungicides in practical use. Therefore experimental antifungal compounds with selective action, such as antibiotics and antimetabolites will also be included in this discussion.

Fungicides may interfere with the metabolism of the fungal cell at various sites. Some samples of fungicidal action and acquired resistance are presented in Table 8.3.

TABLE 8.3 **Sites of action of fungicides and some examples of acquired resistance in fungi**

Site	*Interference with*	*Fungicide*	*Acquired resistance by*
Cell wall	chitin synthesis	Kitazin	*Pyricularia oryzae*[1]
		Hinosan	
		polyoxin D	
Protoplast-membrane	permeability	pimaricin	*Botrytis cinerea*[2]
		nystatin	*Saccharomyces cerevisiae*[3]
		amphotericin B	*Candida* sp.[4]
Mitochondria	energy production	antimycin A	*Venturia inaequalis*[5]
		carboxin	*Ustilago maydis*[6]
		oxycarboxin	*Ustilago hordei*[7]
Ribosomes	protein synthesis	cycloheximide	*Neurospora crassa*[8]
		kasugamycin	*Pyricularia oryzae*[9]
Nucleus	nucleic acid metabolism	6-azauracil	*Cladosporium cucumerinum*[10]
		benomyl	various fungi (see Table 8.2)

[1] De Waard (unpublished). [2] Bollen (unpublished). [3] Patel and Johnson (1968). [4] Winner and Athar (1970). [5] Leben *et al.* (1955). [6] Georgopoulos and Sisler (1970). [7] Ben-Yefet and Dinoor, pers. comm. [8] Vomvoyanni *et al.* (1971). [9] Ohmori (1967). [10] Dekker (1968a)

Resistance of a fungus to a fungicide may be brought about in various ways. The fungal cell may be changed in such a way that the fungicide does not reach the site of action. This may be due to a decreased permeability of the protoplast membrane, or to an increased capacity of the fungus to detoxify the fungicide. When the fungicide does reach the site of action, resistance may be due to a decreased affinity at the site of action to the fungicide or to other changes in the fungal metabolism, resulting in a compensation for the inhibiting effect or a circumvention of the blocked site (Table 8.4).

TABLE 8.4 **Types of resistance to fungicides**

Fungicidal activity in vitro	*Decreased permeability*	*Detoxification(a) no conversion into toxic compound(b)*	*Decreased activity at site of action*	*Circumvention(a) compensation(b)*	*Type of resistance*
+	+				I
−	−	+			IIa, IIb
		−	+		III
			−	+	IVa, IVb
				−	no resistance

It seems plausible that resistance to general plasmatoxicants, to which many of the conventional fungicides belong, must be explained by a decreased permeability of the protoplast membrane or by detoxification before the site of action has been reached. Resistance to selective fungicides might be due also to the other mechanisms mentioned above. Examples of each of these resistance mechanisms are given below.

I. Decreased permeability

The antibiotic blasticidin-S, which is used against *Pyricularia oryzae* on rice, inhibits protein synthesis in this fungus at a concentration of 5–10 μg/ml. In laboratory experiments strains of this fungus were obtained which tolerated a concentration of 1000–4000 μg/ml in the medium. In cell-free extracts of the resistant fungi, blasticidin-S did inhibit protein synthesis. From this follows the conclusion that in the intact cell the antibiotic is unable to reach the sites of protein synthesis, possibly due to decreased permeability of the protoplast membrane (Huang *et al.*, 1964). In work with the systemic fungicide 6-azauracil and the fungus *Cladosporium cucumerinum*, a strain of this fungus was obtained with increased tolerance to the toxicant. It appeared that less 6-azauracil was taken up by the tolerant fungus from an aqueous medium (Dekker, 1971a).

IIa. Increase of detoxication

Detoxication of a fungicide may occur by a modification of its molecule with a concomitant loss of fungicidal activity after entrance into the cell. Evidence has been obtained that naturally-occurring insensitivity of a fungus to a fungicide may be due to this phenomenon. *Fusarium oxysporum* f. *lycopersici*, which is resistant to ascochitin, appears to be able to reduce this antibiotic to the less fungitoxic dihydroderivative (Nakanishi and Oku, 1969). When comparing fungi resistant to quintozene (pentachloronitrobenzene), such as *Fusarium oxysporum* f. *niseum* and f. *lycopersici*, with quintozene-sensitive fungi, such as *Rhizoctonia solani*, the same authors (1970) found a positive correlation between quintozene resistance and ability to convert this fungicide into the much less fungitoxic compounds pentachloroaniline and pentachlorothioanisole. That not only a difference in sensitivity between various fungi, but also between strains of one fungus might be explained by this phenomenon, was indicated by the finding that the same correlation was found for two strains of *Botrytis cinerea* (Nakanishi and Oku, 1970). Therefore it seems quite possible that acquired tolerance might also be attributed to detoxication.

IIb. Decrease of conversion into fungitoxic compound

As a counterpart of detoxication the fungus may convert an inactive compound into a fungicide, which is called lethal synthesis. This appeared to be the case with a 6-azauracil (AzU)-resistant strain of *Cladosporium cucumerinum*, obtained after exposing conidia to ultra-violet irradiation. For antifungal activity AzU has first to be converted to a 6-azauridine-5′-phosphate (AzUMP), which is synthesised by this fungus from AzU via 6-azauridine (AzUR) in the following way:

The final product, AzUMP, inhibits one of the enzymes of the pyrimidine biosynthesis *de novo*. The resistant strain was unable to convert AzU into AzUR, so that the toxicant AzUMP was not formed. This type of acquired resistance was apparently due to a loss of activity of the enzyme uridine phosphorylase. No resistance was observed when AzUR or AzUMP were administered to the fungus (Dekker, 1967). A second AzU-resistant strain showed a decreased conversion of AzUR into AzUMP (Dekhuijzen and Dekker, 1971).

III. Decreased affinity at the site of action

When a fungicide reaches the site of action without being detoxified, tolerance of the fungus may be due to lack of affinity for the inhibitor at the reactive site. An example of this is offered by polyene macrolide antibiotics, which are inhibitory to fungi which do not belong to the Phycomycetes. They are known to interfere with ergosterols in the protoplasma membrane, which results in leakage and ultimate death of the fungal cell. The insensitive fungi, such as *Pythium* and *Phytophthora* species, lack ergosterol in their protoplasma membrane (Schloesser, 1965). This example concerns an existing difference in sensitivity between various classes of fungi. Recently, however, acquired resistance to this type of antibiotic has also been observed. Strains of *Botrytis cinerea* have been obtained which were tolerant to pimaricin (Bollen, pers. comm.). It is of interest that strains of *Candida* species with increased tolerance to amphotericin B and nystatin showed a decreased sterol content. The antibiotics probably have less affinity for the membrane, due to its decreased sterol content.

It is known that the antifungal activity of the antibiotic cycloheximide is due to inhibition of protein synthesis on the ribosomes. Cooper *et al.* (1967) found that in tolerant strains of *Saccharomyces cerevisiae*, obtained from sensitive organisms by gene mutation, resistance was located in the ribosomes. It seems possible that, by gene mutation, the composition of the ribosome has been slightly altered in such a way that affinity to cycloheximide has decreased without impeding the capacity of the ribosome to synthesise proteins.

When a fungicide primarily acts upon one particular enzyme, a slight change in this enzyme as a result of gene mutation may be the cause of loss of affinity to the fungicide without loss of its normal capacities, thus conferring resistance. Such a case has been demonstrated by Lewis (1963) in a study with the fungus *Coprinus lagopus* and ethionine. This compound is an antimetabolite of methionine and interferes with the so-called 'methionine-activating' enzyme. In the resistant mutant this enzyme had been changed in such a way that it lacked affinity to ethionine, but not to methionine. This could be explained as a result of a mutation in the structural gene, which forms the messenger RNA coding for this enzyme.

IVa. Circumvention

If a fungicide blocks a reaction at one site in the fungal metabolism, the fungus may adapt to this situation by shifting its metabolism in such a way that the blocked site is bypassed.

It is known that antimycin A acts upon the electron transport in the respiratory chain between cytochromes *b* and *c*. Tolerance of *Ustilago maydis* to this antibiotic could be attributed to a shift in the electron transport at a site preceding cytochrome *b* to an alternate terminal oxidase (Georgopoulos and Sisler, 1970). This alternate pathway is apparently of considerable value to *U. maydis* for growth in the presence of the antibiotic, which is demonstrated by the fact that a mutant, which lacked this alternate pathway, appeared sensitive to antimycin A. This example concerns a mutation from a normally tolerant fungus to a sensitive one, and it is not known whether the reverse might be possible in a normally sensitive fungus. The resistance mechanism of antimycin-A-tolerant strains of *Venturia inaequalis*, obtained by Leben *et al.* (1955) after irradiation of conidia with ultra-violet light, has not been investigated.

It is of interest that the antimycin-A-sensitive mutant of *U. maydis* appeared resistant to the oxathiins carboxin and oxycarboxin. Although it is not yet clear how oxathiin resistance is controlled by the gene which eliminates the antimycin A tolerance, it is assumed that the action of the two compounds is related and that the oxathiins must also affect respiration.

IVb. Compensation

When a particular enzyme is the primary site of attack by a fungicide, increased tolerance may be the result of the ability of the organism to compensate for the effect of the inhibition, for example by an increased production of the inhibited enzyme. An example of this phenomenon is strain R II-1 of *Cladosporium cucumerinum*, which is resistant to 6-azauracil. As has been explained earlier, 6-azauracil, after conversion into its nucleotide 6-azauridine monophosphate, inhibits primarily one enzyme in pyrimidine biosynthesis, namely, orotidine monophosphate decarboxylase. The activity of this enzyme was determined in the resistant strain and the wild type fungus, using carboxyl-labelled orotidine monophosphate, according to a method described earlier (Dekker, 1968b). It appeared that the resistant strains produced at least three times as much of the enzyme than the sensitive fungus (Dekker, 1971b). As a consequence the resistant fungus needs more of the fungicide to be inhibited, i.e. is more tolerant.

Conclusion

From the above-mentioned examples of acquired resistance to antibiotics and antimetabolites, it becomes clear that fungi possess the potential for a wide range of metabolic variation. This provides numerous opportunities for adaptation to fungicides which interfere selectively with the metabolism of fungi.

This, however, does not allow a general conclusion about the probability that resistance will develop to new systemic fungicides. This may vary greatly for different fungicides, depending on the mechanism of resistance and the type of pathogen. Furthermore, as will be outlined below, laboratory experiments cannot provide information about all the factors that contribute to the emergence of a resistance problem in practice.

EMERGENCE OF RESISTANCE IN THE FIELD

Laboratory experiments with or without the use of rays or other mutagenic agents may provide an estimate of the potential of a fungus for mutation towards resistance to a fungicide. They do not, however, answer the question whether resistant mutants will survive and become economically significant under field or greenhouse conditions. Of interest in this connection are the experiments of Ohmori (1967) with *Pyricularia oryzae* and the antibiotic kasugamycin. In the laboratory, isolates of this fungus were obtained which tolerated 100 μg/ml of kasugamycin in the medium, while the wild strain was inhibited at a concentration of only 1 μg/ml. Although kasugamycin has been used against the *Pyricularia* disease of rice during a number of years, no resistance problems have been encountered in practice. Ohmori observed that the kasugamycin-resistant strains of *P. oryzae* were less infectious on rice plants. Moreover he found that, in spite of tolerance of the mycelium, the formation of spores in the resistant strain was inhibited by the antibiotic.

In our work with 6-azauracil and *Cladosporium cucumerinum* two out of fifteen resistant strains had lost their pathogenicity to cucumber plants, while a few others appeared less virulent. Most of the resistant strains, however, were as virulent as the wild type fungus. Tolerance to pentachloronitrobenzene in *Sclerotium rolfsii* was not associated with loss in pathogenicity (Weigun Yang and Lung-Chi Wu, 1971). Wolfe (1971) concludes that most of the recorded examples of resistance do not seem to be associated with changes in pathogenicity, assessed in laboratory studies, and that the potential might therefore be said to exist for the evolution of forms which possess the combinations of tolerance and pathogenicity necessary for successful competition with sensitive forms of the fungus.

Although virulence and other factors which contribute to survival and frequency of resistant mutants in the field, such as infection chance, speed of colonisation and reproduction rate, can be investigated in laboratory experiments, these do not provide all the information necessary to evaluate the competitive ability of a resistant strain in the field. This is argued by Wolfe (1971) in the following way. 'The theory of genetic homoeostasis indicates that, although particular individuals with a characteristic such as fungicide tolerance may be selected, the whole population does not easily shift in this direction since there are many characteristics in the population which are held in a complex balance in the existing environmental situation. Thus, if the new selection pressure is relaxed, or incomplete, the population may tend to shift back to its previous state of balanced adaptation. It is only when the tolerant strains have a large and continuous selective advantage that the population gradually becomes fully adapted.'

As systemic fungicides have come into practical use only recently, studies on the epidemiology of resistant strains in the presence or absence of these fungicides, have not yet been made. It is important that this should be done in the future.

AVOIDANCE OF FUNGICIDE RESISTANCE

Search for new fungicides

It has been argued above that the frequency with which fungicide-tolerant strains will emerge depends among other factors on the number of mutations required for a certain level of resistance. If this holds true, search for new systemic fungicides should preferably be directed at those toxicants which interfere with the fungal metabolism at more than one site, so that for the emergence of a resistant strain several mutations may be required. This, however, is rather a theoretical possibility than a practical guideline in the search for new fungicides, since hardly any information is available about chemical structures required to obtain such multiple effects. Moreover it seems questionable whether such a compound would be sufficiently selective with respect to the combination of host and parasite. Another possibility is the search for compounds, which though not fungicidal *in vitro*, nevertheless enhance the resistance of the host plant indirectly. This artificially induced resistance may impose a lower degree of selection pressure on the fungus population than the more highly selective systemic fungicides. Possibly this might be compared to the so called 'horizontal' or polygenic resistance of a crop to a fungal pathogen, which is not easily broken down by a new strain. It has been shown that various experimental systemic compounds may protect plants in this way (see Chapter 6). An example is procaine hydrochloride, which after introduction into cucumber plants provides protection against powdery mildew in spite of the fact that it is not a fungicide *in vitro* (Niemann and Dekker, 1966a). It scarcely inhibits germination of conidia of *Sphaerotheca fuliginea*, nor elongation of germ tubes *in vitro* (Van't Land and Dekker, 1972). At present, however, no information about the potential of fungi to acquire resistance to this type of systemic compounds is yet available.

Effect on resistance of pattern of application of fungicides

When using systemic fungicides, to which resistance in fungi may develop, it should be realized that a high and continuous selection pressure tends to enhance the resistant population. Selection pressure will increase, when:

(*a*) One fungicide or closely related fungicides are applied repeatedly.
(*b*) A fungicide level, lethal or sublethal to the wild type fungus, is maintained in the plant continuously. This may be the case when the fungicide is applied frequently or when a continuous supply of fungicide is maintained by administering it as a soil application, or to the seeds or the planting material.
(*c*) The fungicide is used in an isolated area or a more or less isolated space (greenhouse), or in such a large area, that the wild type population and the competition it provides is virtually eliminated.

In order to reduce the emergence of resistance, preferably more than one fungicide, with different mechanisms of action, should be applied, either combined or alternately. The use of the fungicides should be restricted in space and time, as much as economic disease control allows, and the method of application carefully considered, so that selection pressure is not increased unnecessarily. The question remains, however, whether a combination of different chemicals or a restricted use as mentioned above will appear feasible and economical.

Some reserve in the application of selective, systemic fungicides seems advisable for other reasons. Such a fungicide may influence the microbiological balance in the soil or on the plant surface: the activity of plant pathogens may be influenced by the presence of other micro-organisms, which compete for food and space, or produce antibiotic substances. When the latter fungi are eliminated, pathogens which are relatively insensitive to the fungicide may be favoured, so that new disease problems may arise. In a field experiment with rye, a reduction of true eyespot caused by *Cercosporella herpotrichoides* was obtained after application of benomyl, but the incidence of sharp eyespot caused by *Rhizoctonia solani* increased tenfold (Van der Hoeven and Bollen, 1972). These aspects of the use of selective fungicides should be of special concern when fungicides are used which are not broken down readily to non-fungitoxic compounds.

In order to reduce the emergence of fungicide resistance in the field, Wolfe (1971) also advocates the use of systemic fungicides preferably on host varieties which have at least a moderate degree of resistance, and monitoring for resistant strains to detect in an early stage what fungicides are potentially at risk.

The question arises whether, by stopping the application of a systemic fungicide to which resistance has developed, resistance will disappear in the fungal population after a period of time, so that the same fungicide might be used again. As this problem has not yet been studied with respect to fungicide resistance, it is of interest to consider the results obtained in experiments with insects, which have become resistant to new organic insecticides. These led Keiding (1967) to the conclusion that a high level of resistance may persist for a long time and will usually decrease only partially when the selection pressure is relaxed. Even if a population shows an apparently complete reversion of resistance, the potential of a new development is usually much higher than before the pesticide was used, which indicates that resistance, at least against insecticides, is essentially a 'one-way street'. If this should hold true also for fungicides, it should, however, not exclude the possibility of 're-use' of a fungicide as a secondary compound, following the use of a new primary compound, to help reduce the pressure for emergence of isolates tolerant to the new compound.

Acknowledgements The author is indebted to Dr M. S. Wolfe, Plant Breeding Institute, Cambridge, England and to Dr S. G. Georgopoulos, Nuclear Research Centre 'Democritos', Athens, Greece, for critical reading of the manuscript and for valuable suggestions.

9

Methods of application

by E. Evans

As with all fungicide applications the eventual performance of systemics is dependent on.

(*a*) The inherent biological activity of the product;
(*b*) Correct timing of the fungicide treatment;
(*c*) Accurate placement of the fungicide.

Correct placement implies knowing whether the fungus concerned is spread in seed, in soil or through the atmosphere and making maximum use of the compound's systemic properties. Systemicity generally leads to an improved distribution of the fungicide and to a better biological performance. However, the dosage must be adjusted with regard to crop safety, persistence and residues if systemic compounds are to be widely used in agriculture.

BIOLOGICAL ACTIVITY AND SYSTEMICITY

The subject of systemicity and the mechanisms of movement within higher plants has been reviewed by Crowdy (Chapter 5) but from the practical point of view the following characteristics are significant:

(i) The uptake and distribution of a systemic compound within a plant differ according to whether the compound is applied to the aerial or subterranean parts.
(ii) The rate of movement is greater in herbaceous than in woody plants

and even then is dependent on the physiological status of the tissues concerned (Shepherd, 1971).

(iii) Movement of all present-day systemics is unidirectionally upward following the transpiration stream, which in turn is affected by environmental conditions during and after treatment (Peterson and Edgington, 1971).

SEED TREATMENT

A seed dressing is defined as a substance applied to seed for the control of pest or disease problems of that species. Traditionally the objectives of fungicide dressings were to destroy pathogenic organisms on the seed, or to protect germinating seedlings from attack by soilborne pathogens. The introduction of systemic fungicides has added two further possibilities.

(*a*) The control of pathogens situated within the seed and previously inaccessible to chemicals.

(*b*) The control of airborne diseases using the dressing as a reservoir of fungicide during the growth of the crop.

The current methods of treating seedborne diseases are summarised in Table 9.1. These were devised to deal with broad-spectrum non-systemic materials, largely based on a variety of organomercurials.

TABLE 9.1 **Summary of seed dressing techniques**

Formulations	*Application methods*	*Examples of machinery*
1. Adhesive dust	Batch or continuous flow processes	Simple drums, high speed dusters, e.g. Robinson and Plantector and in the row seed box treatments
2. Wet slurries	Dribble or spray applications	Panogen and 'Mistomatic' machines (2 and 3)
3. Solutions	Dribble or spray treatments	

These techniques are equally applicable to systemic products and when used as sterilants of seed as well as protectants during germination, seed dressings provide a most economic and convenient method of control. Used in this manner the amount of chemical required is small and the operation is easily mechanised. When prolonged systemic protection is required the dilution effect of growth and the eventual breakdown of the active ingredient demand that larger amounts of chemical have to be applied to the seed. This can present considerable problems of formulation, which is already restricted by the physical nature of the active compound and by the kind of machinery to be used in dressing the seed. Thus dusting machines demand a solid adhesive product while the 'wet' seed dressers use either solutions or slurries. The dressing process is further limited by the retentive capacity of the seed itself, particularly when it has to be stored with unusually high rates of systemic dressings.

As the seed is handled roughly during the planting operation dressing may be lost but sufficient chemical must remain near the seed if it is to be taken up by the growing seedling. As roots are organs of absorption it is reasonable to suggest that this is the main method of entry of systemic dressings. When dealing with large dicotyledonous seeds however it can be shown that cotyledons are also absorptive organs. Some *Phaseolus* beans were soaked for 24 h in distilled water. After stripping the testa from a sample, excess water was removed with dry filter paper and the naked seed dressed with 0·25 per cent by weight of carboxin. A sample of intact beans was then treated in a similar manner applying the fungicide to the testa. A third series of seedlings were grown in this manner devoid of chemical treatment. The beans were then grown in a humid chamber but not in direct contact with water. Within 14 days plants were produced which were inoculated with the bean rust pathogen *Uromyces phaseoli* and disease assessment made 10 days later still. The results are shown in Table 9.2.

TABLE 9.2 **Per cent disease control and phytotoxicity index after various seed treatments**

Treatments	*Mean % disease control†*	*Mean phytotoxicity index*
1. 0·25% w/w carboxin applied to testa	11	2·0
2. 0·25% w/w carboxin applied to naked seed	100	2·3
3. Untreated seedlings	0	0·3

† Bean rust pathogen *U. phaseoli* used as assay organism

The barrier to absorption set up by the testa is obvious from the small amount of protection afforded in treatment 1, while the excellent performance of treatment 2 reflects the good absorptive nature of the cotyledons. Phytotoxicity as recorded in this experiment consisted of the typical marginal necrosis described by Snel and Edgington (1970).

The success of the use of systemic fungicides as seed dressings is now clearly established and the oxathiin, carboxin, is widely used for the control of the deep-seated loose smut diseases of wheat and barley (Kuiper, 1968) caused by *Ustilago tritici* and *Ustilago nuda* respectively, diseases which were previously only controlled by the hot water treatment of seed. The first practical control of an airborne disease by seed treatment was that of barley mildew *(Erysiphe graminis)* with the pyrimidine, ethirimol (Brooks, 1971a).

SOIL TREATMENT

Until the recent introduction of systemic products, fungicides were only applied to soils for

(*a*) The protection of germinating seed crops from soilborne organisms.
(*b*) The eradication of soilborne parasites prior to planting a new crop.

The highly selective character of present-day systemic fungicides does however enable crops to be treated *in situ* for the control of soilborne organisms and for the control of airborne pathogens following root uptake. The methods of application currently used for treating soil with fungicides depend primarily on the physical characteristics of the products and degree of sophistication of the machinery used. Gases, such as the general sterilant methyl bromide, are injected at set points from pressurised containers and similar 'point' treatments of liquids that liberate gases such as carbon disulphide and formaldehyde are also made. Solids, however, require thorough incorporation in soil. Liquids or suspensions of active ingredient in water including most of the current systemics are frequently used as overall drenches. Thus benomyl and thiabendazole drenches are under test on a variety of crops for the control of typical soilborne pathogens such as *Verticillium* wilt and *Phymatotrichum* rots of cotton (Ranney, 1971; Lyda and Burnett, 1970). This technique is also recommended for a number of systemic products such as thiophanate-methyl, benomyl and ethirimol used against cucumber mildew, an airborne disease of major importance in glasshouse crops (Aelbers, 1970).

A novel modification of the drenching technique has been developed for use in Japan of the fungicide kitazin against rice blast *(Pyricularia oryzae)*. Here the product is applied to the irrigation water as granules containing 17 per cent a.i. and since the active ingredient has only a limited degree of solubility (500 ppm) in water such granules were persistent for three weeks under field conditions (Yoshinaga, 1969).

As with any other chemicals applied to soil, systemic fungicides are potentially subject to inactivation. This may be due to the process of physical adsorption with a consequent lack of availability and reduced biological activity. Chemical and biological interference may also occur, leading to a breakdown of the active ingredient. The rate of breakdown is obviously of considerable importance in determining the fungicidal efficiency and residue status of treated soils. Uptake by plants is greatly affected by a variety of factors which are as yet not fully understood. Hock *et al.* (1970) found that the uptake of benomyl by the roots of elm seedlings was much less in plants grown in potting compost than with those grown in sand cultures. Whether this is a matter of root structure or availability of fungicide was not specified. Similarly the uptake of substances dependent on water solubility might well be affected by drought, and their use correspondingly restricted to situations where irrigation or any other watering procedure is readily available. The complicated nature of this adsorptive phenomenon and its significance to the uptake of the pyrimidine fungicides by higher plants was discussed in some detail by Graham-Bryce and Coutts (1971). They showed that in this instance the process of adsorption was largely reversible if sufficient time was allowed. They also found that the distribution and consequent uptake from localised soil treatments was affected by various structural and textural characteristics of soils.

The activity of certain benzimidazole fungicides has even been influenced by formulating the active ingredient with certain surfactants (Pitblado and Edgington, 1971) although the benefit of such an effect has not yet been established in practice. Stipes and Oderwald (1971) and Biehn and Dimond (1970) investigated the effect of benomyl on Dutch elm disease when injected as a suspension of active ingredient into the feeding root zone of young elm trees with and without added surfactant. The former concluded that Tween 20 improved the performance of benomyl used in this manner.

LEAF AND STEM TREATMENTS

As airborne epidemics develop much more rapidly than seed or soilborne diseases they must be tackled with urgency by means of spray or dust treatments of the aerial parts of the crop. The introduction of systemic products as seed or soil treatments provides an alternative approach for the control of certain airborne organisms, but most fungicides are still applied by spraying or dusting techniques. The field of application technology has received considerable attention from both physicists and biologists alike and was reviewed by Evans (1968). Table 9.3 is taken from this review and summarises the variety of techniques most commonly employed for the application of fungicides to the aerial parts of plants.

TABLE 9.3 **Summary of most commonly employed commercial techniques of applying foliar fungicides (Evans, 1968)†**

Growth form of crop	*Technique of interest*		
	High-volume‡	*Low-volume§*	*Dusting*
Ground or field crop	Hydraulic sprayer with horizontal spray gear	Aircraft with conventional spray bar and nozzles or rotary atomisers	Single or multi-outlet row crop dusters, or aircraft dusting operations
Bush or orchard plantation	Hydraulic sprayers with vertical spray bars or hand lances	Airblast machines and occasionally aircraft, e.g. for control of downy mildew of vines	Power-operated dusters
Closed-canopy plantations	Hydraulic sprayers with hand lances	Airblast machines from below, and aircraft from above the canopy	Power dusters from below canopy, and dust applications from aircraft

† Taken from *Plant Diseases and their Chemical Control* (1968) by E. Evans, published by Blackwell Scientific Publications, Oxford

‡ High volume – any spray volume which leads to the coalescence of spray droplets and a consequent 'drip pattern' on sprayed foliage

§ Low volume – spray volumes where coalescence of spray droplets does not occur

In order to prevent the development of disease symptoms, a series of treatments is usually applied according to a pre-determined schedule designed to cover all potentially infective periods. Thus the efficiency of any treatment will depend on the persistence of the active ingredient between one treatment and the next and the distribution of the ingredient during these infective periods. Non-systemic products often have to withstand severe weathering as surface deposits while systemics are generally much less dependent on weather because they are able to penetrate host tissues. Thus a 21-day schedule of benomyl on apples gave a level of control of scab *(Venturia inaequalis)* equal to a 14-day schedule of the non-systemic captan standard (Evans *et al.*, 1971).

Another major factor governing the efficiency of any spray deposit is the distribution of the active product on or in the host plant. Final distribution of non-systemics is largely dependent on the method of application used and the redistribution of the deposits as solutions, suspensions or vapours on the surface of the crop (Evans, 1968). Tissue penetration and internal movement of the fungicide is implicit in the term systemicity and hence systemic distribution is usually superior to that of non-systemics. Morgan (1952) showed how a given quantity of a copper fungicide gave vastly different biological results dependent on the distribution pattern of that deposit. This effect is even more dramatic when the procedure is repeated with a systemic product. Commencing with a spray concentration of 0·025 per cent dinocap and benomyl respectively a series of five suspensions were prepared by repeated 1:1 dilutions of the previously higher concentration. By means of a micropipette droplets of approximately 0·05 ml in volume were placed on young cucumber foliage and allowed to dry prior to incubation for a period of 14 days in an atmosphere heavily loaded with spores of the mildew pathogen *Sphaerotheca fuliginea*. By placing equal amounts of active ingedient/leaf as 1, 2, 4, 8 or 16 droplets of the appropriate concentration, protectant action was then estimated as a percentage of the leaf area free from disease. The mean data recorded in this manner are summarised in Table 9.4.

TABLE 9.4 **The effect of distribution on the activity of dinocap and benomyl deposits (mg of a.i. applied/leaf)**

Concentration of applied droplets, %	*Number of drops/leaf*	*% disease control*	
		Dinocap	*Benomyl*
0·025	1	5	20
0·0125	2	10	40
0·0067	4	20	100
0·0033	8	55	100
0·0017	16	100	100

Since the final pattern of results of both compounds gave almost 100 per cent control it follows that the amount of active ingredient was not limiting the activity of either dinocap or benomyl and that the overall result was largely dependent on the distribution pattern. The superiority of the benomyl treatment was obvious throughout and its apoplastic movement in the transpiration stream clearly shown in the wedge-shaped pattern of control associated with single drops of benomyl, particularly when placed adjacent to main veins.

Ultimately the efficiency of any fungicide depends on the amount of active compound that arrives at the site of action. The active compound may not be the substance supplied as the formulated product, as shown for benomyl by Clemons and Sisler (1969) and for thiophanate-methyl by Selling *et al.* (1970).

As long ago as 1956 Gray was able to show that the uptake of streptomycin by leaves and flowers could be increased by the addition of glycerol to the spray liquid. This effect is generally attributed to the humectant properties of the glycerol with a corresponding delay in the rate of drying and greater uptake of the surface deposit. Surfactants have traditionally been added to improve biological efficiency of most fungicides in two ways (*a*) by enhancing the wetting properties of sprays and possibly by improving distribution and (*b*) by increasing the solubility of the active ingredient as shown by the effect of Tween 20 on triarimol (Brown and Hall, 1971). These improvements apply equally well to both systemic and non-systemic products. Adhesive supplements were added to benomyl by Epstein (1969) who showed that improved control of apple scab was achieved by so doing. Thus although generally some improvement in disease control has been achieved by the addition of various supplements, the overall superiority of the systemics is largely dependent on better distribution and a more regulated persistence largely dependent on the applied dosage. The uptake, distribution, accumulation and persistence of toxicant by a given tissue is eventually responsible for its biological performance, but the level of toxicant which will control one pathogen may not be sufficient for another. This was clearly demonstrated by Frick (1971) who showed that while apple scab was well controlled by systemically-moved benomyl in half leaf tests, this level of toxicant was insufficient to control the apple mildew pathogen *(Podosphaera leucotricha)*.

While foliar spray treatments are convenient for the control of airborne pathogens of either annuals or perennials this is not a convenient technique for tackling systemically distributed pathogens of woody tissue such as Dutch elm disease (caused by *Ceratostomella ulmi*) or silver leaf disease of stone fruit (caused by *Stereum purpureum*). Stem treatments with or without wounding are most commonly used on large perennials (Schreiber, 1970) but various other techniques have been applied to a variety of crops for experimental purposes (Wain and Carter, 1967; North, 1962). In practice stem injection has been used for the correction of mineral deficiencies particularly that of calcium-induced iron deficiency in top fruit and citrus. Tangential holes are bored into the trunks and solid or encapsulated doses of active ingredients implanted. Adhesive bands enclosing absorbent substrates containing the active ingredients are also used, but all such implantation techniques involve the use of locally high concentrations of chemical or else large reservoirs of dilute solutions. From the practical point of view the latter technique is cumbersome while the use of concentrated preparations tends to cause local necrosis and eventual failure to take up the toxicant from solution.

POST-HARVEST TREATMENTS

Harvest in this sense implies that part of the crop which is to be marketed as:

(*a*) Raw material for manufacturing or processing
(*b*) Food for human or animal consumption
(*c*) Propagating material for crop maintenance purposes.

As systemic fungicides generally are very active there has been a tendency of late to use them in preference to the traditional non-systemics but perhaps their greatest asset is their ability to penetrate host tissues, enabling them to eradicate many of the latent infections which might have occurred prior to harvest. Late sprays and post-harvest dips of benomyl greatly reduced the incidence of *Gloeosporium* fruit rot in stored apples (Pleisch *et al.*, 1971) and many boll rots of cotton (Ranney, 1971) at harvest. Dip treatments of benomyl at harvest also reduced the incidence of various vegetable rots in storage (Derbyshire and Crisp, 1971).

The category designated as propagating material includes seed pieces for potato growing, bulbs or ornamental cuttings, all of which have been treated with broad-spectrum fungicides such as benomyl in order to rid the planting material of diseases which might have affected the crop. Hide *et al.* (1969) used a 10 per cent benomyl dust for the treatment of potatoes and readily eradicated several potential pathogens from the seed tubers. Various benzimidazole fungicides have similarly been tested as pre-planting treatment of bulbs with the same object in mind (Rooy, 1969).

Although the details of these processes would vary from crop to crop the basic principle of application remains, as the dust, sprays and dip treatments are aimed at attaining a uniform distribution of active ingredient on the potentially infected surface in order to destroy the pathogen *in situ*, and possibly in the case of propagating material to protect the crop during its establishment period.

TIMING SYSTEMIC FUNGICIDE TREATMENTS

The overall efficiency of any fungicide treatment is eventually dependent on accurately timing the application in a manner dictated by the biology of the pathogen. With non-systemic products the effect of timing can be extremely critical (Evans *et al.*, 1965) but the advent of systemics has in some instances made this requirement much less important due to tissue penetration giving improved distribution and weathering characteristics and to the curative nature of most present-day systemic products.

On the principle that a pathogen is most easily controlled when its inoculum potential is minimal it is generally agreed that treatment should be most effective when (*a*) the pathogen population is lowest, (*b*) the host least susceptible to the fungus and (*c*) the environmental conditions least congenial to infection. This entails treating seed with chemical dressings for the control of seedborne organisms and treating soils for the control of soilborne pathogens prior to planting (although this is not usually possible with perennial crops). It is more difficult to apply this principle to airborne diseases generally but some success has been achieved with post-harvest treatments of apple orchards for the control of scab with both non-systemic (Burchill and Hutton, 1965) and systemic fungicides (Darpoux, 1970). The objective here is to destroy the pathogen while the inoculum potential is declining naturally. Unfortunately however airborne pathogens tend to be extremely mobile and re-infection from outside the treated area could well nullify the effect of such treatments.

Accurate disease forecasting is now widespread for a variety of crops throughout the world, but the application of the chemical itself is often more of a practical difficulty than the decision to spray. For this reason there is a tendency to follow routine schedules throughout the normally infective season and it would seem that 'negative forecasting' could prove to be economically most desirable (Visser *et al.*, 1962). Negative forecasting is here taken to imply forecasts of when a given treatment could safely be omitted from the standard schedule of either systemic or non-systemic products. The actual interval recommended between successive treatments depends on the growth rate of the crop as well as the dosage applied on each occasion. The growth rate and the physiological status of the host tissues may affect the susceptibility of certain crops to specific pathogens and this may complicate the issue even further. However the internal transfer of systemics means that previously unsprayed tissues may receive toxicant and theoretically this should improve its biological performance *vis-à-vis* non-systemics. Growth in this sense may simply imply expansion of a single organ such as a leaf or fruit which had previously received some spray deposit; more commonly however new structures such as leaves are produced between treatments and in order to gain a specific advantage in this case a systemic product would have to be redistributed from older sprayed organs to the newly developed structures. The first of these effects can frequently be seen in half-leaf tests on rapidly growing foliage on young laboratory grown crops used for fungicide screening tests. However there is as yet no evidence to suggest that any of the present systemic fungicides are redistributed to the terminal buds in the manner of the aminotriazole and TBA herbicides which tend to accumulate in these areas, and until such products are available, the advantages of systemic products in controlling such pathogens as apple mildew (caused by *Podosphaera leucotricha*) will be minimal. Tissue penetration is essential to curative action of any fungicides as well as being the prelude to systemic movement and as all present-day systemics are curative in the early stages of infection their use should be less dependent on absolute precision timing than on their equivalent non-systemic standards. This generalisation seems to be particularly true of diseases that are essentially internal parasites but care should be taken when applying this philosophy to the superficial powdery mildew pathogens where even non-systemic products may have a considerable curative effect on the pathogen at any time (Evans *et al.*, 1970).

10.I

Results in practice – I. Cereals

by D. H. Brooks

INTRODUCTION

The main world cereal crops are wheat, maize, rice and barley. Cereals such as sorghum, millet, rye, oats and buckwheat are grown on a smaller scale, and are not covered in this review. The cereals are relatively low-value crops, and the highly-priced systemic fungicides have not found widespread acceptance. The notable exceptions are for the control of rice blast – particularly in Japan where sophisticated products were already used on a wide scale – and more surprisingly for the control of barley mildew, mainly in Europe. Systemic fungicides are also used for the control of loose smuts, mainly on crops grown for seed production.

The importance of barley mildew in Europe has only been fully recognised in recent years, and a very large amount of effort has been put into investigating its control with systemic fungicides. So successful has this work been that in the second year of availability of systemic fungicides, more than one-quarter of the barley crop in England was treated, and market penetration was even more rapid in Northern Germany. Not surprisingly, consideration of the control of powdery mildew forms a major part of this review.

BARLEY AND WHEAT

Powdery mildew

The extent to which fungicides for control of mildew have been developed depends, amongst other things, on the regularity with which mildew occurs, its severity and the cash return to the farmer of the cereal crop.

Erysiphe graminis DC. is far less dependent on weather conditions than most cereal pathogens, and it is found wherever suitable hosts are grown. Its severity depends on such factors as summer temperature, rainfall and humidity, the presence of bridging crops overlapping with the main crops, and fertiliser usage. It is probably the most important disease attacking barley in West and East Europe, particularly in the UK, Denmark, West and East Germany, parts of Holland and Belgium, Northern France and Hungary. Outside Europe, it is sometimes important in New Zealand, West Australia, Canada (Ontario), and parts of the USA (particularly the eastern and western seaboards and south of the Great Lakes). Definitive information on the importance of mildew on wheat is more scarce – it occurs wherever wheat is grown, but is possibly of most importance in eastern European countries such as Hungary and Yugoslavia where the very susceptible cultivar Bezostaja is widely grown. Oats are severely attacked by mildew in many countries, but are not widely enough grown to have attracted much attention from those concerned in developing fungicides.

Because of the low returns to the farmer in the Americas and Australasia, most attention to fungicide development has been centred in West Europe on barley (and to a lesser extent wheat) and East Europe on wheat (and to a lesser extent barley), and commercial usage is almost entirely limited to these areas. By far the majority of the results available have been obtained on barley in Europe.

With the exception of the acid anilides (carboxin, pyracarbolid (2-methyl-5,6-dihydro-4H-pyran-3-carboxylic acid anilide), etc.) almost all of the synthetic systemic fungicides are active against *E. graminis*, and most of them (HOE 2873 is an exception) give some control from root uptake.

Soil application The attraction of this method is that the soil acts as a reservoir for the chemical, and the plant continues to absorb and translocate the fungicide over relatively long periods. The period of control obtained in this way depends on the length of time the chemical remains available in the region of root absorption and this depends, amongst other things, on the physical properties of the molecule. Although promising experimental results have been obtained with benomyl, triarimol, ethirimol and thiophanate methyl, only ethirimol is used commercially in this way, and most of the available results have been obtained with this compound. Since the cereal crop is one of the few situations where systemic properties have been fully exploited, some discussion of this usage is appropriate here.

Method of application Numerous methods of application have been used, including seed dressing, pelleted seed, liquid sprayed into the furrow, broadcast granules and combine-drilled granules. With ethirimol, the chemical leaches only very slowly in most soil types, and broadcast granules have given variable and generally poor disease control. The other methods have all been used successfully. Where the chemical is placed apart from the seed, the degree of disease control is affected by the evenness of application; for example, where fertiliser granules coated with ethirimol were combine-drilled in an accurate experimental drill, good results were obtained, but when a range of farmer drills were used (most drills dispense fertiliser very unevenly) variable results were obtained. Always the most consistent results have been obtained with ethirimol used as a seed dressing.

One of the major problems with seed dressing is that many of the commercial seed-treating machines are not capable, without modification, of handling the relatively large quantity of chemical required (600–1000 g formulation/100 kg) at the same time as the standard fungicide/insecticide dressing (usually 200 g/100 kg); it has therefore been necessary to carry out costly conversions to machinery. Using converted machinery, very good adhesion of both compounds to the seed is obtained.

Factors affecting activity The duration and rate of uptake, and rate of translocation of a soil-applied systemic fungicide depend on a large number of factors, including those which affect foliage-applied systemic fungicides (transpiration rate, etc., see Chapter 5).

The chemical can only be taken up if it is in an available form in the root region. The mobility of the chemical depends on adsorption, and the degree to which it is reversible, and on structural and textural features of the soil (Graham-Bryce and Coutts, 1971). With ethirimol, adsorption increases with decreasing pH and increasing organic matter content; for this reason ethirimol is not very effective as a seed treatment in acid peat soils. In light, sandy soils with little tendency to form aggregates, the chemical leaches in a roughly conical pattern beneath the seed. In heavier-textured soils which are usually well aggregated, leaching is much reduced (Graham-Bryce and Coutts, 1971). In practice, excess leaching does not seem to be a factor limiting uptake in most situations.

Free water in the soil is necessary for uptake of fungicides. In pot experiments with ethirimol incorporated into the soil, uptake did not cease until wilting occurred. In the field, however, the plant may get its water from deep in the soil whilst the surface soil containing the fungicide is dry. In such conditions uptake will cease. This does not appear to limit uptake in most areas, but in experiments with ethirimol in South Australia, where the water table receded very rapidly after sowing, disease control persisted for approximately four weeks. In parts of West Germany in the severe drought of June 1970, uptake also appeared to cease for a while, although it was resumed when the drought broke.

In practice in the field, uptake of ethirimol by the barley plant appears to cease whilst there is still available chemical in the surface layers; probably the roots remaining in the treated surface layers of the soil are no longer capable of uptake.

Spray application The economic return from mildew control does not normally justify more than one spray application. In most of Europe, mildew generally begins to appear in the spring barley crop towards the end of tillering, although in some years it can appear earlier (particularly in the vicinity of winter barley) or later. With the xylem-translocated fungicides currently available, leaves formed after the time of spraying receive little or no direct chemical protection. The degree of mildew control achieved, therefore, depends not only on the inherent activity, rate of application and persistence of the fungicide, but also, most important, on the timing of application and the inoculum pressure on the crop. When applied as sprays, the systemic fungicides are superior to non-systemic fungicides such as drazoxolon and quinomethionate in their limited movement within the leaves to which they are applied, and in their powerful eradicant activity.

The fungicides have normally been applied by ground equipment in 200–400 l water/ha. Aerial application, including ultra-low-volume application, has also been used experimentally with some success.

Results on spring barley The duration and degree of control of mildew achieved with ethirimol seed dressing depends, not only on the factors already outlined, but also on such factors as inoculum pressure and, obviously, rate of application. There is, however, a relatively flat dosage–response curve, possibly because root distribution is one of the main factors limiting uptake. Mildew usually appears at the same time on treated and untreated crops, but the speed of development is greatly reduced by treatment. When high rates of application are used (675–2000 g/100 kg) almost complete control of disease can be expected for 15 weeks or more (Brooks, 1971a; Little and Doodson, 1971). At lower (commercial) rates (370–500 g/100 kg) the degree of control is somewhat less although the persistence does not appear to be altered appreciably. The precise time of breakdown of control is difficult to pinpoint since environmental factors and increased host resistance frequently cause the rate of mildew spread to decline in later (post-flowering) growth stages. In a series of trials carried out in 1968 (Brooks, 1971a) seed treatment apparently kept the disease in check throughout the life of the crop; there was, however, little mildew spread on to the upper leaves even in untreated plots.

In 1969, when the disease continued unchecked, disease started to build up in the treated plots during the grain-filling stages. Probably, therefore, under commercial conditions the effectiveness of ethirimol drops off very sharply after flowering. The degree to which the crop becomes infected at this stage depends on environmental conditions and the intensity of inoculum; re-invasion certainly occurs more rapidly on small plots surrounded by untreated areas than in large treated blocks. In trials in Northern France, crops have sometimes become heavily infected in the late growth stages in spite of excellent control at flowering. The average mildew control at growth stage 10·5 is usually about 90 per cent.

On spring barley the persistence of ethirimol seems to be unaffected by sowing date; there is no evidence that early-drilled crops lose their protection before those drilled later. One complicating factor is that the degree of control achieved with ethirimol depends not only on the load of chemical on the seed, but also on the density of sowing of the seed. Consequently, if very low seed rates are used, disease control will be less good than where very high seed rates are used. Within the normal range of rates used in any one country, however, the differences are probably not significant.

Yield responses from seed treatment in large-scale plots have varied from slight decreases to increases of up to 80 per cent, with not always a very close correlation with the amount of mildew. Average increases from large numbers of trials in England and West Germany in 1969, 1970 and 1971 were 7, 12 and 13 per cent respectively in replicated trials, and 10, 10 and 18 per cent in 'farmer' trials.

Of the sprays, the most widely tested and used material is tridemorph. In over 200 trials carried out in Germany in the period 1962–68, an average increase in yield of 10 per cent was obtained with an application rate of 0·5–0·6 kg a.i./ha; similar results have been obtained in England and many other countries (Anon., 1971a). Mildew is normally controlled for four to five weeks following application, and the effect of this delay on further epidemic build-up is frequently seen until harvest. In general the yield response has been well correlated with mildew control (Kradel *et al.*, 1969) but on some occasions yield increases have been obtained at very low mildew levels. When tridemorph is applied to an established infection it rapidly kills the mildew, and the crop sometimes appears to regain its green colour, giving what has been called a cosmetic effect.

Ethirimol has also been widely used both experimentally and commercially as a foliar spray at a rate of about 0·5 kg/ha. Average yield response for a series of trials carried out in England and Germany in 1970 and 1971 varied from 5 to 12 per cent.

The only other fungicide to have been widely used commercially is chloraniformethan [1-(3,4-dichloranilino)-1-formylamino-2,2,2-trichloro-ethane], usually at a rate of 290 g/ha. In England yield increases in replicated trials in the years 1970 and 1971 averaged 8 and 6 per cent (Davies *et al.*, 1971).

Triforine, benomyl, thiophanate methyl, chlorquinox (5,6,7,8-tetra-chloroquinoxoline) and triarimol have all given promising results in large-scale field evaluation. Triarimol is notable for its extremely low rate of application, 30–75 g/ha, which in trials in England in 1971 gave yield increases mostly in the range of 5–10 per cent (Rea and Brown, 1971).

Thus a large number of new systemic compounds have given good control of mildew on spring barley. All of these compounds have a similar systemic action when applied as sprays, and all have eradicant as well as protectant properties. Since timing of an application has a very marked effect on the performance of a spray (see below) it is almost impossible to make valid comparisons between sets of data on individual compounds obtained in separate trials, and very few trials have been carried out in which all the compounds have been compared with one another under similar conditions. Even in these trials, the results are not always very useful, because the rates and formulations used have not always been those recommended commercially. Usually tridemorph, ethirimol and triarimol have been slightly more effective in controlling mildew than thiophanate methyl, chloraniformethan and chlorquinox, although the results are by no means consistent. In terms of yield the broad-spectrum compounds have sometimes given bigger yield response than expected from mildew control, presumably due to control of other pathogens. Almost always, spray applications have been inferior to ethirimol seed dressing.

From a comparison made of the results obtained with sprays at different times, and of sprays with seed dressings, it is obvious that mildew affects the barley plant in different ways depending on the time of attack. In large numbers of trials carried out on spring barley in 1969 and 1970, when mildew attack was relatively late, yield increases could almost always be accounted for in terms of increase in grain size, whether resulting from seed dressing or spray application (Brooks, 1971b). In 1968, and more so in 1971, mildew attacked early and caused a decrease in tiller number, a marked reduction in root growth and a small reduction in straw length (Brooks, 1971b). In these trials yield increases usually resulted from a combination of more tillers and plumper grain although in a few cases grain size was actually reduced, presumably because of increased competition between the tillers.

Several workers have noticed that different varieties of barley respond to different extents to mildew control. For example, the variety Julia responded far more than was expected to control of mildew with ethirimol (Doodson and Saunders, 1969). With tridemorph it has been possible to divide both susceptible and resistant varieties into two groups, one of which responded well to treatment (even with little mildew), and one which did not respond well (Kradel *et al.*, 1969).

Time of spray application Much work has been carried out on the timing of spray application, mainly by the commercial companies. The time has sometimes been expressed as calendar dates, sometimes in terms of growth stage of the crop and sometimes in relation to mildew attack.

From trials carried out in 1968 and 1969, Rosser (1969) concluded that spraying should be carried out in May or June when the mean temperature had exceeded 20°C for seven days (i.e. conditions suitable for rapid epidemic development); the correct dates were approximately June 5th in 1968 and June 18th in 1969, and these represented relatively late growth stages in the crop. In 1971, attempts to determine the correct time of spraying in relation to 'Rosser Periods' were unsuccessful since although mildew was early and severe, the first 'Rosser Period' did not occur until July (Evans and Hawkins, 1971; Gilmour, 1971). Gilmour suggested that a better forecasting system could be based on the accumulated day-degrees in excess of 15°C from May 1st. From trials carried out with tridemorph in 1970, Lester (1971) concluded that the optimum time for spray application in that year was at growth stage 8 (Feekes Scale – roughly mid-June).

The experience of the chemical companies has generally been that somewhat earlier sprays (G.S. 5–6) give optimum yield response (Anon., 1971c; Jung and Bedford, 1971; Rea and Brown, 1971; Brooks, 1971a). For example, in 58 trials carried out in 1961–64 in Western Germany, sprays of tridemorph at G.S. 5–6 increased yield by an average 419 kg/ha, compared to 339 kg/ha at G.S. 8–9. A similar pattern was found in England in 1970, although in a series of trials in 1968 sprays at G.S. 6 were markedly superior to sprays at G.S. 4 (Anon., 1971a). Frequently, although the later sprays have given maximum disease control (when assessed at G.S. 10–11) earlier sprays have given greater yield response. For example, Evans and Hawkins (1971) found that in 1971 progressively later sprays gave increased mildew control up until early June, but that sprays in mid-May gave the biggest yield response.

It is clear from the data available that it would be quite wrong to standardise spray timing recommendations in terms of time of year or stage of growth of crop, since these will vary from area to area and from year to year. The new chemicals have powerful eradicant activity, and the optimum time to spray is probably at the beginning of the rapid phase of epidemic build-up, erring on the early side rather than the late. In

Western Europe it is normally between G.S. 4 and 8, mid-May to mid-June. Where there is abundant inoculum coming into the field from adjacent infected crops, the spray could be made at the expected time of appearance of infection, bearing in mind the weather conditions needed (Rosser, 1969); where there is not an adjacent source, it would probably be better to wait until the appearance of mildew in the crop, although marginally lower yields might be expected.

The discrepancy between maximum yield and maximum mildew control makes it very difficult for the farmer to judge the correct timing of application, and it is here that he needs expert advice. It is important therefore that further results are obtained and made available to him via the extension services.

Results on winter barley In Europe, winter barley is generally drilled in September or October. It may become infected with mildew immediately after emergence before cold weather prevents the rapid spread of mildew, or in spring when the temperature rises, or both. Even following a severe attack in the autumn, there is not always an appreciable amount of mildew in the spring, but the yield potential may already have been reduced (Brooks, 1971b). The problem of spray timing, and whether more than one spray is justified is therefore far more complex on winter barley than on spring barley.

Single sprays of tridemorph applied in October have caused increased tillering and root growth, and have greatly improved the vigour of crops (Lester, 1971). Yield response has, however, not always been obtained (Jung and Bedford, 1971).

Sprays of tridemorph (Kradel and Pommer, 1967; Jung and Bedford, 1971), triarimol (James *et al.*, 1971) and ethirimol applied in the spring (usually at G.S. 5–6) have also given marked disease control and have led to significant yield increases. For example, in two trials sprayed with tridemorph at G.S. 5–6 in 1970, excellent mildew control resulted for four weeks, and yield was increased by 15 per cent.

As would be expected from the results on spring barley, ethirimol as a seed dressing gives good control of mildew after emergence until weather prevents further spread of the disease. In autumn 1969 almost complete control was achieved during this period in trials in Europe and Western Germany (Hall, 1971); in 1970, infection continued later in the autumn and disease started to appear in the treated plots (Hall, 1971; Yarham *et al.*, 1971). In both years mildew control was resumed in the spring and, particularly in 1970, the disease was largely checked until after ear emergence (Hall, 1971). Where comparisons were made, disease control achieved in the spring following seed dressing application was as good as that obtained with a spray applied in the spring with a hormone weed killer. As with the autumn spray of tridemorph, the seed dressing of ethirimol had a profound effect on the vigour of the crop. In one trial in 1970 (assessed in November) tiller no./plant was increased from 3·0 to 5·4 and tiller weight from 0·05 to 0·14 g; mean adventitious root length was increased from 1·5 mm to 77·5 mm, and the number of adventitious roots per plant from 0·1 to 1·8 (Finney and Hall, 1971). In 1969 mildewed plants suffered much more from winter kill and were much less vigorous in the spring than seed-treated ones; in 1970 there was less winter kill and the effects in spring were less striking.

The yield response of winter barley is difficult to interpret. In 1969, ethirimol applied as a spring spray gave an average yield increase of 6 per cent, but the seed dressing, which controlled autumn mildew and gave moderate control of mildew in the spring, increased yield by 12 per cent. The extra increase obtained from control of autumn mildew was associated with an increase (16 per cent) in the number of tillers at harvest. In 1970 there was an equally marked increase of tillering. In some trials this resulted in large yield increases. In other trials the treated plants produced fewer grains per ear, and less plump grain, and this largely offset the effect of increased tiller numbers. In one trial where there was a 15 per cent yield increase, tiller number was increased by 37 per cent but 1000-grain weight was decreased by 7 per cent and the number of grains per ear by 12 per cent (Finney and Hall, 1971). The average yield response varied from 20 per cent increase (early, severe attack in autumn) to 5 per cent (late, mild attack in autumn) (Hall, 1971). A comparison of the two sets of trials suggests that in a year where autumn attack is followed by winter kill and perhaps by adverse growing conditions in the spring (e.g. the drought of 1970 or low soil fertility) the increase in tiller number will result in a direct yield benefit. In a year with little winter kill and good growing conditions, the crop attacked by mildew in autumn can compensate to such an extent that the increased tiller number resulting from chemical treatment can result in less productivity per tiller. These results were obtained with ethirimol seed dressing, but a similar pattern of results has been obtained with sprays of tridemorph (Jung and Bedford, 1971). Clearly more work is necessary to establish the best method of controlling mildew on winter barley.

Effect on other diseases of barley Barley is attacked by a number of other foliar diseases, the most important of which in Western Europe is brown rust *(Puccinia hordei)*. The benzimidazoles, triarimol, triforine and tridemorph all have some activity against brown rust, but since the attacks of brown rust usually occur much later than the optimum time for mildew sprays, it is not normally possible to exploit this activity. There has, in fact, been speculation that the control of mildew could lead to an increase in brown rust, particularly with specific compounds like ethirimol. This does not, however, seem to be the case.

In three winter barley trials in 1969/70, ethirimol seed dressing apparently significantly reduced brown rust attack (Brooks, 1971b) and similar observations were made in 1971 (Yarham *et al.*, 1971). In spring barley trials in 1970, in three out of the 14 trials infected with brown rust there was an apparent slight increase on ethirimol-treated plots, (Brooks, 1971b). Where the differences are relatively small, it is difficult to judge the reliability of the results because of the difficulty of accurately assessing brown rust in the presence of mildew on the untreated plots (it is easy to underestimate brown rust on mildewed plants). In trials carried out by the National Institute of Agricultural Botany in 1971 (Little and Doodson, 1971; Little, 1971 – personal communication), taking the trials as a whole, there was a slight tendency for brown rust to be worse on ethirimol-treated crops. On individual trials, there was an apparently large increase in brown rust on seven of the 64 variety/site combinations, but an apparent decrease in two. These differences were not always connected with the control of mildew. For example, on the highly mildew-resistant variety Feronia, there was a large increase in brown rust at one site, but no increase at the remaining seven sites. Thus this change in level of brown rust was not related to the increased clean leaf presented by the control of mildew.

There is thus little evidence that the control of mildew will lead to an increase in the levels of brown rust. Nevertheless, rust can limit yield potential gained by controlling mildew, and an integrated approach using rust-resistant varieties should be adopted.

Results on wheat Curiously the first two fungicides to achieve prominence, tridemorph and ethirimol, are less active against wheat powdery mildew than against barley powdery mildew, and tridemorph is also phytotoxic to certain cultivars. This, taken with the fact that much less is known about the importance of wheat mildew, has meant that far less work on chemical control has been carried out on this crop.

With ethirimol, the seed dressing has usually given only 30–50 per cent disease control when assessed at growth stage 10·5, and sprays applied in the spring were no better. Yield increases have rarely exceeded 5 per cent but increases of 10–20 per cent were obtained on the very susceptible Russian cultivar, Bezostaja, in trials in Hungary. On soft and hard spring wheats the seed dressing has been more effective and has given worthwhile yield increases in some trials.

Yield results with tridemorph on winter wheat have also been disappointing, and less than would have been expected from equivalent mildew levels on spring barley; the size of the response has varied with variety. For example, in a trial in Germany in 1970, the yield of Cappelle was decreased by a programme of sprays but that of Hanno was increased by over 20 per cent (Anon., 1971a). Sprays applied as late as G.S. 10 have sometimes given optimum response.

Chloraniformethan and triforine have given results on wheat roughly comparable with those of tridemorph. Benomyl is active on wheat both as a spray and a seed/soil treatment (Johnston, 1970) but it has not been exploited in this outlet.

Thus, although there exists a potential market for a fungicide on wheat in West and particularly East Europe, the fungicides currently available do not seem ideal for powdery mildew control.

Rusts

Wheat and barley are attacked by a number of rust diseases. The rusts are far more dependent than mildew on suitable weather conditions, and in a highly variable climate like that of West Europe, rust intensity varies greatly from year to year. Thus on winter wheat, yellow (stripe) rust (*Puccinia striiformis* West.) can cause losses of 20 per cent or more (Batts and Elliott, 1952) but is sometimes quite insignificant. Brown rust of barley (*P. hordei* Otth.) is common in Europe, but usually occurs only late in the season, so losses are not normally very great. Black (stem) rust (*P. graminis* Pers.) and leaf (brown) rust of wheat (*P. recondita* Rob. ex Desm.) are important in some areas in some years. In the United States, severe rust epidemics occur more frequently, and black rust and leaf rust are particularly damaging: in some years almost complete losses can occur locally (Rowell, 1968a).

Chemical control The compounds most widely evaluated for the control of rust are oxycarboxin (von Schmeling and Kulka, 1966) and RH-124 (4-*n*-butyl-1,2,4-triazole) (von Meyer *et al.*, 1970). Both chemicals are freely translocated within the xylem, and both can be applied either as spray or soil treatment. Oxycarboxin is active against all the rusts, but curiously, RH-124 is active only against leaf rust (von Meyer *et al.*, 1970). Other chemicals being evaluated against rust include the benzimidazole derivatives, triarimol, mebenil, pyracarbolid, triforine, G-676 (2,4-dimethyl-5-carboxanilido thiazole) and BAS 3170 (2-iodo-benzanilide).

Seed treatment Following seed treatment with carboxin at 2630 g/100 kg, almost complete control of yellow rust on wheat was obtained for 31 days, and after a further 30 days (when the plants were in 'boot') only 5 per cent of the foliage was infected, compared to 25 per cent on untreated plants (Powelson and Shaner, 1966). There was no impairment of germination at that high rate, but no yield measurements were made.

At the much lower rate of 600 g/100 kg, Rowell (1967) noted a reduction in germination of about 20 per cent, and partial disease control of leaf rust lasting for about 60 days. Application of a further spray to the seed-treated plots provided control as good as that obtained with two sprays. In further trials (Rowell, 1968b), applications of granules combine-drilled with the seed to give 1·12 kg a.i./ha followed by two sprays at 1·68 kg/ha again illustrated the importance of control of early and mid-season stages of rust, although in this trial there was a relatively late attack of rust. Poorer results were achieved by Hagborg (1970) when using granules drilled between the seed rows. Combinations of seed treatment and sprays have also been used with some success against black rust and leaf rust of wheat in India (Pathak and Joshi, 1970).

RH-124 appears to be more active from root uptake than carboxin (von Meyer *et al.*, 1970), giving complete control after 14 days compared to no control with carboxin, both at the rate of 1 ppm in the soil. Field trial results following soil applications are not available.

Thus, at least where rust attacks early in the life of the crop, seed dressings provide some useful protection. In Europe, where attack normally comes later, it is doubtful whether useful control could be obtained with the chemicals currently available.

Spray application Numbers of trials have been reported where single or multiple spray programmes have been evaluated for the control of various rusts, but it is very difficult to make a valid comparison between them because of the wide range of rates and conditions used. Furthermore, interpretation of yield data is complicated by the fact that artificial inoculation was frequently used.

In the United States, single sprays of oxycarboxin have not, in general, given sufficient control, due partly to the short half-life of the chemical giving insufficient persistence in early sprays, and insufficient eradicant activity in late sprays (Rowell, 1967). Nevertheless significant yield increases have sometimes been obtained, particularly with late sprays. For example, Powelson and Beaver (1970) obtained a significant increase from a single spray of 2·2 kg/ha against yellow rust on wheat applied just before flag-leaf emergence. In Europe, where inoculum pressure is often less severe, single sprays of oxycarboxin at 3 kg/ha can cause a prolonged delay in epidemic build up (Rapilly, 1970).

Where two or three sprays have been applied, good control of black and leaf rusts has generally been obtained with consequent yield increases. For example, Hagborg (1970) obtained good results with three sprays totalling 3·2 kg/ha or two sprays on any two of three dates. In general, oxycarboxin has not proved markedly superior to nickel/maneb sprays; recent formulation improvements have, however, increased its activity somewhat (Hagborg, 1971).

RH-124 has been less extensively evaluated, but preliminary results suggest that on leaf rust it is appreciably more effective than oxycarboxin (Rowell, 1969), and control until late into the season has been obtained on spring wheat from single sprays of 0·11 and 0·45 kg/ha (Buchenau, 1970). A single spray of 1·12 kg/ha was at least as effective as three sprays of oxycarboxin each at 2·24 kg/ha. RH-124 is inactive against other rusts.

Against yellow rust in Europe, mebenil at 1·87 kg/ha applied once during jointing gave an average yield increase of 17 per cent in three trials on wheat; when a second application was made 10–12 days later, the increase was 20 per cent. On spring barley, an average increase of 8 per cent was obtained with a single spray of 2·25 kg/ha at the beginning of jointing (Pommer and Kradel, 1969). The authors concluded that where there is an early attack, it would probably be necessary to spray during jointing and again shortly before or during ear formation.

The more recent compound, BAS 3170, has given better results than mebenil against both brown rust on barley and yellow rust on wheat, and single sprays have led to yield increases of about 10 per cent. The addition of the mildew fungicide tridemorph as a tank mix has further enhanced rust control (Frost, 1971 – personal communication).

Triarimol is active against rust, but again has been less widely evaluated than oxycarboxin. In Minnesota, a single spray of 0·56 kg/ha was markedly inferior to RH-124 (1·12 kg) for control of leaf rust, but two sprays at this rate gave results equal to three sprays of oxycarboxin (each of 2·4 kg) or the single spray of RH-124. Against black rust, triarimol gave quite good control from a single early application of 0·56 kg (Rowell, 1970).

Benomyl (2 kg/ha) was only moderately effective against stem and leaf rust on wheat (Rowell, 1970; Buchenau, 1970) and against brown rust of barley (at 1·7 kg) (Evans *et al.*, 1969).

Triforine gives good eradicant and protectant activity against leaf rust (Schicke and Veen, 1969; Ebenebe *et al.*, 1971) but has not been widely field-evaluated: some control of brown rust of barley and yellow rust of wheat and rye has been noted in field plots (Schicke *et al.*, 1971).

Discussion Thus, although there are several systemic compounds active against the various rusts of wheat and barley, the difficulty of exploiting this in commercial practice hinges on the difficulty of timing the spray correctly in relation to epidemic build-up, particularly since rust frequently attacks in the later growth stages.

In conditions suitable for rust development, there is at first a slow build-up of disease while inoculum is limiting; in the later stages there is abundant incoulum and the disease builds up rapidly on individual plants (van der Plank, 1963). With the older fungicides, control of the later stage alone was almost impossible, and repeated sprays were necessary. With the new systemic fungicides which have more powerful eradicant activity, later applications alone can be effective. The earlier rust attacks a plant, the greater the damage done (Batts and Elliott, 1952); ideally, therefore, a spray should be applied as soon as possible after rust appears (Rapilly, 1970). In Europe, promising results have been obtained using this approach (Pommer and Kradel, 1969), although a second spray is necessary during a prolonged attack. In the United States, where rust can cause premature death of the crop before the grain starts to fill, a spray applied to protect the crop during the last 30 days might be essential (Rowell, 1968a). In such severe conditions, more than one application is economic (Hagborg, 1970, 1971) and this can be a programme of sprays, or seed treatment followed by spray application (Rowell, 1968b). The seed treatment, if used widely, could be important in restricting inoculum and thus prolonging the slow stage of disease build-up. Its use also makes the timing of subsequent sprays less critical (Rowell, 1968b).

In Europe, when triforine, triarimol, benomyl and thiophanate methyl are being developed for commercial control of powdery mildew on wheat and barley, their activity against rust could give an added bonus. In fact, because the optimum time for spraying against mildew is generally much earlier than that for spraying against rust, control of rust is not normally achieved in practice.

Other foliar diseases

Systemic fungicides have not been used commercially against diseases other than mildew and rust. Experimentally, good control of glume blotch of wheat (*Septoria nodorum* Berk.) has been obtained with programmes of benomyl sprays.

Seed and soil-borne diseases

Loose smuts Since the loose smuts of barley (*Ustilago nuda* (Jens.) Rostr.) and wheat (*U. tritici* (Pers.) Rostr.) are borne internally in the seed, control is not possible with non-systemic fungicides. The most usual method hitherto has been the use of a 'hot' water steep which eradicates the fungus without causing major damage to the seed. This method is, however, costly and difficult to control, and the advent of suitable systemic fungicides has greatly increased the ease and efficiency of disease control.

Successful control of *U. nuda* with carboxin was first reported by von Schmeling *et al.* (1966). This has since been confirmed and expanded by numerous workers in several countries (early results were collated by Moseman, 1968). The rates used vary very widely within the range 25 g to 600 g a.i./100 kg seed. On barley, rates of 100 g/100 kg and above almost always give complete disease control. At lower rates, small numbers of smutted ears were found, but even at 25 g/100 kg infection was reduced from 293 to 2 smutted plants per 55 000 seeds (Maude and Shuring, 1969).

Extremely good results have also been obtained with carboxin on *U. tritici*, but here higher rates of chemical are required to give complete control (Jones *et al.*, 1968; Maude and Shuring, 1969). Some workers have found almost complete control at rates above 100 g/100 kg (e.g. Hansing, 1967); other workers have failed to get complete control even at rates as high as 375 g/100 kg. (Maude and Shuring, 1969; Kuiper, 1968). Nevertheless, at the rate of 110 g/100 kg, carboxin gives excellent control on both wheat and barley in commercial usage (Macer *et al.*, 1969). Because of its relatively narrow spectrum of disease control, carboxin is normally used in mixtures with other fungicides such as copper oxyquinolate (with which it has some synergistic effect) (Richard and Vallier, 1969) or mercury (Macer, 1969).

Carboxin appears to be without phytotoxicity except at very high rates (Reinbergs and Edgington, 1968); furthermore, in some conditions it appears to be directly beneficial to the plant (von Schmeling and Clark, 1970). Yield increases are not normally obtained unless very high levels of disease are present; Stoker and Dewey (1970) however, found yield differences at smut levels as low as 5 per cent.

Pyracarbolid (closely related chemically to carboxin) has a similar range of activity to carboxin, but is slightly more potent (Stingl *et al.*, 1970; Jank and Grossmann, 1971). For example, in field trials against *U. nuda*, pyracarbolid reduced infection from 13·3 to 0·1 per cent, compared to 1·9 per cent with carboxin when both compounds were applied at 25 g a.i./100 kg (Stingl *et al.*, 1970). BAS 3191 (2,5-dimethylfuran-3-carboxylic acid anilide) gives excellent control of the smuts, at least comparable with that obtained with carboxin (Pommer, 1970; Ventura *et al.*, 1970). It has, however, been replaced by BAS 3270 (2,5-dimethylfuran-3-carboxylic acid cyclohexylamide) which has a similar spectrum of activity but is less phytotoxic (Pommer *et al.*, 1971).

Thiabendazole gives only moderate control of loose smut. For example, an application rate of 200 g a.i/100 kg only reduced infection from 16·0 to 6·3 per cent, and there was slight reduction in germination at this high rate (Darpoux *et al.*, 1968).

Benomyl can be used to obtain complete control of loose smut on wheat and barley, but on barley at rates much higher than those needed with carboxin. At very low inoculum levels, Mankin (1969) obtained good control on barley at 62·5 g/100 kg, but in other trials at higher inoculum levels, 250 g was inferior to 94 g of carboxin (Mankin, 1970). Benomyl appears to be more active against wheat smut than against barley smut, complete control against the former being obtained at 100 g/100 kg (Ventura *et al.*, 1970). NF48 [2-(3-methoxycarbonyl-2-thioureido)aniline] is identical in behaviour to benomyl. Triarimol (at the low rate of 8 g/100 kg) was slightly inferior to benomyl.

Covered smuts, bunts The main diseases in this group are wheat bunt (*Tilletia caries* (DC.) Tul and *T. foetida* (Wallr.) Liro), dwarf bunt (*T. contraversa* Kühn), covered smut of barley (*Ustilago hordei* (Pers.) Lagerh.), loose smut of oats (*U. avenae* (Pers.) Rostr.) and covered smut of oats (*U. kolleri* Wille). These diseases are either carried superficially on the seed, or are soil-borne; in general, they have been very well controlled by non-systemic chemicals such as mercury and maneb. Many of the new systemic fungicides also give good or excellent control.

Against covered smut on oats, carboxin gave almost complete control at 150 g/100 kg (Wallace, 1968) and 90 g/100 kg (Kuiper, 1968); Stingl *et al.* (1970) found complete control at all rates down to 40 g, and Ventura *et al.* (1968) down to 30 g. Very similar results were obtained against covered smut of barley (Wallace, 1968; Ventura *et al.*, 1968). Carboxin is somewhat less active against wheat bunt, incomplete control being obtained within the range 150–225 g/100 kg (Wallace, 1968; Ventura *et al.*, 1968; Kuiper, 1968). Against soil-borne *T. foetida* carboxin was considerably inferior to HCB (hexachlorobenzene) on all except HCB-resistant strains (Kuiper, 1968), and was ineffective against dwarf bunt (Hoffmann, 1969).

Pyracarbolid has been reported to be slightly more active than carboxin against covered smut of oats (complete control at 10 g/100 kg) (Stingl *et al.*, 1970). Results are not available against the other diseases.

Thiabendazole gave complete control of seed-borne bunt at 120 g/100 kg (Hoffmann, 1970), and good control at rates as low as 20 g/100 kg (Darpoux *et al.*, 1968). Similar results have been obtained against covered smuts of oats and barley. Like carboxin, thiabendazole is not very effective against soil-borne bunt, but it has given some control of dwarf bunt at high rates (250–480 g/100 kg) (Hoffmann, 1971; Dickens and Oshima, 1971). It is the only chemical to have proved effective against this disease.

Benomyl is highly effective against seed-borne bunts and smuts, giving complete control at 100 g/100 kg on wheat, barley and oats (Ventura *et al.*, 1970; Mankin and Wood, 1969). It is relatively ineffective against soil-borne bunt and dwarf bunt (Hoffmann, 1970). Where NF48 has been compared with benomyl, results have been comparable.

BAS 3191 was as effective as benomyl against bunt and loose smuts of oats, giving complete control at 100 g/100 kg (Ventura *et al.*, 1970), but it has been replaced by BAS 3270 (Pommer *et al.*, 1971).

Flag smut Flag smut of wheat (*Urocystis agropyri* (Preuss) Schroet.) is mainly soil-borne and has not been very well controlled by standard fungicides.

Of the new systemic fungicides, oxycarboxin gives the best control. Almost complete eradication was obtained at 30 g/100 kg: carboxin and pyracarbolid were also extremely effective (control at 75 and 50 g/100 kg respectively): benomyl was only moderately effective and thiabendazole was ineffective (Line, 1970). Metcalfe and Brown (1969) found a similar pattern of results, using artificial inoculation and higher rates of chemical application.

Seedling blights The diseases in this group are seed- or soil-borne and are mainly well controlled by organomercurial or non-mercurial seed dressings such as maneb. Whilst many authors have noted the effect of systemic fungicides on field emergence, few have studied the effect of the chemicals on specific diseases.

Snow mould. *Fusarium nivale* Fr. None of the new fungicides is as effective as the standard compounds. Benomyl, NF48, thiabendazole, BAS 3191, BAS 3270 and carboxin all have some activity at 100 g/100 kg; triarimol is without effect at 8 g/100 kg (Ventura *et al.*, 1970; Darpoux *et al.*, 1968; Pommer *et al.*, 1971). Benomyl is used commercially at 100 g/100 kg as a seed dressing and at 200 g/ha as a spray for the control of snow mould in Finland.

Brown foot rot. *Fusarium culmorum* (Sacc.) Sn. and H. and *F. graminearum* (Schwabe) Sn. and H. Where wheat seed internally infected was used, thiabendazole gave complete control at 160 g/100 kg, whilst maneb and mercury were relatively ineffective (Darpoux *et al.*, 1968). Results are not available with the other compounds.

Leaf stripe. *Helminthosporium gramineum* Rabenh. In trials on spring and winter barley, triarimol, BAS 3191 and BAS 3270 gave slight control, but benomyl and NF48 were ineffective. All treatments were inferior to maneb and mercury (Ventura *et al.*, 1970; Pommer *et al.*, 1971). Poor results were also obtained with thiabendazole (Darpoux *et al.*, 1968). Good control of leaf stripe has been obtained in the field with carboxin applied at a rate of 125 g/100 kg (Kingsland, 1970) and 150 g/100 kg (Macer *et al.*, 1969).

Glume blotch. *Septoria nodorum* Berk. Triarimol (8 g), benomyl (150 g) and NF48 (100 g), all per 100 kg, reduced infection of seedlings, but were inferior to mercury; oxycarboxin and BAS 3191 were ineffective (Ventura *et al.*, 1970). Thiabendazole, however, was found to have good activity, comparable with that of maneb and mercury (Darpoux *et al.*, 1968).

Discussion. The great advantage of the systemic fungicides as seed dressings is that they provide control of pathogens borne within the seed (the loose smuts); these are inaccessible to non-systemic fungicides. None of the new chemicals has a sufficiently broad spectrum to control the wide range of diseases currently controlled by the organomercurials or certain non-mercurial dressings like maneb or copper oxyquinolate. It is therefore necessary to use mixtures of systemic and non-systemic fungicides: because of the high cost of these mixtures they are normally used only on stocks known to be heavily infected with loose smut, or where the crop is being produced for seed.

Eyespot *Cercosporella herpotrichoides* Fron, occurs wherever wheat and barley are grown. It reduces yield by directly reducing the amount of carbohydrate going to the grain, and by causing lodging. Fungicides have not been used commercially for the control of eyespot, but the growth regulant, chlormequat, has been widely used to shorten and stiffen straw, and so make it less prone to lodging.

Many workers have investigated the effectiveness of benomyl in controlling eyespot in glasshouse experiments (Defosse, 1970; Lemaire *et al.*, 1970; Fehrmann, 1970). In these experiments excellent control was achieved by spray, seed dressing and soil applications. Even when applied 30–50 days after inoculation, progress of the disease was halted, indicating a very powerful eradicant activity, (Defosse, 1970; Fehrmann, 1970). Protective sprays were only effective if applied not more than 10 days before infection.

The effectiveness of benomyl has been confirmed in the field, where the eradicant effect has again been established, allowing good control over widely different spray times. In the United States, in artificially inoculated plots of winter wheat, benomyl spray at 5·6 kg/ha in November, March or April gave almost complete disease control and significantly increased yields (Powelson and Cook, 1969). In Germany, however, autumn application was relatively ineffective, but sprays of 240 g/ha in the spring were very effective, a spray in May reducing infection from 95 to 7–10 per cent when assessed six weeks before harvest (Fehrmann, 1970). In severely infected artificially inoculated plots in England, a spray at G.S. 9 (Feekes Scale) reduced infection by 95 per cent and increased yields on average from 6–8 per cent on moderately resistant varieties to 43 per cent on very susceptible varieties (Doodson and Saunders, 1969).

Equally striking results have been obtained in naturally infected wheat in the field using low rates of application. In England, sprays of 1·12 kg/ha at G.S. 3–4 gave excellent control, but slightly increased sharp eyespot (*Rhizoctonia solani* Kühn). A seed treatment at 280 g a.i./ha gave poor control of eyespot (Prew, personal communication). In New Zealand, a single spray of 0·28 kg/ha applied at G.S. 7–8 reduced lodging from 36 per cent to almost nil, and increased yield by 46 per cent (Witchalls and Close, 1971). No doubt the efficiency of benomyl spray depends on correct timing in relation to the main infection period, and Rapilly (1970) has attempted to forecast this using the method devised by Ponchet (1959) based on the mean daily temperature. Benomyl has also proved active against eyespot of rye (Rawlinson and Davies, 1971) and barley (Davies, 1970).

Little work has been reported on other chemicals, but thiabendazole was found to be almost as active as benomyl when applied as a seed dressing (Lemaire *et al.*, 1970).

RICE

Rice is attacked by a number of fungal diseases, the most important of which is blast *(Pyricularia oryzae* Cav.), and this is the only disease on which extensive results with systemic fungicides have been published. Although blast is important in virtually every rice-growing country, it is largely in Japan that progress with fungicidal control has been made.

In the early 1960s, phenylmercury acetate and the antibiotic blasticidin S were widely used for control of blast; both compounds have now been replaced because of their high toxicity to mammals. Blasticidin is absorbed into plants, but is not effective as a systemic fungicide because of its rapid degradation in the plant (Misato *et al.*, 1959). More recently a far safer antibiotic, kasugamycin, has been used for blast control. This compound, obtained from the culture filtrate of *Streptomyces kasugaensis,* is very active when sprayed at 20–30 ppm, and there is a wide margin of safety from phytotoxicity. The chemical is very effective in preventing secondary spread and sporulation, but has little effect until after penetration (Okamoto, 1968). It is therefore rather more effective as an eradicant than as a protectant fungicide. It is absorbed into the leaves and roots and translocated readily into the upper parts of the plant, where further lesion enlargement is prevented (Ishiyama *et al.*, 1965a, b; 1967). In practice, however, it is used only as a foliar spray since excessively high quantities are required for application through the paddy water (Kozaka, 1969).

More recently a number of closely related organophosphorus fungicides have been developed in Japan, including kitazin, kitazin P, hinozan, inezin and cohnen. All of these compounds give effective control of blast in sprays of 4–500 ppm, but above this there is a danger of phytotoxicity. The chemicals have eradicant as well as protectant properties, but are best applied not later than 2–3 days after infection. By far the most readily systemic of these compounds is kitazin P (Kozaka, 1969), and this compound is freely translocated from roots to leaves.

In practice kitazin P can be applied in a programme of 2–3 sprays at 450 ppm or it can be applied to paddy water at a rate of 5–10 kg/ha (Yoshinaga, 1969). The granules appear to become active about three days after application, to reach their maximum effectiveness after five to seven days but to retain marked activity for at least 30 days (Kozaka, 1969). For leaf blight control, granule application at the time blast was observed, or one week earlier or later, gave equivalent disease control. With neck and node blast, best yield results were achieved with an application about one week before heading, although later applications gave slightly better disease control. Even at the lowest rate of application, granules were markedly more effective than a programme of three sprays (Yoshinaga, 1969).

Benomyl has marked activity against blast, although this does not appear to have been widely exploited. In trials in Louisiana, two sprays of benomyl each at 560 g/ha gave control of neck blast equivalent to that obtained with phenylmercury acetate, and superior to blasticidin S (Lindberg, 1969). In the Philippines, two sprays of benomyl at 200 g/ha were markedly superior to kasumin for control of neck rot (kasumin is reported to be relatively ineffective in the tropics) (Anon., 1970).

Benomyl has also been used as a seed treatment to control seedling blast. In box tests, benomyl at 400 g/100 kg of seed gave almost complete control during the seedling stage, whilst untreated plants were almost killed. Soil treatment with benomyl also gave good control of leaf and neck blast (superior to kitazin) but only at rates unlikely to be economic (Anon., 1970).

MAIZE

Most maize diseases are not amenable to control with the systemic fungicides, and work has largely been restricted to trials against head smut, *Sphacelotheca reiliana* (Kühn) Clint. Moderate control of this disease with carboxin was obtained in naturally infested field plots at a rate of 250g/100 kg (Simpson and Fenwick, 1971) and in seedlings transplanted from artificially inoculated soil in boxes to the field at 400 g/100 kg (Jooste, 1969).

Carboxin is also reported to have some activity against southern leaf blight (*Helminthosporium maydis*, Nisikado and Miyake) when used as a seed dressing.

Acknowledgements My thanks are due to many colleagues at Jealott's Hill, notably Dr M. C. Shephard and Dr T. J. Purnell whose results I have quoted without reference. My thanks are also due to Mrs S. E. Handysides for help in the preparation of the manuscript.

10.II

Results in practice – II. Glasshouse crops

by D. M. Spencer

INTRODUCTION

The glasshouse crops industry exists almost entirely in the northern hemisphere. Indeed, it is mainly confined to the temperate areas of Europe and North America with less extensive areas in Japan and the USSR. Consequently the range of crops with which the glasshouse industry is concerned is somewhat limited. In Europe, for example, tomatoes, cucumbers and lettuces constitute the bulk of the industry with flowers and other decorative plants making up most of the remainder.

Holland has a larger glasshouse acreage than any other country (Table 10.II.1) and of this the major part is devoted to vegetable production. The same is true for the British Isles where flower production accounts for less than one-quarter of the glasshouse acreage. In France and Germany on the other hand a relatively larger proportion, though still less than half of the glasshouse space, is occupied by decorative crops, whereas in the USA three-quarters of the glasshouse acreage is used for flower crops.

These proportions have remained fairly stable in recent years though the acreage of ornamental crops in northern Europe has been increasing.

Italy and Japan have relatively small areas of traditional glasshouses, but both countries have rapidly increasing areas of crops grown under plastic structures. It is against this background of an established range of crops with a corresponding range of diseases that the practical application of systemic fungicides must be considered.

Crops in glasshouses are usually grown without rotation and thus tend

TABLE 10.II.1 **Area and function of glasshouses in major glasshouse-crop-producing countries**

	Acreage	*% Acreage*	
		Vegetables	*Flowers*
Holland	17 166	84	16
USA	6 570	25	75
France	2 500	60	40
Germany	4 500	62	38
British Isles:			
(*a*) England	4 210	74	26
(*b*) Scotland	307	†	†
(*c*) N. Ireland	56	†	†
(*d*) Guernsey	1131	87	13
(*e*) Jersey	109	†	†
(*f*) Eire	393	†	†

† Figures not available; mainly vegetables

to suffer from diseases which can very quickly become epidemic in one house or on one holding. Within the protection of the glasshouse, however, the grower does have the facility for a measure of control over the environment in which his crop is growing and hence, by careful manipulation of this control the experienced grower can minimise the losses caused by disease. For example, by the judicious use of his ventilators the astute tomato-grower will so control the relative humidity of the atmosphere within his crop as to reduce the likelihood of spores of leaf-invading pathogens such as *Botrytis cinerea* and *Cladosporium fulvum* being able to germinate and to invade their host. Even with the wilt organism *Fusarium oxysporum*, the inoculum of which may be present in the deeper soil layers where steam or chemical soil partial-sterilants cannot penetrate, the experienced grower can accommodate the infection that may develop and can grow his crop with the minimum of loss through wilt. In considering the use of chemical crop-protectants, such as systemic fungicides, in the glasshouse it is therefore important also to take account of the non-chemical methods of disease control available to the grower and as far as possible to integrate the various methods of disease control.

The 1971 edition of the Agricultural Chemicals Approval Scheme (Ministry of Agriculture, 1971) which lists the 834 proprietary products officially approved for use in horticulture and agriculture in the United Kingdom, mentions only four systemic fungicides – benomyl, carboxin, ethirimol and tridemorph. Dodemorph has appeared in a supplementary list and this compound, together with benomyl, are, as yet, the only systemic fungicides recommended for use on glasshouse crops.

DISEASE CONTROL IN THE GLASSHOUSE

Special requirements The two most important glasshouse vegetable crops, tomatoes and cucumbers each have a long cropping season. From the time that the first fruit is harvested to the time that the plants are pulled out, may be as long as eight months and for some of this time fruit may be picked almost daily. Thus, any compound will be unacceptable for use on such crops unless it can be shown that the compound is either non-toxic to those who eat the fruit or that such toxicity is lost very soon after the compound has been applied. Benomyl, for example, has been cleared for use in the United Kingdom on all edible crops with no time interval between application and harvest. Similarly there is no restriction on the time between application of dimethirimol and picking the cucumber fruit. No other systemic fungicides are yet at this advanced stage of registration though there seems good reason to expect some of them to be useful where there is a tolerably short time interval between application and harvest.

Because of the short interval between successive harvests those protectant fungicides which have to be applied as sprays at frequent intervals to protect new growth and which leave noticeable deposits on the fruit will also be unacceptable. Furthermore, glasshouse crops are grown intensively and therefore bulky machinery cannot be used without the danger of damaging tall-growing crops. It follows, therefore that a material which has to be applied at short intervals by hand spraying will be less attractive to the grower than one which can be applied automatically and at longer intervals.

The currently available systemic fungicides which can be applied as drenches need to be used only monthly, whereas sprays are applied fortnightly. Furthermore, when applied to the soil and taken up by the roots these compounds will not cause blemishing deposits on the outside of the crop. For these reasons systemic fungicides have been long-awaited by the glasshouse industry and it is, perhaps, not surprising that growers have shown great enterprise in making full and enthusiastic use of benomyl, the first wide-spectrum systemic fungicide to become available to them.

In the past the glasshouse industry has been obliged to make use of crop-protectant compounds primarily developed for other agricultural applications. Dimethirimol is a notable exception in that it was developed specifically for use against powdery mildews of cucurbits and therefore had direct applicability to the glasshouse cucumber crop.

USE OF SELECTED SYSTEMIC FUNGICIDES

Tomato *Grey mould (Botrytis cinerea)* This organism can cause death of the whole plant if lesions girdle the lower stem. It also causes loss through reduced yield and reduced value of 'ghost-spotted' fruits. Primary infections usually occur on leaves and flowers (Smith, 1970), and infection spreads to other parts either where infected flowers fall or through the distribution of airborne spores.

Though Bollen and Fuchs (1970) showed the organism to be very sensitive to benomyl none of a group of eight systemic fungicides was able to control ghost-spotting of the fruit (Smith, 1971a). Indeed, though an *in vitro* bioassay method, using *Cladosporium fulvum* and *B. cinerea* as test organisms indicated the presence of fungitoxic materials in the leaves, sepals and petals of plants treated with benomyl or with two thiophanate compounds, the method failed to demonstrate fungitoxicity in fruits (Smith and Spencer, 1971). Several compounds, applied as sprays or drenches were able to inhibit the spread of stem and leaf lesions, suggesting that if they were to be applied as a protectant treatment to the entire crop in a glasshouse they could prevent the build-up of conidial inoculum and thus reduce ghost-spotting of the fruit.

Spray applications of benomyl, 8 oz (a.i.) per 100 gal at 14-day intervals or one pint per plant of a drench containing 4 oz (a.i.) per 100 gal applied at monthly intervals, have been used for the control of *B. cinerea* but in practice growers seem to have adopted a range of concentrations and a variety of methods of application.

Leaf mould (Cladosporium fulvum) This disease is associated with high temperatures and high humidity and can often be controlled by manipulation of the environment, where this is possible. The causal organism is sensitive to low concentrations of thiabendazole, benomyl, and thiophanate compounds, less so to triarimol and triforine. It has been used in tests *in vitro* for the presence of fungitoxic materials in treated plants (Smith and Spencer, 1971). In small-scale glasshouse tests, six systemic fungicides were found to control the disease for at least six weeks after a single drench treatment (Smith, 1971a).

At present benomyl is alone in having been cleared for use on all edible crops in the United Kingdom and has been recommended for use as a spray for the control of *C. fulvum* infection in the same way as it is used for the control of grey mould *(B. cinerea)*.

Stem rot (Didymella lycopersici) This disease is much worse in some years than in others and usually attacks first at soil level. The sunken dark brown lesions with barely visible black pycnidia may appear on other parts of the plant later in the season, often at the site of injury but not necessarily so, and usually cause the death of those parts of the plant above the infection.

The ED_{50} of benomyl against both *D. lycopersici* and *B. cinerea* is 1–5 μg/ml (Bollen and Fuchs, 1970) and the symptoms of stem rot *(D. lycopersici)* have much in common with those of grey mould *(B. cinerea)*. Smith (1971a) found that benomyl applied as root drench or spray, though retarding the growth of stem lesions caused by *B. cinerea*, was not completely inhibitory and results obtained for control of *D. lycopersici* by benomyl have been variable. However it has been found that benomyl applied monthly as a whole-stem spray or as a stem-base drench can reduce the level of disease caused by *D. lycopersici*.

Wilt diseases (Verticillium albo-atrum, Verticillium dahliae and Fusarium oxysporum) These important diseases are soil-borne and therefore efficient soil-sterilisation is essential where they are a problem. The losses they cause can, to some extent, be mitigated by careful control of the environment: with *Fusarium* wilt the temperature needs to be kept down and with *Verticillium* wilt, higher temperatures are required. Until the advent of systemic fungicides the growers' only alternative to adequate environmental control was to pull out all infected plants as they appeared.

Bollen and Fuchs (1970) found benomyl to be highly active against *V. albo-atrum* and *V. dahliae* but less active against *F. oxysporum*. Fuchs *et al.* (1970) found the last-named fungus to be inhibited by 1 μg/ml benomyl *in vitro* and when they supplied the compound as a daily drench to potted seedlings which had been inoculated with *F. oxysporum*, symptoms of wilt were suppressed. Weinke *et al.* (1969) showed thiabendazole to be active *in vitro* against *Verticillium* and *Fusarium* species and were able to detect some movement of the compound into roots of tomato plants after foliar applications.

Soil drenches of benomyl and thiabendazole applied to potted tomato plants previously inoculated with *V. albo-atrum* controlled the disease for 14 weeks, though the pathogen could be isolated from stem segments after 15 weeks (Ebben, 1970a).

Soil drenches of benomyl were found to control *Fusarium* wilt of potted tomato plants by Thanassoulopoulos *et al.* (1970) who also noted that spraying the leaves with an aqueous suspension had no effect on the disease and that the chemical had no apparent phytotoxic effect when applied as a soil drench at rates up to 10 000 ppm (a.i.). In the writer's laboratory potted tomato plants treated with high levels of benomyl have shown leaf necrosis, particularly on younger leaves. In field-grown tomatoes, Matta and Garibaldi (1970) found benomyl, thiophanate and thiophanate-methyl highly effective when applied as soil drenches in reducing *Verticillium* wilt, whilst Jones *et al.* (1971) achieved wilt control for over four months when benomyl was applied at the same time as the soil fumigant 1,2-dibromoethane.

Benomyl is used as a soil drench, at one pint per plant, of a suspension containing 4 oz (a.i.) per 100 gal applied at four-week intervals for the control of *Verticillium* wilt and the compound is being used at this concentration in Jersey for the control of *Fusarium* wilt.

None of the other systemic fungicides has reached this stage of commercial use against wilt diseases.

Brown root rot (Pyrenochaeta lycopersici) This disease reduces plant vigour and crop production by destroying part of the root system. Normally the disease is kept under control by annual or twice-yearly soil sterilisation with steam or methyl bromide. Ebben (1971) found that benomyl applied at 40 lb/acre as a whole-plot drench or applied dry and raked into the soil surface, reduced brown root rot and increased tomato yields in soils known to be heavily infested with *P. lycopersici*. Thiabendazole was phytotoxic at the concentration used (40 lb (a.i.) per acre) and though it brought about a decrease in brown root rot the yields from plots treated with this compound were less than those from untreated plots. The cost of materials for this treatment would be about £350 per acre and, as Ebben (1971) points out, more than one treatment would be needed in the season to control brown root rot. In addition, benomyl is ineffective against some soil-borne pathogens and so the compound cannot be considered in any way as a substitute for effective soil sterilisation. It may be that post-planting drenches could be used to control brown root rot in patches where soil sterilisation had not been efficient.

Cucumber *Powdery mildew (Sphaerotheca fuliginea)* Powdery mildew is the most important cucumber disease. It spreads very quickly in hot dry conditions and its control by standard protectant fungicides is complicated by the very rapid growth of the cucumber plant accompanied by a rapid proliferation of new and therefore unprotected leaves.

Dimethirimol is a systemic fungicide which is highly active against certain powdery mildews. It was shown by Elias *et al.* (1968) to be particularly effective and to give long-lasting protection against powdery mildews of cucurbits when applied to the soil. Bent (1970) found dimethirimol to be toxic to the spores of *S. fuliginea in vitro* and *in vivo* and was able to correlate the disease control provided by this compound with its direct fungitoxic action. The compound was widely used on the British cucumber crop in 1968 and gave spectacular control of powdery mildew. Since 1969 the material has been used widely in Holland but in both these countries its continued use has been complicated by the emergence of strains of the fungus which are tolerant to dimethirimol (see Chapter 8). In 1971 it was sold for the first time in Eastern Europe and at present it is also in extensive use in Israel where strains of *S. fuliginea* tolerant to benomyl have appeared.

Benomyl is highly effective against cucumber powdery mildew and is widely used in the United Kingdom both as a spray at 4 oz (a.i.) per 100 gal, or soil drench at 3 oz (a.i.) per 100 gal, each applied at monthly intervals. At these concentrations the compound also gives effective control of *B. cinerea* and the spray application has also been found to give a measure of control of *Mycosphaerella melonis*. Cole *et al.* (1970) found that soil amendment with benomyl at concentrations of 40 ppm gave complete control of powdery mildew of cucumber.

Other systemic fungicides provide control of *S. fuliginea*. Triforine was shown to be systemically active against cucumber powdery mildew when applied as a pre- or as a post-infectional treatment (Fuchs *et al.*, 1971). Thiophanate and thiophanate methyl have also given protection of the same order as that provided by benomyl in laboratory and greenhouse tests (Mizukami, 1970; Van Assche and Vanachter, 1970).

Smith (1971a) examined a group of 14 compounds for systemic fungicidal activity in tests on tomato and cucumber plants growing in pots. Many of the compounds gave a measure of control over the diseases and in particular benomyl, triarimol, thiophanate and thiophanate methyl were effective against cucumber powdery mildew when applied as sprays and drenches and in addition to these, triforine and dodemorph also prevented infection but only when applied as drenches. Further tests with some of these compounds on cucumber crops have confirmed their usefulness against powdery mildew.

Black root rot (Phomopsis sclerotioides) Where this disease is a problem, soil sterilisation is essential though root lesions have been found on plants from steamed plots after about four months (Ebben, 1970b). More powdery mildew was found in plants with black root rot than on those with healthy roots and the application of dimethirimol gave poorer control of mildew on plants with root rot.

Ebben and Last (1969) grew plants on plots infested and non-infested with *P. sclerotioides* and controlled both black root rot and powdery mildew *(S. fuliginea)* with benomyl drenches. Where no benomyl was applied yields were very much reduced by mildew on plots not infested with *P. sclerotioides* but were even lower in plots where root rot also occurred.

Cruger (1969) noted in his experiments on the control of cucumber mildew by benomyl and dimethirimol that the former compound applied as a drench at 500 mg a.i. per plant gave some control of *P. sclerotioides*.

Stem rot (Mycosphaerella melonis) In tests on two-week-old seedlings growing in pots, Ebben (1971) found that benomyl, thiophanate and thiophanate methyl prevented the development of lesions when petioles were inoculated with mycelium from growing cultures of *M. melonis*. By contrast triarimol and thiabendazole did not prevent lesion development even when they were used at phytotoxic concentrations. In older cropping plants, benomyl sprays were only able to retard the development of *M. melonis* lesions.

The use of systemic fungicides in sprays and drenches on cucumbers for the control of mildew, grey mould and stem rot has to be integrated with any scheme for the biological control of insects and mites.

Lettuce *Grey mould (Botrytis cinerea) and downy mildew (Bremia lactucae)* These are the two most important diseases of lettuce: both are associated with adverse growing conditions, both occur first on lower leaves and can cause loss from unsaleable heads, when, for example, the disease appears on the heart leaves.

Sprays and dusts with thiram or zineb have given adequate protection and, as yet, there seems no real indication that systemic fungicides have any particular advantages. *Botrytis cinerea* is sensitive to benomyl and this compound is therefore useful for the control of grey mould when applied as a spray at 4 oz (a.i.) per 100 gal.

There have been reports of the compound causing a slight check to growth when applied at a higher concentration as a spray or as a drench. Furthermore, benomyl is ineffective against *Bremia lactucae,* a member of the Phycomycetes. None of the other systemic fungicides has been reported to be useful for the control of lettuce diseases.

Carnation *Fusarium wilt (Fusarium oxysporum) and Verticillium wilt (Verticillium cinerescens)* These are both serious soil-borne diseases which can be introduced on planting material. Furthermore, it has been considered as almost inevitable that wilt would sooner or later appear on any commercial carnation holding. *F. oxysporum* can be distributed by air-borne spores. The usual recommendations of soil sterilisation and careful attention to hygiene have given only a partial control of the diseases on some holdings. The introduction of systemic fungicides has therefore been particularly welcomed by carnation growers who have been quick to develop their own control measures using the new compounds.

Benomyl has been widely used for the control of these diseases since it was demonstrated to give control of *V. cinerescens* infection in trials in the Lea Valley (Knight, 1971).

Garibaldi (1968) reported on the efficacy of thiabendazole and carboxin used as a drench of 0·5–1·0 g/m^2 every one or two weeks in controlling *Verticillium* wilt. He obtained only partial control with carboxin, which was phytotoxic and other workers have since found thiabendazole occasionally to cause phytotoxicity.

In her earlier experiments, Ebben (1970a), though achieving control of *V. cinerescens* infection in pot tests, found neither benomyl nor thiabendazole applied at 0·5 g and 1·0 g a.i. per plant to control the disease when applied, before planting, to plots infested with the disease organism. When she applied the compounds to a growing crop in which *Verticillium* wilt was spreading Ebben (1971) found the rate of spread to be greatly reduced in the succeeding weeks. Flower yields were reduced in the thiabendazole-treated plots.

Work in France (Tramier and Antonini, 1971) has shown that soaking cuttings in a 5 000 μg/ml (a.i.) suspension of benomyl before planting in soil infested with *V. cinerescens* can delay infection but that better protection is obtained by applying 1 g/m^2 every 8–15 days in the irrigation water. In order to control established infection a first treatment with 5–10 g/m^2 is required, followed by the fortnightly application of 1 g/m^2. *F. oxysporum* is more resistant to benomyl than is *V. cinerescens* and therefore requires 2–3 g/m^2 applied fortnightly. These rates seem to be high, and leading to a high soil level of the systemic fungicide when compared with makers' recommendations for its use in the British Isles. Here a rate of $\frac{3}{4}$ lb Benlate (50 per cent benomyl) in 100 gal applied as a drench at 2 gal/yd^2 is repeated two weeks later and is then not applied again within three months of the second drench. These rates approximate to 75 g (a.i.)/100 l applied at 3 l/m^2, i.e. just over 2 g/m^2. However, though there have been occasional reports of slight checks to growth by applications of benomyl it seems that the carnation crop can withstand high levels of benomyl without damage (Fletcher, 1971).

Rust diseases of carnation (*Uromyces caryophyllinus* and *Puccinia arenaria)* are controlled by sprays containing 0·075 per cent triforine whilst carboxin was shown by Umgelter (1969) to control rusts of several ornamentals including carnation.

Rose

It is a feature of the systemic fungicides currently available that none of them is translocated to any great extent in woody tissues. Therefore, whereas drench treatments can be applied to other crops, spraying at the present time is the only useful method of application of systemic fungicides to rose bushes. The compounds are thus used mainly as protectants though eradicant activity has been demonstrated for some of them.

Powdery mildew *(Sphaerotheca pannosa)*, black spot *(Diplocarpon rosae)* and rust *(Phragmidium* spp.*)* are the three most important diseases of cultivated roses and, of these, powdery mildew is the more troublesome under glass. Dodemorph has been shown to be effective in controlling powdery mildews in a wide range of ornamental plants (Kradel *et al.*, 1969) and used as a spray at 0·1 per cent a.i. at 10-day intervals controlled the disease on Hybrid Tea roses in the glasshouse (Frost and Pattisson, 1971).

Triforine, shown by Fuchs *et al.* (1971) to be fungitoxic to a range of fungi *in vitro* has effectively controlled black spot and rust as well as powdery mildew when used at 150–200 ppm (a.i.) as a spray at 7–14 day intervals (Schicke and Arndt, 1971).

Benomyl is registered in most European countries and has been given clearance in North America for use on roses at a rate of 200–1000 ppm (a.i.).

In trials of five systemic fungicides for the control of black spot on roses, Kavanagh (1970) found triforine to give better results than triarimol and benomyl, thiophanate was less effective and thiabendazole gave poorest results as measured by the extent of defoliation at the end of the season.

Engelhard (1969) compared benomyl, fuberidazole and thiabendazole with a zinc-plus-maneb formulation for the control of *D. rosae*. Benomyl provided residual protection for over three months and was the only effective fungicide of the three benzimidazole compounds.

Other ornamentals

A wide range of ornamentals is produced in the glasshouse as bedding plants, pot plants or as cut flowers. Such plants include cyclamen, chrysanthemum, begonia, Saintpaulia, poinsettia, hydrangea, pelargonium, kalanchoe and azalea. These species each suffer from specific diseases which commonly include damping off and foot rots, caused by members of the Phycomycetes, powdery mildews, grey mould *(B. cinerea)* and diseases caused by *Fusarium* and *Rhizoctonia* species.

None of the currently available systemic fungicides seems to give control of diseases caused by Phycomycetes. Powdery mildews and rusts of many ornamentals are well controlled by triforine as a spray at 0·075 per cent in water (Schicke and Veen, 1969).

Thiabendazole and benomyl are both very effective against grey mould and powdery mildews and the latter compound is registered in Europe and North America as a standard treatment as a spray or pot drench for the control of these diseases.

Thielaviopsis basicola on cyclamen (Stalder and Barben, 1970) and on poinsettia (Raabe and Hurliman, 1971) is controlled by benzimidazole compounds.

Benomyl and thiabendazole were shown by Crane and Jackson (1969) to give adequate control of *Ascochyta chrysanthemi* which, like *B. cinerea*, causes losses of chrysanthemum cuttings, whilst white rust of chrysanthemum *(Puccinia horiana)*, pelargonium rust *(P. pelargonii-zonalis)* and several other rusts of flowering plants were shown by Umgelter (1969) to be controlled by carboxin.

Bulbs and corms

Most bulbs and corms of commercially grown crops are used to produce stocks for forced flowers in the glasshouse. They suffer from a number of diseases of which the most serious are caused by species of *Fusarium* and *Botrytis*. These diseases have been subjected to experimental control by systemic fungicides, particularly by thiabendazole and benomyl, with encouraging results and these compounds are now used routinely for the control of certain of the diseases.

Tulip fire (Botrytis tulipae) This serious disease has an underground phase in which infection is transferred from mother to daughter bulbs and is also carried-over from season to season by sclerotia remaining on bulbs and in the soil. Price (1970) showed that thiabendazole and benomyl were toxic to sclerotia of *B. tulipae* and that bulbs dipped in suspensions of the fungicides remained toxic to the pathogen for several weeks. De Rooy and Vink (1969) showed that a dip of 0·5 per cent of 50 per cent benomyl or 0·5 per cent of 60 per cent thiabendazole gave good control of *B. tulipae* in forcing bulbs.

Fusarium basal rots (Fusarium oxysporum) *Fusarium* rots of tulip, narcissus, iris, gladiolus, lily and freesia, grown for cut-flower production under glass, have each been treated successfully with thiabendazole and benomyl (Magie, 1970; Gould and Miller, 1970, 1971). Better results are obtained when bulbs are dipped in the fungicide suspension within two days of lifting, though with bulbs intended for forcing, the dip can be applied just before cold treatment and for protection during growth, bulbs and corms are dipped immediately before planting.

The benzimidazole fungicides have also been used to control *Penicillium* spp in various bulbs, and Smith (1971b) found that benomyl, applied as a bulb dip or post-emergence foliar spray could effectively decrease the incidence of primary leaf infection by *Stagonospora curtisii* (Narcissus leaf scorch), as well as subsequent spread and thus could increase bulb weight. This compound is now registered for use on ornamental crops, including bulbs in Europe and North America.

Mushroom

Though not widely grown in the glasshouse nowadays, the mushroom *(Agaricus bisporus)*, being a crop of the protected environment, is included in this chapter. This crop has shown outstanding annual increase in output. In 1950 the world annual production of mushrooms was about 112 million lb, whereas now the annual output is of the order of 650 million lb. Of the major-producing countries, USA heads the list with 175 million lb annually, France 150 million lb, Taiwan 115 million lb and Holland 60 million lb. In the British Isles the annual output of about 100 million lb, worth some £13 million, is now second in value to that of the tomato crop at £14 million per annum.

The composted animal manure on which mushrooms are usually grown provides a selective medium for the growth of mushroom mycelium, though under adverse conditions other fungi can colonise the medium and compete directly with the mushroom for space and nutrients. In addition, the cultivated mushroom, like any other crop, is attacked by insect pests and suffers from diseases caused by viruses, bacteria and fungi. Of the latter, the diseases known as dry bubble, wet bubble and cobweb caused respectively by *Verticillium malthousei*, *Mycogone perniciosa* and *Dactylium dendroides* are the three most serious. Each of these seems to have been important at various times in the past but, at present, *V. malthousei* is most frequently encountered on mushroom farms and can cause serious crop losses wherever *A. bisporus* is grown (Cross and Jacobs, 1968).

Chemical control of the disease poses problems because both host and parasite are fungi. In the past, zineb has been widely used but the variable results obtained with this and other compounds led Gandy (1971a) to evaluate more recently available compounds including systemic fungicides for use against *V. malthousei*. She found that benomyl (1·0 g (a.i.)/m^2) applied to the casing, a layer of peat and chalk which must be placed on the compost to induce fruiting, gave a considerable degree of control without any apparent toxic effect on the crop. Since the disease develops very rapidly the compound must be applied before cropping begins, otherwise one or two flushes of mushrooms may be lost before the fungicide becomes fully effective (Gandy, 1972).

Snel and Fletcher (1971) showed that whereas the ED_{50} values of benomyl and thiabendazole against mycelial growth of *A. bisporus* were 48 and 100 ppm the ED_{50} values for both compounds against *V. malthousei, D. dendroides* and *M. perniciosa* were all less than 1 ppm. When the two fungicides were incorporated in the casing layer of mushrooms grown in plastic pots both gave complete control of *M. perniciosa* infection.

Further work by Gandy (1972) has shown that thiophanate and thiophanate methyl are also active against *V. malthousei*. There is, as yet, little information on uptake and translocation of systemic fungicides in *A. bisporus* though Drakes and Fletcher (1971) using bioassay methods were able to demonstrate fungitoxic action by sporophore material during six weeks of cropping following the application of benomyl to the casing layer.

METHODS OF APPLICATION IN GLASSHOUSES

The systemic fungicides at present available for practical application to the growing crop move in the plant only in the direction of the transpiration stream. Accordingly, the grower can apply them to the soil and thereby make use of their systemic properties, always assuming that there is efficient root uptake, or he can apply them to the above-ground parts of the plant. In this case there will be little uptake and movement and, consequently no protection of new growth formed after the treatment has been applied so, in effect, the compounds are used as orthodox protectant fungicides, though in some cases they can eradicate established infections.

There is no doubt that systemic fungicides have been welcomed by glasshouse growers who are now applying benomyl by almost any means at their disposal.

It is likely that growers of those crops in which it would be more satisfactory to apply systemic fungicides as sprays will have suitable equipment for this purpose and there is little need here to discuss the merits of the various types of spraying equipment. There are circumstances however, in which fixed equipment in the glasshouse, used for example for overhead watering or for the application of foliar feed could conceivably be used for the application of systemic fungicides. It is difficult to get uniform distribution from such equipment particularly in tall-growing crops and, on very large holdings, this is accentuated by the large volume of liquid needed to fill the system and to be purged from it at the end of treatment.

The use of low-level fixed irrigation lines to apply systemic fungicides as soil drenches is less costly in labour than high-volume spraying though at present some compounds such as triarimol are not recommended for soil application. Trickle irrigation, by which plants in pots or larger plants such as tomatoes and cucumbers are each watered from a single nozzle, has been used for the application of systemic fungicides to the soil. So too has the drip-watering system in which the resistance required to give a standard dosage to each plant is provided by lengths of small-bore plastic tubes on the ends of the distribution lines. The outlets in both of these systems are very liable to become blocked by dirt, pipe scale, hard-water scale or, when used for systemic fungicides, by settlement of improperly mixed wettable powder formulations.

Plants such as carnations, grown in beds, are watered by a modification of the low-level trickle system. In this a harness of lay-flat polythene tubes, punched with small holes are used to apply the water. This is a low-pressure system and thus, if used for the application of systemic fungicides, it must be borne in mind that a large amount of suspension will drain to the lower end of the system each time the supply is turned off.

None of these systems is ideal; all of them suffer to a greater or lesser extent from lack of uniformity in delivery. Some degree of uneven application of systemic fungicides may be acceptable but overdosage could lead to phytotoxicity and undesirably high residues. Blocked outlets on the other hand could mean that some plants were not treated, thus leading to the false impression that the compound had failed to control the disease or, more ominously, that a strain of the pathogen resistant to the compound had made its appearance.

It would be safer, where possible, to apply a measured dose of the compound to each plant. Dimethirimol is recommended to be applied directly to the base of the plant from a special lancet applicator which delivers 20 ml of the diluted liquid from a knapsack container. Benomyl is also recommended to be applied as a soil drench at 1 pint per plant which could well be provided by a suitable applicator supplied from the sprayer tank.

RESISTANCE

The development of resistance of pests and diseases to the action of crop protectants is a problem of long standing. Insects quickly develop tolerance to insecticides and strains of disease organisms quickly appear to infect the resistant varieties of crop plants produced by the plant breeder. Therefore, when systemic fungicides became commercially available, it was feared that tolerant strains of disease organisms might not be long in making their appearance. This is more likely under the selection pressure exerted by the use of these specific compounds which effectively protect the whole of the plant and the whole of the crop.

Indeed strains of organisms tolerant to the systemic fungicides currently available have appeared very quickly (Woodcock, 1971). This subject is dealt with at length elsewhere in this volume (Dekker, pp. 156–174). In addition, Gandy (1971a) has found that several fungi, including *Cladosporium fulvum, Mycogone perniciosa, Dactylium dendroides* and *Verticillium malthousei*, develop slightly increased tolerance to benomyl when 'trained' on successive media containing higher levels of the compound but, again, this tolerance to concentrations of less than 10 ppm should not present a problem in the field.

It was suggested long ago for insecticides and has now been suggested for systemic fungicides that to ring the changes, by using different compounds in succession would prolong the effective life of each compound (Anon., 1971). There would seem to be more likelihood of success if a mixture of compounds with different modes of action were to be used in that the pathogen would be less likely to be able to circumvent more than one metabolic block at one time (Crowdy, 1970). In the present circumstances, however, there is very little choice of systemic fungicides with different modes of action. For example, for the control of *Botrytis cinerea* in tomato or cucumber the very active compounds are all benzimidazole derivatives, probably having the same mode of action. We may have to think of the alternate use of protectant and systemic fungicides.

Any attempt to draw parallels between the development of resistance by insects to insecticides and the development of resistance by fungi to fungicides should take account of the fact that though resistance has now been demonstrated for several fungi and that Hastie (1970) found benomyl to induce instability in diploids of *Aspergillus nidulans* it has not been suggested that the systemic fungicides have themselves induced mutations among pathogens of glasshouse crops. Indeed, it seems that the tolerant strains were already in existence and have simply been selected out by the applied treatment.

The experience at the Glasshouse Crops Research Institute is of interest here. When tolerance of cucumber powdery mildew to dimethirimol had appeared it was decided to arrange for a period of six weeks in autumn 1970 when no cucurbit plants would be grown at the Institute. At the end of this time, in order to re-start work on the control of this disease, infected plants were introduced from another site some 100 km distant. The new strain was susceptible to dimethirimol and during 1971 there was no indication of the failure of the compound to control the disease.

NON-FUNGICIDAL EFFECTS OF BENZIMIDAZOLE COMPOUNDS

It is possible, experimentally, to produce phytotoxic symptoms such as stunting and leaf necrosis with drench applications of benzimidazole derivatives at high concentrations. There have been reports of other physiological effects of some of these compounds. Cole *et al.* (1970) found benomyl applied to the soil at levels above 20 ppm to bring about a progressive reduction in plant weights but at 10 ppm the compound caused an increase in fresh and dry weight of cucumber plants.

Lettuce growers, using benomyl for the control of *B. cinerea* have noticed a more intense green coloration of the crop and plants which have a mature appearance when young. Such plants may be heavier than those not treated with the fungicide. Benomyl-treated celery has been found to be crisp and the green parts still fresh after winter storage whereas the untreated crop becomes yellow in store. Tomato plants which have been treated with benomyl, thiabendazole or thiophanate derivatives are usually a darker shade of green, particularly in the younger leaves, than are the untreated controls. Producers of chrysanthemum cuttings who have used benomyl to control *Verticillium albo-atrum* infection of stock plants have commented on the darker green of treated plants.

These responses, in which chlorophyll retention is prolonged, are all of the type which could be associated with the activity of cytokinins. Compounds with cytokinin-like activity are related to adenine (I) (Rothwell and Wright, 1967). The similarity in structure between this compound and methyl-2-benzimidazole carbamate (II) which is the systemically fungi-

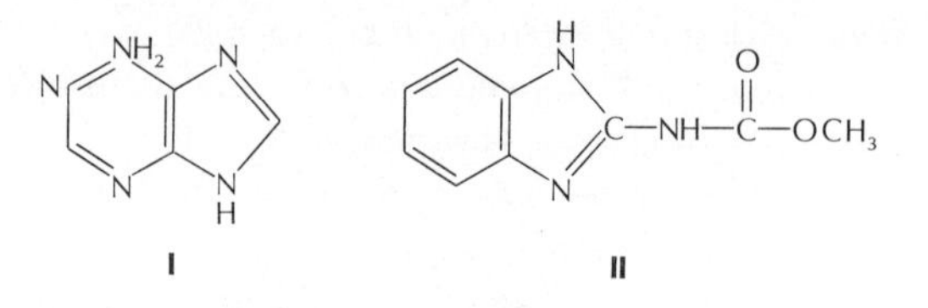

toxic moiety derived from benomyl, supports the possibility that the benomyl-derivative could have cytokinin-like activity.

Indeed, when detached wheat leaves were floated on benzimidazole solutions not only was breakdown of proteins and chlorophyll prevented but the leaves were resistant to rust infection (Person *et al.*, 1957; Samborski *et al.*, 1958).

BIOLOGICAL CONTROL

An aspect of the use of systemic fungicides in the glasshouse which may well command increased attention in the future concerns their integration into a programme which includes the use of insect predators to control pests.

The biological control of red spider mite and white fly on cucumber is now in commercial use but for this method of pest control to be successful the balance between host and predator must be maintained and thus the use of chemicals can only be tolerated if they do not seriously upset this balance. The introduction of systemic fungicides has been of benefit to the development of biological control as a means of reducing the losses caused by pests of glasshouse crops. The chemicals used previously, such as sulphur, quinomethionate and dinocap, though effective against cucumber powdery mildew were also phytotoxic and acaricidal.

Dimethirimol was one of the first systemic fungicides to be used in conjunction with biological control and proved to be completely compatible. When cucumber powdery mildew became tolerant to this compound, benomyl used in a spray programme was found to be less satisfactory (Parr and Binns, 1970). This compound when applied as a spray was found to reduce the fecundity and egg-hatch of the red spider mite and thus to upset the chain by removing the food supply of the predator. However, when benomyl is applied as a drench to cucumber plants it controls powdery mildew and its effect on the biological control of red spider mite is much less serious.

The more recently available compounds, thiophanate and thiophanate methyl, seem to behave in a manner similar to that of benomyl. As high-volume sprays at 0·01–0·05 per cent concentrations they are both strongly acaricidal and harmful to *Phytoseiulus* and would probably be better used as drenches (Parr, 1971).

PRESENT TRENDS: FUTURE NEEDS

At the 6th British Insecticides and Fungicides Conference in Brighton, 1971, several speakers, in considering the problem of tolerance of pathogens to systemic fungicides, emphasised the need to avoid applying compounds at sub-lethal doses. If this is to be done and if the materials are to be used as sprays then the grower will need to make sure that he covers the entire crop with an adequate amount of the fungicide; this will be difficult to achieve in practice. Furthermore, the new growth produced between applications will be inadequately protected and it is on such tissues that tolerant strains may appear and develop.

The tolerances so far reported seem to be at two quite distinct levels: that of the strain of *B. cinerea* which is tolerant to benomyl at over 1000 ppm compared with the non-tolerant strain which is susceptible to 1 ppm (Bollen and Scholten, 1971) and that of other fungi tolerant to systemic fungicides at levels little above those which are toxic to non-tolerant strains, for example, the strain of *Sphaerotheca fuliginea* normally encountered at Jeallott's Hill Research Station is susceptible to less than 1 ppm dimethirimol whereas other strains are regarded as tolerant if they can withstand levels of up to 5 ppm (Bent *et al.*, 1971).

When used as protectant sprays the systemic properties of the compounds are not being utilised at all and it would seem that future development of systemic fungicides could be aimed at the use of materials which can be applied to the roots and be retained within the plant at a level well above that which is toxic to the pathogen.

None of the systemic fungicides currently in use is active against Phycomycetes. This, in itself, is a drawback which must be remedied by the future introduction of compounds active against this group. At present, however, there seem to be indications of further hazards resulting from the use of selective compounds active against organisms other than Phycomycetes. The inference is that the systemic fungicides not only control disease organisms but also kill soil-borne organisms which are themselves antagonists of such Phycomycetes as *Phytophthora* and *Pythium*. These latter are thus liable to cause disease when systemic fungicides are used, though they may not have done so in the absence of the chemical. An example of this was demonstrated by Dekker (1971a) who found that aster seedlings were killed by *Phytophthora parasitica* in the presence of benomyl but were unharmed when benomyl was not used. It is interesting to speculate here on the relative potential for antagonism of soil-borne fungi as compared with bacteria and actinomycetes, which are much less affected by benomyl and which therefore may well continue to antagonise the Phycomycetes in the presence of the systemic fungicide.

It is likely that the selective effect of the systemic fungicides extends beyond the Phycomycetes and their antagonists. Indeed the soil microflora is likely to be considerably affected by drenches of the systemic compounds. It is said that the half-life of benomyl in soil is approximately six months. During this time it will be subject to the action of a wide range of micro-organisms, some of which will increase in number because of the presence of the compound. An example of this is shown by the production of nitrite in soils to which benomyl has been applied (Spencer, 1971). Isolates from soils treated with the chemical include organisms which, in pure culture, can use methyl benzimidazole carbamate, the active moiety derived from benomyl, as their sole source of carbon and thus degrade the compound. If such organisms were not operating, the build-up of systemic fungicides in soils over a long period could well lead to problems of phytotoxicity, bearing in mind that the benzimidazole fungicides seem to be unaffected by steam sterilisation. The long-term effects of the use of systemic fungicides in the protected environment of the glasshouse are worthy of further study.

The requirements for future systemic fungicides can now be summarised to include compounds with different and preferably wider spectra of activity and different modes of action from those presently available. Compounds active against Phycomycetes will be valuable and, in view of the activity of present compounds against antagonistic fungi, there could conceivably be an increase in bacterial diseases, and it would be an advantage if materials with antibacterial properties were available. In order to be truly systemic the compounds should be translocated downwards as well as upwards so that effective methods of application specifically for glasshouses, can be used. Compounds applied by spray or drench should move freely, even in woody tissues and be retained within the plants at levels well above those required to control the disease, so that the development of tolerant strains can be avoided.

It must be borne in mind that, unlike the protectant fungicides which, in the main, stay on the surface, the systemic fungicides are intended to be retained within the tissues of the plant. Therefore the compounds themselves as well as all possible degradation products within the plant, must be free from any danger to those who eat the crop.

10.III

Results in practice – III. Vegetable crops

by R. B. Maude

INTRODUCTION

The role of conventional fungicides in the treatment of vegetable diseases has been mainly that of protection rather than cure. Compounds were applied to seeds, soil and foliage to form local barriers against fungal invasion. Such treatments were most successful when used before the onset of infection and little disease control was obtained after fungi had become established in the plant tissues. This situation has been radically altered by the discovery and development of systemic fungicides which are mobile and fungitoxic within the plant system. Their use has made possible the therapy of established infections and the internal protection of the roots and leaves of plants from invasion by external fungal pathogens.

The potential of such compounds for the improved control of vegetable diseases has been assessed in practice by comparing their performance with that of the non-systemic fungicides. They have been applied in a variety of ways for the control of seed-borne, soil-borne, leaf and post-harvest diseases of vegetables. The effectiveness of these treatments is discussed here.

SEED-BORNE DISEASES

Seed dressings with fungicidal dusts, whilst effective against superficial pathogens, have in the past given relatively little control of fungi which are deep in the seed tissues and these have been treated either by steeping the seeds in hot water (50°C for 25 min – Bant and Storey, 1952) or soaking them in water containing 0·2 per cent a.i. thiram for 24 h at 30°C (Maude *et al.*, 1969).

The latter method, which is effective against many internal seed pathogens, is a systemic treatment, involving the penetration of the seed by thiram. This only occurs, however, under the very special conditions of the treatment and requires that the seed shall become fully imbibed in the constant presence of 20–30 ppm thiram in a large volume of water. Because of the soaking and subsequent drying of seeds necessary in the thiram soak treatment, the method has proved of commercial use mainly against the pathogens of small, high-value seeds such as *Septoria apiicola* on celery, and was not practicable for the treatment of bulky seeds such as peas and beans. For such seeds there was obviously a need for systemic fungicides which could be applied to the seeds as a dry powder or slurry and then be absorbed during the imbibition of the seeds in the soil.

The possibility that such fungicides might be found was opened up by the discovery by von Schmeling and Kulka (1966) that dry seed dressings of the systemic oxathiin fungicide, carboxin, controlled deep-seated infection by *Ustilago nuda* in barley seed.

The narrow spectrum of activity of the oxathiin compounds (effective mainly against Basidiomycete fungi – Edgington *et al.*, 1966) severely limited their potential value for the treatment of seed-borne vegetable diseases, the majority of which are caused by Ascomycetes or Fungi Imperfecti. The situation was considerably changed however, by the introduction of the benzimidazole compounds, particularly benomyl (Delp and Klopping, 1968) and thiabendazole (Weinker *et al.*, 1969) which are toxic to many fungi (Bollen and Fuchs, 1970; Edgington *et al.*, 1971).

It was shown by Maude and Kyle (1970a) that slurry (1·25 g a.i./kg seed) or dry (1·875 g a.i./kg seed) dressings of benomyl applied to pea seeds eliminated internal infection by *Ascochyta pisi*. Thiabendazole dressings were equally effective in this respect (Maude and Kyle, 1971a) though carboxin dressings failed to control the disease. Benomyl and thiabendazole seed-dressings also controlled *A. fabae* (leaf and pod spot) of field beans and *Colletotrichum lindemuthianum* (anthracnose) of French beans (Maude and Kyle, 1971a) and their use represents the first practical and effective treatment of these diseases of the larger vegetable seeds. Benlate T, a commercial seed dressing containing 30 per cent benomyl and 30 per cent thiram is now marketed for the treatment of the seed-borne *Ascochyta* diseases of peas and field beans.

Although the benzimidazoles are designated 'broad-spectrum fungicides' there are certain seed-borne fungi against which they have little toxicity in culture, e.g. *Alternaria* and *Stemphylium* spp. (Edgington *et al.*, 1971), *Phoma betae* (Bollen and Fuchs, 1970), and *Colletotrichum spinaceae* (Maude and Kyle, 1971a).

In addition, the use of dry dressings of a systemic fungicide demonstrably toxic to a fungus in culture has sometimes failed to give effective control of the same pathogen on the seed. Thus dry dressings of benomyl failed to eliminate *Septoria apiicola* from celery seed or to rid cabbage seeds of *Phoma lingam* (Maude and Kyle, 1970b). Possibly in these cases lethal concentrations of the fungicide failed to reach the deeper seated mycelium of the fungi. When maximum penetration of fungicide was ensured by 24 h soaks at 30°C of celery seeds in 0·2 per cent benomyl (Maude and Kyle, 1970b) or cabbage seeds in 0·2 per cent benomyl or thiabendazole (Jacobsen and Williams, 1971) both seed stocks were completely disinfected. It is obvious that the search must continue for other systemic fungicides which are effective against these and other fungi and that each pathogen may require separate treatment. Other compounds are available but several, e.g. tridemorph, are rather phytotoxic as seed dressings and although this compound inhibited the growth of *A. brassicicola* and *P. betae* in culture (Maude and Kyle, 1971a) it depressed seed germination when applied at fungitoxic rates.

The selectivity of systemic fungicides is also illustrated in the treatment of tuber-borne diseases of seed potatoes. Here the latent fungi such as *Oospora pustulans* (skin spot), *Helminthosporium solani* (silver scurf) and *Rhizoctonia solani* (black scurf) are normally controlled by tuber dips in an organomercury fungicide soon after lifting (Western, 1971). Recently, Hide *et al.* (1969) demonstrated that tuber dips in thiabendazole or tuber dusting with benomyl effectively reduced the incidence of the first two fungi on seed tubers in store and on potatoes raised in the field from this seed material. These fungicides were not similarly effective however, against the seed-borne Basidiomycete *R. solani*, better control of which was obtained with dusts of carboxin or oxycarboxin.

Seed dusts with benomyl or thiabendazole (1–10 per cent a.i. in kaolin, applied at 5 g dust/kg seed) are now preferred to dips in thiabendazole for the control of skin spot and silver scurf on seed potato tubers (Hide and Hirst, 1971) and a prototype machine capable of applying systemic dust treatments to small quantities of potato tubers has been developed (Hirst, 1971). In America where tuber infection by *Verticillium albo-atrum* has been reported in certified seed stocks (Beckman *et al.*, 1969) 3-min dips in benomyl (12 g a.i./l) gave some control of the disease on the harvested crops (Biehn, 1970).

Thus where formerly thiram soak or organomercurial dip treatments were effective against a wide range of seed- and tuber-borne diseases of vegetables, the same latitude of use is not equally possible with the systemic compounds which will have to be carefully selected for use against specific pathogens. Nonetheless, the simplicity of their application coupled with their effectiveness against deep-seated fungi makes their use as seed dressings in commercial practice almost certain.

SOIL-BORNE DISEASES

The problems of treating field soil with conventional fungicides to control soil-borne fungi are considerable (Torgeson, 1967). The most effective and economic control has usually been achieved by localised treatments using heavy seed dressings or row or furrow applications. Under these circumstances the fungicides can sometimes provide protective barriers near the young plant.

While similar restrictions may also apply to the placement of systemic fungicides their uptake by the seed tissues or by the roots of plants affords additional protection against invasion by soil-borne fungi.

In practice both seed and soil applications have been used to protect vegetables from attack by these pathogens.

Seed and soil applications of systemic fungicides have given considerable control of *Rhizoctonia solani*, which infects the hypocotyls of seedling vegetables causing damping-off or wire-stem. Thus thiabendazole seed dressings (1·9 g a.i./kg seed) considerably reduced damping-off of peas due to *Rhizoctonia* in field soil (Crosier *et al.*, 1970) and seed slurries, seed steeps and soil drenches with benomyl prevented pea seedlings from becoming infected in artificially contaminated garden soil (Jhooty and Behar, 1970). Neither fungicide, however, was as successful as the oxathiin compounds (carboxin and oxycarboxin) in controlling *Rhizoctonia* infections. Thus carboxin dressings more effectively reduced damping-off of peas than benomyl, and both oxathiin fungicides gave superior control of seedling losses in French beans due to the fungus (Beyries, 1969). In fact, benomyl slurries (5 g a.i./kg seed) were less effective than similar slurries of quintozene (commonly applied to soil to control *Rhizoctonia*) and neither fungicide gave the same measure of control as carboxin (5 g a.i./kg seed) which completely prevented seedling infection. Seed dressings of this fungicide also prevented seedling losses due to the fungus in *Phaseolus* beans *(P. mungo* var. *aureas)* and at application rates of 20 g a.i./kg seed gave almost complete control of *Rhizoctonia* infection of cabbage seedlings. Both carboxin and oxycarboxin, when dusted on to potato tubers (10 per cent dusts – 1 g a.i./kg seed) have given considerable control of stem canker infection due to *R. solani* (Edgington and Busch, 1967) and benomyl similarly applied proved more effective than broadcast applications of quintozene (25 lb PCNB/ac) (Davis *et al.*, 1971a) against the same disease.

It is apparent that the hypocotyls and stem bases of plants below soil level can be effectively protected from invasion by *R. solani* by soil applications but more particularly by seed applications of systemic fungicides. It has not been shown however, whether the compounds are acting as external protectants by inhibiting the growth of the soil fungus in the soil or whether by their presence in the vascular system of seedlings and plants they are preventing fungal penetration of the hypocotyl and stem base tissues. Possibly a combination of both modes of action occurs. In one instance, however, ineffectual internal protection of the hypocotyl region may be responsible for the failure of a systemic fungicide to control field outbreaks of brown stem rot *(Cephalosporium gregatum)* on soybeans. Thus although soil drenches of thiabendazole (Gray and Sinclair, 1971) and benomyl (Thapliyal, 1970) in glasshouse tests restricted stem invasion by the fungus, furrow applications (1 or 2 g a.i./10 kg soil) of thiabendazole were not effective (Gray and Sinclair, 1971). Since thiabendazole was found to accumulate in the root and epicotyl but not in the hypocotyl tissues of the soybean the authors suggested that possibly insufficient fungicide was present in this region to prevent invasion by the fungus.

Seedling infection of onions by the onion smut fungus *(Urocystis cepulae)* occurs on the first leaves of plants at or slightly below soil level, only during emergence. Some protection during this susceptible period is afforded by heavy seed dressings of thiram applied at 50 per cent or more of the seed weight (Cruickshank and Jacks, 1953). Far more successful control of the disease has recently been reported by Harrow (1970) using seed dressings of benomyl (12·5–50 per cent (a.i.) applied w/w seed) which virtually eliminated smut in onion seedlings. Seed dressings with thiram (40–80 per cent a.i. applied w/w seed) were much less effective than this and dressings of carboxin and oxycarboxin proved too toxic to onion seeds for practical use.

A more prolonged period of protection is required against the onion white rot fungus *(Sclerotium cepivorum)* which may attack salad, bulb onions and garlic at any stage in the growth of these crops. In France, Lafon and Bugaret (1968) found that slurry dressings of benomyl (2·0 g a.i./kg cloves) applied to garlic cloves considerably reduced the incidence of white rot in heavily contaminated field soil and increased the numbers and weight of healthy plants at harvest. Recently, Ryan *et al.* (1971) also reported that larger numbers of Rijnsburger bulb onions survived to harvest if the sets from which they were raised were dressed with benomyl at rates ranging above and below 2·0 g a.i./kg set. They also stated that benomyl (62·5–500 g a.i. fungicide/kg seed) used on White Lisbon onion seed gave almost as effective control of onion white rot in artificially contaminated field soil as calomel used at 500–1000 g a.i./kg seed.

Thus systemic fungicides have been applied to soil, seeds, seed tubers and bulbs to protect extremely local areas on the developing plants, e.g. hypocotyls, epicotyls or the fairly restricted root systems of onion, from infection. In these circumstances considerable control of several diseases has been achieved by their use. Where, however, a much wider area of plant root systems, e.g. stolons and tubers which develop from and are at some distance to the original seed potato, require protection then systemic seed dressings have not been effective though some measure of control has been achieved by soil treatments with these compounds.

Dusts of carboxin and oxycarboxin (Edgington and Busch, 1967), dry dressings of benomyl (Davis and Callihan, 1970) and 3-min dips in benomyl or dusts of carboxin (Biehn, 1969) applied to seed potatoes failed to reduce the incidence of black scurf *(R. solani)* on tubers harvested from naturally or artificially contaminated field soil. It is presumed that effective control of infection on the developing tubers would only have occurred had sufficient systemic fungicide been supplied to them via the stolons from the treated tubers during the growth of the crop. This apparently was not the case. Slightly more effective control of tuber infection occurred when seed piece treatments with benomyl were followed by soil drenches with the same fungicide (Biehn, 1969) possibly because of the more effective placement of benomyl in a readily available form in relation to the developing stolons and tubers. Furrow applications of granules of (5 per cent a.i.) carboxin and oxycarboxin (4·5 kg a.i./ha), however, failed to reduce the incidence of *R. solani* on harvested tubers (Edgington and Busch, 1967). Similarly, seed tuber dressings of benomyl have given little reduction in the incidence of *Verticillium dahliae* (potato wilt) propagules in the soil (Corbett and Hide, 1971) but some control of vascular infection of harvested tubers was obtained where seed piece dips in benomyl were followed by soil drenches of the same fungicide applied two (6·5 lb/ac) and nine (16 lb/ac) days after planting (Biehn, 1970). The rotavating of benomyl into the soil (20 lb a.i./ac) also effectively decreased the number of *Verticillium* propagules present in the soil at harvest and increased the yield of potatoes (Corbett and Hide, 1971).

Thus where fungi are dispersed through the soil the task of protecting fairly extensive plant root systems by means of systemic fungicides is practically difficult and economically may not be feasible.

The problem of adequately protecting the root system also arises in the use of conventional fungicides for the control of brassica clubroot *(Plasmodiophora brassicae)*. Seed pelleting with calomel is unreliable and offers ineffective protection to brassica root systems in direct-drilled crops, though a practical, if not completely effective control, on the transplanted crop is achieved by seed-bed treatments and the dipping of transplants in calomel. The use of systemic seed dressing for the control of clubroot has not as yet been tested, but from the work of Jacobsen and Williams (1970) it is apparent that benomyl mixed with clubroot-infected pot soil prevented the roots of cabbage seedlings from becoming infected, apparently by preventing zoosporangial invasion of the root zones. Control of the disease on cauliflower seedlings was also achieved by van Assche and Vanachter (1970) with benomyl, thiophanate or thiophanate methyl mixed with pot soil. Findlayson and Campbell (1971) found however, that soil type was critical in determining the effectiveness of benomyl. Thus in naturally infested peat soil benomyl (10 g a.i./l soil) failed to control clubroot on cauliflower seedlings although this was effectively treated by much smaller applications of calomel (2·5 g a.i./l soil). In mineral soil however, benomyl was more effective by comparison at lower rates of application.

Thus the possibility exists of using systemic fungicides as alternatives to calomel for the treatment of seedlings in mineral soil prior to transplanting them into the field. Van Assche and Vanachter (1970) consider that such treatments may give sufficient initial protection to young plants to ensure that these give rise, when planted in the field, to a completely marketable if slightly infected crop. The fact remains that in the limited soil system of a pot there may be little to choose between the protection afforded to root systems by conventional and systemic compounds, and more effective use of the systemic action of benomyl for the control of clubroot might be made by using it as a soil drench treatment on a limited scale in the fields as is suggested by Jacobsen and Williams (1970).

Furrow applications of small amounts of benomyl or thiabendazole in conjunction with crop rotations are suggested by Papavizas and Lewis (1971) as potentially feasible measures for the control of black root rot of beans *(Thielaviopsis basicola)* which was considerably reduced in pot soil by drenches of these fungicides.

One important group of soil-borne diseases which cannot be controlled by the available systemic fungicides is those attributable to attack by Phycomycetes, which usually cause damping-off of seedlings or root rot of older plants. Thus soil applications of benomyl and thiabendazole failed to control root infection of peas by *Aphanomyces euteiches* (Papavizas and Lewis, 1970) and seed dressings of both fungicides were ineffective against the Phycomycete fungi causing pre-emergence losses in peas (Harper, 1968; Crosier *et al.*, 1970). Increases in emergence of peas and in their fresh weight at harvest were achieved, however, if the seeds were dressed with captan (to prevent seedling losses by Phycomycete fungi) and benomyl was added to the soil or the seeds to control hypocotyl infections by *Fusarium solani* var. *pisi* and *Rhizoctonia solani*. Similar results were obtained by Natti and Crosier (1971) who found that furrow sprays of benomyl, benomyl plus chloroneb or thiabendazole provided effective control of bean *(Phaseolus vulgaris)* root rot *(Fusarium solani* f. *phaseoli)* during the early part of the growing season. Dual-function dressings of captan plus benomyl were also used by Maude and Kyle (1971) on early-sown peas. The addition of captan to the mixture substantially improved the emergence of the peas and benomyl successfully eliminated seed-borne *Ascochyta pisi*. Seed dressing mixtures of captan + thiabendazole proved equally effective, as did Benlate T (a mixture of thiram with benomyl) which is now recommended for use as a seed dressing for the control of seed-borne *Ascochyta* and soil-borne pathogens *(Pythium, Fusarium* and *Rhizoctonia* spp.) of peas and field beans.

LEAF DISEASES

Historically, conventional fungicides probably achieved their greatest use as foliar sprays. In the vegetable crop the spraying of potato foliage for the control of blight (*Phytophthora infestans*) is an outstanding example of this. Although no systemic fungicide yet produced is effective against this fungus, many offer improved control of other leaf pathogens, in some cases by penetration and limited movement within the leaf tissues; in others by their direct protectant action. Additional uses, not possible with non-systemic compounds, are the control of leaf diseases by seed or soil applications. In concept, the treatment of root-invading fungi by foliar sprays of systemic fungicides, though highly desirable, is at present unrealistic since active transport of these chemicals so far, occurs only in an upwards direction in the plant tissues.

Improved control of disease by the use of sprays of a systemic fungicide was first reported by Froyd and Johnson (1966) who found that two high-volume sprays of thiabendazole (6 oz a.i./ac) gave more effective control of leaf spot of sugar beet than five applications of much greater concentrations of captan. Later, Delp and Klopping (1968) reported the superiority of benomyl over maneb in controlling the same disease. Erwin (1970) suggested that the improved disease control obtained with thiabendazole may have been due to local systemic movement of the fungicide within the leaf tissues, and the same may apply to benomyl. Solel (1970a) demonstrated that thiabendazole, benomyl and thiophanate, when sprayed onto the lower surfaces of sugar beet leaves, prevented infection of the upper surfaces of the leaves inoculated with *Cercospora beticola* spores, indicating translaminar movement of the fungicides. Such an effect did not occur if the non-systemic fungicide Brestan 60 (fentin acetate with maneb) was used instead. Recently (1971a), Solel has shown that the vapour of benomyl but not that of thiabendazole is toxic to *C. beticola* spores. Possibly both modes of action contribute to the effectiveness of foliar sprays of this fungicide against leaf spot of sugar beet. Certainly, in field spray trials (Solel, 1970b) three benomyl sprays (6 oz a.i./ac each) at three-weekly intervals completely controlled the disease and were superior to similar thiabendazole or fentin acetate applications. Under conditions conducive to disease expression (i.e. successive applications of spores to plants and frequent sprinkler irrigation) Paulus *et al.* (1970b) also found that benomyl gave superior disease control to thiabendazole when both fungicides were sprayed at 18- to 22-day intervals. Recently low-volume aerial sprays of benomyl (2·4 oz a.i./ac) have also given effective control of this disease (Solel, 1971b).

Weekly high-volume sprays of benomyl (0·25–0·5 lb a.i./ac) gave more effective control of late *(Septoria apiicola)* and early *(Cercospora apii)* blights of celery than similar applications with the non-systemic fungicide maneb (Delp and Klopping, 1968). Benomyl sprays, however, proved less efficient against celery late blight if applied at fortnightly intervals. Similar results were obtained by Paulus *et al.* (1970a), who found that fortnightly sprays of benomyl (0·25 lb a.i./ac) were no more effective than those of several non-systemic compounds when infection in the crop was severe. When sprays were applied as soon as the disease appeared in the crop more successful disease control was achieved with fortnightly sprays though thiabendazole proved slightly superior to benomyl. Ryan and Kavanagh (1971) reported however, that four sprays of benomyl (0·4 lb a.i./ac) at similar intervals gave more effective control of the disease under conditions of moderate and severe infection than a number of conventional fungicides. Benomyl in these trials was inferior only to the tin fungicides. Similar results were obtained by Maude and Shuring (1970) and Maude and Moule (1971a) who found benomyl to be superior to several non-systemic fungicides but slightly less effective than the tin fungicide Brestan 60 (fentin acetate with maneb) in reducing the severity of the disease on the foliage. Both fungicides (0·5 lb a.i./ac) however, gave comparable reductions in petiole infection when sprayed at fortnightly intervals. Thiabendazole was less effective than benomyl. No evidence was found by Maude and Moule (1971a) that either of these fungicides moved within the plant in sufficient concentrations to give any effect against the fungus other than that which could be ascribed to their direct protectant action. More effective control of ringspot (*Mycosphaerella brassicicola*) of Brussels spouts was achieved by Welch *et al.* (1969) with four fortnightly benomyl sprays (1·5 lb a.i./ac) than with a similar number of sprays of dithiocarbamate, copper and other fungicides. Infection of the seed inflorescences of cabbages by *Alternaria brassicicola* (dark leaf spot) was not prevented, however, by seven weekly sprays of tridemorph (1 lb a.i./ac) although this fungicide was inhibitory at low concentrations to the growth of the fungus in agar culture (Maude and Moule, 1971b).

The timing of the application of sprays of systemic fungicides to coincide with a period of maximum susceptibility of crop to infection has sometimes given very effective control of disease. Thus *Sclerotinia sclerotiorum* (white mould of snapbean) which infects the flowers of beans, was controlled by Natti (1967) with one (full bloom) or two (full- and post-bloom) high-volume sprays of benomyl (0·75 or 1·5 lb a.i./ac) which effectively prevented foliage and pod infection for a further 22 days. Although similar sprays of thiabendazole (0·5, 1·0 and 2·0 lb a.i./ac) and dicloran (1 or 2 lb a.i./ac) reduced infection initially, neither maintained a high level of disease control. Since none of the fungicides effectively controlled the disease when applied as single post-bloom sprays, the timing of such applications is clearly critical and early spraying is presumably necessary to prevent the establishment of the fungus in the flower tissues.

Although a similar control of *Botrytis cinerea* on the pods of French beans at harvest with two spray applications (full- and post-bloom) of benomyl (0·5 lb a.i./ac on each occasion) was initially reported by King (1971), single full-bloom applications (0·5 lb a.i./ac) were later shown to be just as effective (King, 1971, Ryan *et al.*, 1971). In trials, sprays of thiophanate methyl and benomyl gave equal control of *Botrytis* and these fungicides were superior in this respect to thiabendazole and several non-systemic fungicides. A single high-volume spray of benomyl (0·5 a.i./ac) at full bloom is now recommended by the manufacturers of Benlate (Du Pont and Co.) for the control of this disease in the French-bean crop in Britain.

In controlling both diseases on bean pods at harvest by applications made to the flowers the effect of benomyl when compared with that of thiabendazole appears to be far greater than that which can be ascribed to protectant action alone. Its additional effect of reducing anthracnose (*Colletotrichum lindemuthianum*) on the pods of French beans (King, 1971, Ryan *et al.*, 1971) is furthermore difficult to understand since this fungus normally causes pod infections by the transfer of spores from stem lesions. Possibly the effect is one of reduction of inoculum potential in this case.

Considerable control of bean rust (*Uromyces phaseoli*) has also been obtained with one, or at most, two sprays of systemic fungicides applied, at high volume, at the first sign of the disease. In America one spray (0·5 or 1 lb a.i./ac) (Baldwin, 1970) and in Tanzania two sprays (2·2 lb a.i./ac) (Hudson and Jaffer, 1970) of oxycarboxin gave excellent control of the disease. In both countries non-systemic fungicides failed to give effective control, though in Tanzania sprays of other systemic fungicides, triarimol, BAS3050F (mebenil) and HOE6952 also considerably reduced infection.

The application of systemic fungicides to seeds can result in some protection of the cotyledons and first leaves of some vegetables from attack by airborne fungi. Von Schmeling and Kulka (1966) were the first to demonstrate this effect by showing that infection of the unifoliate and trifoliate leaves of beans by *Uromyces phaseoli* (bean rust) was largely prevented by dusting the seeds with carboxin or oxycarboxin. Natti (1970) reported that the incidence of *Erysiphe polygoni* (powdery mildew) on kidney bean cotyledons was reduced if the seeds from which they were grown were dressed with benomyl. These effects, however, possibly may be too transient to represent a practical control method under field conditions of infection and a more lasting control of leaf pathogens has been obtained by incorporating systemic fungicides with field soil. Thus soil drenches of benomyl controlled *Erysiphe cichoracearum* (powdery mildew) on the foliage of muskmelons (Wensley and Huang, 1970; Netzer and Dishon, 1970). In this respect one benomyl drench (0·2–0·5 g a.i. in 100 ml/plant) was as effective as four fortnightly sprays of the same fungicide (800 g a.i./ha) (Netzer and Dishon, 1970). Similarly Derbyshire (Anon., 1971b) found that a single benomyl drench applied to the soil (0·25 or 0·35 g a.i./l per m^2) of nursery beds containing young celery plants infected with *Septoria apiicola* five days before transplanting them into field soil prevented the development of the disease in the main crop. The treatment was practised in some celery nurseries in 1971. Control of the disease on infected celery seedlings was also obtained by transplanting them into compost containing benomyl (4·2 g a.i./m^2) to a depth of 5 cm. When these plants were later grown to maturity they remained healthy and the disappearance of the disease was correlated with the natural loss of affected foliage associated with the complete protection afforded to the new foliage by the systemic uptake of benomyl from the soil and its presence in the plant tissues (Maude and Moule, unpublished results).

POST-HARVEST DISEASES

Systemic fungicides already have a well established use as post-harvest treatments for the control of several storage rots of bananas and citrus fruits (Eckert, 1969). Thus the dipping of bananas and citrus in suspensions of thiabendazole to prevent outbreaks of blue and green mould *(Penicillium italicum* and *P. digitatum)* during storage had been practised in America since 1966 where the treatment has been cleared for commercial use (Erwin, 1970). More recently benomyl dips have been used for the same purpose.

By comparison the post-harvest treatment of vegetable produce with systemic fungicides for the control of storage rots has received little attention.

Recently, however, 2-min dips in, and the dusting of tubers with benomyl or thiabendazole were shown by Leach (1970) to prevent the development of dry rot (*Fusarium roseum*) during a period of five months' storage of artificially inoculated potato tubers. Similar treatments with non-systemic fungicides were not successful. Ten-second dips in benomyl, thiabendazole and in the non-systemic fungicide dicloran (1·2 g a.i. of the fungicide/l) were used by McMillan (1970) to control outbreaks of watery soft rot *(Sclerotinia sclerotiorum)* which frequently occur during the long-distance transport of pole beans to market. Although the benzimidazole fungicides considerably reduced outbreaks of the disease, it was completely eliminated by the dicloran treatment which is therefore preferred for this use. Dipping or soaking (15 min) of carrot roots or celery butts in benomyl (1 g a.i./l) more effectively controlled rotting by *Centrospora acerina* (liquorice rot of carrots, crown rot of celery) during storage than did similar treatments in dicloran (0·5 g a.i./l) (Derbyshire and Crisp, 1971). So effective were both benomyl treatments that after seven months' storage, carrot root rotting, due to the fungus, was reduced by two-thirds and the marketability of the roots was doubled. Likewise, decay of stored celery sticks by *C. acerina* was prevented for 12 weeks by the treatment. Since deterioration in stored carrot roots is often complete after 3–4 months and celery *Centrospora* may produce losses of about 30 per cent after eight weeks' storage the extensions of the shelf-life of both carrots and celery by these treatments may have considerable application in practice.

It is apparent that in their effectiveness against the storage diseases mentioned here the benzimidazoles exhibit activity against some of the main genera of fungi which cause post-harvest decay, and may therefore hold considerable promise for the treatment of mixed populations of spoilage fungi on stored vegetable produce.

10.IV

Results in practice – IV. Fruit crops

by R. J. W. Byrde

INTRODUCTION

An account of the performance of systemic fungicides on fruit crops – particularly on top fruits – is somewhat a contradiction in terms. Evidence to date suggests that fungicides which are systemic in some herbaceous plants, such as cucumbers or cereals, show little or no movement in top fruits, particularly when applied as foliage sprays. Nevertheless, these compounds are powerful fungicides in their own right; with even the very limited movement that can be achieved – perhaps aided by an increased tenacity (Doma *et al.*, 1971) – they have opened up new possibilities for the control of diseases which have hitherto proved intractable.

The poor translocation in top fruits is well illustrated by the work of Cavell *et al.* (1971). Autoradiographs of young apple seedlings treated at the root with ^{14}C-ethirimol show clearly the localisation of the compound or its breakdown products within the vascular tissue, whereas this compound is fully systemic in cereals. However, benomyl has occasionally been effective as a root application to young plants for the control of apple powdery mildew caused by *Podosphaera leucotricha*. Cimanowski *et al.* (1970) obtained control of the disease on one-year-old rootstocks growing in soil, but failed to do so when they were growing in Perlite: clearly soil texture can be an important factor. Gilpatrick (1969) obtained good control on very small seedlings, but Hammett (1968) was unsuccessful with one-year-old seedlings growing in John Innes compost. Lewis (1968) recorded the failure of soil applications of benomyl to control scab (caused by *Venturia inaequalis*) on apple trees 8–10 ft high, even by root injection, and Byrde *et al.* (1969) did not obtain a significant reduction of the disease by application of benomyl as a spray to the grass sward, or as a slurry to the collar of the tree or in bore holes in the trunk. When applied as a foliar spray to apples, the compounds do not appear to move from one leaf to another, at least in concentrations sufficient to control diseases; moreover, Mercer (1971) reported little or no movement of thiophanate methyl even within a treated leaf, and Doma *et al.* (1971) found only triarimol capable of giving a marked zone of mildew protection on a leaf from a single deposit.

On cherries, there is a single report of effective control of powdery mildew (caused by *P. oxyacanthae*) following root application of benomyl (Gilpatrick, 1969). On strawberries, there is some evidence to suggest a more systemic distribution (see p. 249).

This review is not exhaustive: references to trial reports, such as those in *Fungicide–Nematicide Tests* or Annual Reports, have been included only when giving information apparently not yet described in scientific papers.

TEMPERATE FRUIT

Top fruit *Apples* The most important diseases of apples are scab and powdery mildew (Table 10.IV.1). Although lime-sulphur was formerly used to control both diseases, the superior fruit quality and disease control obtained with organic fungicides led to the adoption of the latter for routine spraying during the 1950s. Where both diseases are severe, a separate fungicide is used for each, generally at 7–10 day intervals – e.g. captan, dithianon or dodine against scab, and dinocap or binapacryl against mildew. Since most of the new systemic fungicides are active against both fungi, to a varying degree, a single compound can again be used for disease control.

The control of apple scab by benomyl, first described by Delp and Klöpping (1968) is so well known that a comprehensive reference list is not given: many examples are cited by Hickey (1971a). The compound has often given good control even when used at extended intervals, e.g. up to 28 days (Burchill and Williamson, 1971; Byrde *et al.*, 1969; Catling, 1971b; Delp and Klöpping, 1968; Evans *et al.*, 1971; Kavanagh and O'Kennedy, 1971; Moller *et al.*, 1969). Benomyl owes its success partly to its activity as a curative fungicide (e.g. Catling, 1971b; Kavanagh and O'Kennedy, 1971; Zehr, 1971a). Szkolnik (1969) considered that its post-infectional activity was intermediate between those of captan and dodine.

TABLE 10.IV.1 **Control of apple scab (*Venturia inaequalis*) and powdery mildew (*Podosphaera leucotricha*)†**

Disease	*Compound*	*References*
Scab	Benomyl	Delp and Klöpping (1968), and see text
	Thiophanate, thiophanate methyl	Borecka *et al.* (1971); Cole *et al.* (1971); Doma *et al.* (1971); Douchet *et al.* (1971); Formigoni *et al.* (1971); Ishii (1970)
	Triarimol	Arnold *et al.* (1971); Borecka *et al.* (1971); Burchill and Williamson (1971); Doma *et al.* (1971); Gramlich *et al.* (1969); Kavanagh and O'Kennedy (1971)
	Triforine	Adlung and Drandarevski (1971); Harper and Byrde (1972); Schicke and Veen (1969)
Mildew	Azepine derivative	Byrde *et al.* (1969); Schwinn (1969)
	Benomyl	Bourdin *et al.* (1969); Burchill and Williamson (1971); Catling (1971b); Cole *et al.* (1971); Covey (1971); Coyier (1971); Delp and Klöpping (1968); Doma *et al.* (1971); Evans *et al.* (1971)
	Pyrazophos	Hay (1971); Woodward (1971a)
	Thiophanate, thiophanate methyl	Aelbers (1971); Burchill and Williamson (1971); Cole *et al.* (1971); Doma *et al.* (1971); Douchet *et al.* (1971); Formigoni *et al.* (1971); Woodward (1971a)
	Triarimol	Arnold *et al.* (1971); Bryant and Hitchman (1971); Burchill and Williamson (1971); Butt (1971); Covey (1971); Coyier (1971); Doma *et al.* (1971); Gramlich *et al.* (1969); Wiggell and Simpson (1970)
	Triforine	Adlung and Drandarevski (1971); Harper and Byrde (1972); Schicke and Veen (1969)

† See also Hickey (1971a)

The compound also shows great promise for use as an autumn (pre-leaf-fall) spray to prevent perithecial formation on fallen leaves (Borecka *et al.*, 1971; Borecki and Nowacka, 1969; Burchill and Swait, 1971; Connor and Heuberger, 1968; McIntosh, 1969; Paisley and Heuberger, 1971; Vojvodic, 1970), so bringing about a drastic reduction in inoculum the following season and so probably allowing a reduced spray programme. It has also been used in the spring to prevent ascospore discharge (Miller, 1970). Even when a normal spray programme is used, finishing in July, a residual effect on scab the following season has been observed (Byrde *et al.*, 1972).

Although almost always highly effective against scab, a few exceptions have been recorded and seem worthy of special mention. Thus Fridrich and Gilpatrick (1971a) found unsatisfactory control of leaf scab (but good fruit scab control) in the Hudson Valley, and Palmiter (1970), in the same area, noted leaf infection earlier in the season: most of the lesions were abnormal and produced few if any spores. The addition of Tween 20 spreader improved the control of foliage infection. Klos *et al.* (1971), in Michigan, similarly recorded a high incidence of flower cluster leaf scab under severe infection conditions although, again, conidiophores and conidia were absent from the lesions, which soon stopped developing. Under extremely severe infection conditions in Northern Ireland in 1971, benomyl gave an unsatisfactory control of the disease, and proved inferior to dithianon and dodine (J. Cartwright, priv. comm.). There is no reason to think that resistant races of *Venturia inaequalis* are responsible for the above results.

Benomyl is also effective against powdery mildew, though not to the same degree: thus, it has not been satisfactory when used at extended intervals (e.g. Burchill and Williamson, 1971; Evans *et al.*, 1971). One of its attributes is its marked effect on primary mildew, both in 'cleaning up' blossom truss infections in the spring and in controlling overwintering bud infections, as recorded the following season (e.g. Catling, 1971b). From greenhouse tests its curative action against mildew appears greater than that of dinocap and binapacryl (Frick, 1971).

Use of benomyl as a routine spray for scab and mildew control has given rise to several side-effects. Thus, an increased level of fruit russet (by comparison with standard fungicides) has sometimes been encountered, e.g. on Golden and Red Delicious (Gilpatrick, 1970) and on Cox's Orange Pippin (Byrde, 1972; Kirby, 1971). A mixture with lead arsenate proved highly phytotoxic (Ross and Sanford, 1971). The compound sometimes appears to stimulate vegetative growth (e.g. Evans *et al.*, 1971), and yield increases, sometimes statistically significant, have been recorded (Byrde *et al.*, 1970; Evans *et al.*, 1971; Kirby, 1971). Although the basis of this effect is not clear, it may well reflect increased fruit set, and it is significant that the compound, or its breakdown products, has been detected by bioassay in extracts of floral parts including the stigma (Corke *et al.*, 1972).

Triarimol has also proved highly effective against the two diseases (Table 10.IV.1). Against scab, its curative properties seem particularly strong: Brown *et al.* (1970) showed that it was completely effective up to at least 72 h after infection; Frick (1971) rated it superior to benomyl in this respect, and Zehr (1971a) ascribed its good performance under severe infection conditions to this property. The material has proved particularly effective when used as an aerial spray (Szkolnik and Gilpatrick, 1971). However, there have been occasions when the control of fruit scab has not been commensurate with the compound's effectiveness on leaves (e.g.

Borecka *et al.*, 1971; Doma *et al.*, 1971). Against powdery mildew the compound has been outstandingly effective (e.g. Hickey, 1971a): its powerful curative action (Frick, 1971), and its ability to control the disease over a wide leaf area from a small deposit (e.g. Doma *et al.*, 1971) are undoubtedly important factors. Even as an aerial spray, the compound, unlike other mildew fungicides thus applied, gave some control of the disease (Szkolnik and Gilpatrick, 1971). As with benomyl, russet has been recorded – e.g. on Golden Delicious (e.g. Gilpatrick, 1970) and on Cox's Orange Pippin (e.g. Byrde, 1972; Kirby, 1971). The compound seems more erratic than benomyl when used to suppress perithecial development (e.g. Burchill and Swait, 1971; Paisley and Heuberger, 1971).

So far tested on a less extensive scale, the thiophanate fungicides have given similar results to benomyl. Thiophanate methyl has been found to have excellent curative action against apple scab (Formigoni *et al.*, 1971). When applied as a routine spray thiabendazole, in contrast, has given erratic results, well illustrated in the reports edited by Hickey (1971a) although it has proved moderately effective as a spring spray to reduce ascospore inoculum (Miller, 1970). Triforine, while giving good scab and mildew control, has also resulted in improved foliage, albeit with slight fruit russet (Adlung and Drandarevski, 1971). Pyrazophos, while active against powdery mildew, had little effect on scab in greenhouse tests (Frick, 1971). Dimethirimol, though not normally effective against apple mildew, has shown some effect as a drench to apple seedlings in the greenhouse (E. C. Hislop, priv. comm.).

Systemic fungicides have been useful against other apple diseases (Table 10.IV.2) and several have shown antisporulant activity against *Nectria galligena*, the apple canker pathogen. Thus Bennett (1971) showed a marked and prolonged inhibition of spore production in the orchard following benomyl treatment: thiabendazole was less effective. Cole *et al.* (1971) demonstrated the antisporulant activity of a canker paint based on thiophanate methyl. Benomyl (McDonnell, 1970) was also superior to thiabendazole (McDonnell, 1971) for Nectria fruit rot control, and Manners and Corke (1971) showed that Nectria spores were much more resistant to thiabendazole than to benomyl. Spore output from cankers due to *Gloeosporium perennans* was also affected more by benomyl than by thiabendazole (Corke *et al.*, 1971). Routine benomyl applications for scab and mildew control result in the accumulation of toxic residues in the bark, which may account for the residual effect in controlling canker (Byrde *et al.*, 1972). The use of benomyl, thiabendazole and thiophanate methyl as sprays or post-harvest dips against Gloeosporium rots represents a greatly improved control measure for this serious disease.

Diseases not readily controlled by benomyl include Alternaria rot

TABLE 10.IV.2 **Control of apple diseases**

Causal fungus	*Disease*	*Compound*	*References*
Gloeodes pomigena	Sooty blotch	Benomyl	Delp and Klöpping (1968)
Gloeosporium spp.	Bitter rot	Benomyl	Bompeix and Morgat (1969); Borecka *et al.* (1971); Catling (1971b); Cole *et al.* (1971); Edney (1970)
		Thiabendazole	Bompeix and Morgat (1969); Bondoux (1967); Edney (1970); Pleisch *et al.* (1971); Pratella and Tonini (1968)
		Thiophanate, thiophanate methyl	Borecka *et al.* (1971); Cole *et al.* (1971); Formigoni *et al.* (1971); Ishii (1970)
Glomerella cingulata	Necrotic leaf blotch	Benomyl	Taylor (1971)
Gymnosporangium juniperi-virginianae	Cedar-apple rust	Triarimol	Fridrich and Gilpatrick (1971b); Lewis (1971)
Microthyriella rubi	Fly speck	Benomyl	Delp and Klöpping (1968)
Nectria galligena	Canker	Benomyl	Byrde *et al.* (1972); Catling (1971b); Corbin (1971a)
	Fruit rot	Benomyl	McDonnell (1970)
		Thiabendazole	Bondoux (1967)
Penicillium expansum	Blue mould	Benomyl	Cargo and Dewey (1970); Daines and Snee (1969); Spalding *et al.* (1969); Spalding and Hardenburg (1971)
		Thiabendazole	Blanpied and Purnasiri (1968, 1970); Cargo and Dewey (1970); Daines and Snee (1969); Pierson (1966); Pleisch *et al.* (1971); Spalding *et al.* (1969); Spalding and Hardenburg (1971)
Pezicula (Gloeosporium) spp.	Canker	Benomyl, thiophanate	Borecki *et al.* (1970)
Physalospora obtusa	Frog-eye leaf spot	Triarimol	Lade (1970)
	Black rot	Benomyl	Delp and Klöpping (1968)
Podosphaera leucotricha	Powdery mildew	See Table 10.IV.1	
Sclerotinia fructigena	Brown rot	Benomyl	Cole *et al.* (1971)
		Thiophanate methyl	Cole *et al.* (1971)
S. laxa f. *mali*	Blossom wilt	Benomyl	Byrde and Melville (1971); Catling (1971b)
Trichoseptoria fructigena	Trichoseptoria rot	Benomyl	Bompeix and Morgat (1969)
		Thiabendazole	Bompeix and Morgat (1969); Bondoux (1967)
Venturia inaequalis	Scab	See Table 10.IV.1	

(Spalding, 1970), which is also not controlled by thiabendazole (Blanpied and Purnasiri, 1970), cedar apple rust (Delp and Klöpping, 1968) and collar rot caused by *Phytophthora cactorum* and *P. syringae*. A tendency for a disease caused by a *Phytophthora* sp. to increase when benomyl was applied has even been reported (van den Berg and Bollen, 1971).

Pears The most important of the diseases of pears (Table 10.IV.3) is scab, which can cause severe fruit disfigurement and also persistent lesions on the wood. By contrast with apple scab, triarimol has given somewhat erratic results against pear scab (Gramlich *et al.*, 1969; Ross, 1971; Shabi, 1971), as has thiabendazole (Darpoux *et al.*, 1966; Shabi, 1971). Russeting of pear by benomyl and triarimol has been recorded by Kirby (1971).

TABLE 10.IV.3 **Control of pear diseases**

Causal fungus	*Disease*	*Compound*	*References*
Botrytis cinerea	Storage rots	Benomyl	Beattie and Outbred (1970); Coyier (1970)
		Thiabendazole	Beattie and Outbred (1970)
Entomosporium maculatum	Leaf spot	Benomyl	Kantzes and Weaver (1971)
Penicillium expansum	Fruit rot	Benomyl	Maas and MacSwan (1970); Spalding (1970)
		Thiabendazole	Maas and MacSwan (1970); Scott and Roberts (1970); Spalding (1970)
Venturia nashicola	Scab	Benomyl	Kishi (1971)
V. pirina	Scab	Benomyl	Bourdin *et al.* (1969); Kirby *et al.* (1970); Ross (1971); Shabi (1971)
		Thiophanate, thiophanate methyl	Aelbers (1971); Cole *et al.* (1971); Douchet *et al.* (1971); Formigoni *et al.* (1971); Ishii (1970); Shabi (1971)
		Triforine	Schicke and Veen (1969)

Stone fruit The most important group of diseases are those caused by the brown rot fungi which attack the fruit both in the orchard and after harvest, and also the blossom trusses. Three species of *Monilinia (Sclerotinia)* are responsible, with different geographical distributions (Wormald, 1954). Although difficult to control by conventional fungicides, with the possible exception of the very recent introduction 3-(3,5-dichlorophenyl)-5,5-dimethyl-2,4-oxazolidinedione (Sclex) (Hickey, 1971b), the advent of

TABLE 10.IV.4 **Control of stone fruit diseases**

Host	*Causal fungus*	*Disease*	*Compound*	*References*
Apricot	*Eutypa armeniacae*	Dieback	Benomyl	Moller and Carter (1969)
	Monilia (Sclerotinia) fructicola	Brown rot	Benomyl, thiabendazole	Beattie and Outbred (1970)
	Monilia sp.	Blossom blight	Benomyl	Waffelaert (1969)
	Oidium sp.	Mildew	Benomyl	Waffelaert (1969)
Cherry	*Coccomyces hiemalis*	Leaf spot	Benomyl	Cimanowski *et al.* (1970); Delp and Klöpping (1968)
	M. fructicola	Brown rot	Benomyl, thiabendazole	Beattie and Outbred (1970)
	M. fructicola and *M. laxa*	Blossom blight	Benomyl	Ogawa *et al.* (1968a)
	Podosphaera oxyacanthae	Mildew	Benomyl	Gilpatrick (1969)
			Triarimol	Lade and Christensen (1971)
Nectarine	*M. fructicola* and *M. laxa*	Blossom wilt and fruit rot	Benomyl	Wells and Gerdts (1971)
Peach	*Taphrina (Exoascus) deformans*	Leaf curl	Triforine	Ost *et al.* (1971)
	Cladosporium carpophilum	Scab	Benomyl	Delp and Klöpping (1968)
	Cytospora (Valsa) cincta	Cytospora canker	Cycloheximide semicarbazone	Helton and Rohrbach (1967)
			Benomyl, thiabendazole	Springer (1970)
	M. fructicola	Brown rot	Benomyl	Chandler (1968); Daines (1970); Delp and Klöpping (1968); Kable (1970); Ogawa *et al.* (1968a); Smith (1971); Wells (1971)
			Thiabendazole	Daines (1970); Fripp and Dettman (1969)
		Blossom blight	Benomyl, thiophanate methyl	Corbin (1971b)
	Monilinia spp.	Brown rot	Thiophanate methyl	Ishii (1970)
	Stigmina carpophila	'Stigmina'	Benomyl	Corbin and Kitchener (1970)
Plum	*Cytospora cincta*	Cytospora canker	Cycloheximide thiosemicarbazone	Helton and Kochan (1967)
	M. fructigena and *M. laxa*	Blossom blight and brown rot	Benomyl, thiophanate methyl	Renaud (1970)
			Thiabendazole	Renaud (1968, 1970)

benomyl has resulted in a wide measure of control (Table 10.IV.4). As well as acting as a prophylactic spray on living tissue, the compound (at the high rate of 0·1 per cent a.i.) has been shown to suppress sporulation of the fungus on infected peduncles and mummified fruits of peach (Kable, 1970). One instance of less effective brown rot control (on prune) has been recorded (MacSwan, 1971a). Its action (and that of thiabendazole) as fruit dips has been found to be enhanced by higher water temperatures (Daines, 1970; Smith, 1971). Benomyl stimulated the leaf size of cherry seedlings (Cimanowski *et al.*, 1970). There is also evidence that a pre-blossom application can advance the flowering date of plums and somewhat increase the effective pollination period (Stott and Jefferies, 1972).

Although also controlling several other diseases of stone fruit (Table 10.IV.4), benomyl is generally less effective against peach leaf curl (e.g. Corbin and Kitchener, 1970) and Rhizopus rot of peach (e.g. Hickey, 1971b). Often the latter disease is controlled by the addition of dicloran to post-harvest dips (e.g. Wells, 1971). Hickey's survey of 1971 results shows that, of other fungicides tested against stone fruit diseases, triarimol has given an erratic performance against peach scab, cherry leaf spot and the brown rot diseases, and that thiophanate methyl (NF 44) and thiophanate (NF 35) were comparatively ineffective against peach brown rot in one trial.

Soft fruit *Bush and cane fruit* The principal diseases of black currant and gooseberry are grey mould, leaf spot and American gooseberry mildew. Control of these three diseases can be achieved by the use of three separate conventional fungicides: dichlofluanid is particularly valuable against grey mould, the dithiocarbamates have been extensively used against leaf spot, and quinomethionate has brought mildew under control. The advent of broad-spectrum systemic fungicides has opened the way to control of all three diseases by a single compound, and Table 10.IV.5 shows that benomyl can achieve this. On occasions triarimol and thiabendazole (O'Riordain *et al.*, 1971) and thiophanate methyl (Borecki *et al.*, 1971) have proved somewhat inferior against *Pseudopeziza ribis*. Ingram (1971) found benomyl (without a wetting agent), pyrazophos and thiophanate were relatively ineffective against *Sphaerotheca mors-uvae* on black currant, though Clifford and Hislop (1971) showed that the latter compound together with thiophanate methyl had a strong vapour action against this fungus *in vitro*.

Although less information is available on the control of raspberry diseases, of which grey mould, cane spot and spur blight are the most common, benomyl has given adequate control of two of these: however, control of spur blight with this compound has been erratic (O'Riordain *et al.*, 1971).

TABLE 10.IV.5 **Control of bush and cane fruit diseases**

Host	*Causal fungus*	*Disease*	*Compound*	*References*
Black currant	*Botrytis cinerea*	Grey mould	Benomyl	Tapley *et al.* (1969)
	Cronartium ribicola	Rust	Benomyl, triarimol, thiophanate, thiophanate methyl	Borecki *et al.* (1971)
	Microsphaera grossulariae	European gooseberry mildew	Benomyl, triarimol	Borecki *et al.* (1971)
	Pseudopeziza ribis	Leaf spot	Benomyl	Borecki *et al.* (1971); Gilchrist and Cole (1971); O'Riordain *et al.* (1971)
			Thiabendazole	Ayers and Gunary (1971)
			Thiophanate methyl	Gilchrist and Cole (1971)
			Triforine	O'Riordain *et al.* (1971)
	Sphaerotheca mors-uvae	American gooseberry mildew	Benomyl	Gilchrist and Cole (1971)
			Thiophanate methyl	Gilchrist and Cole (1971); Ingram (1971)
			Triarimol	Ingram (1971)
Gooseberry	*Botrytis cinerea*	Grey mould	Benomyl	Tapley *et al.* (1969)
	Puccinia pringsheimiana	Rust	Triarimol	O'Riordain *et al.* (1971)
	Pseudopeziza ribis	Leaf spot	Benomyl	O'Riordain *et al.* (1971); Tapley *et al.* (1969); Woodward (1971b)
			Thiophanate, thiophanate methyl	Woodward (1971b)
			Triarimol, triforine	O'Riordain *et al.* (1971)

	Sphaerotheca mors-uvae	American gooseberry mildew	Benomyl, triarimol	O'Riordain *et al.* (1971); Woodward (1971b)
			Pyrazophos, thiophanate, thiophanate methyl	Woodward (1971b)
			Triforine	O'Riordain *et al.* (1971)
Raspberry	*Botrytis cinerea*	Grey mould (fruit)	Benomyl	Freeman and Pepin (1967)
		Grey mould (stem)	Thiabendazole	O'Riordain *et al.* (1971)
	Elsinoe veneta	Cane spot	Benomyl, thiabendazole	O'Riordain *et al.* (1971)
	Nectria sp.	Cane death	Benomyl	O'Riordain *et al.* (1971)
Blackberry	Four pathogens	Storage rot	Benomyl, thiabendazole	MacSwan (1971b)
Blueberry	*Botrytis cinerea*	Grey mould	Benomyl	Delp and Klöpping (1968)
	Exobasidium vaccinii	Red leaf	Carboxin	Lockhart (1971)
	Monilinia vaccinii-corymbosi	Twig and blossom blight	Benomyl	Pepin and Ormrod (1968)

TABLE 10.IV.6 **Control of strawberry diseases**

Causal fungus	*Disease*	*Compound*	*References*
Botrytis cinerea	Grey mould	Benomyl	Freeman and Pepin (1967, 1968); Gilchrist and Cole (1971); Kavanagh and O'Callaghan (1969); Maas (1970, 1971); Müller (1970); Paulus *et al.* (1969); Tapley *et al.* (1969)
		Thiabendazole	Maas (1970)
		Thiophanate, thiophanate methyl	Aelbers (1971); Formigoni *et al.* (1971); Gilchrist and Cole (1971)
Botrytis cinerea, Rhizoctonia sp.	Storage rots (runners)	Benomyl	Maas and Scott (1971)
Colletotrichum fragariae	Anthracnose	Benomyl	Howard (1971a)
Dendrophoma, Diplocarpon, Mycosphaerella	Leaf spots	Benomyl	Howard (1971b); Pessanha *et al.* (1970)
		Thiophanate	Pessanha *et al.* (1970)
Sphaerotheca macularis	Powdery mildew	Benomyl	Freeman and Pepin (1968); Jordan (1971); Maas (1970); Scott *et al.* (1970)†
		Triarimol	Jordan (1971)
Verticillium dahliae	Wilt	Benomyl	Jordan (1972a); Lockhart *et al.* (1969)
		Thiabendazole	Jordan (1972a)
		Thiophanate methyl	Jordan (1972b)

† This reference describes a trial on greenhouse-grown plants

Benomyl has proved useful against two blueberry diseases, but it was relatively ineffective against powdery mildew on this host (caused by *Microsphaera penicillata* var. *vaccinii*) (Hilborn, 1969), and neither benomyl nor thiabendazole successfully eradicated *Botryosphaeria corticis* in dormant cutting wood (Beute and Milholland, 1970).

Strawberries The chief fungal diseases of strawberries are grey mould, powdery mildew, *Verticillium* wilt and red core (caused by the Phycomycete *Phytophthora fragariae*). Good control of grey mould can now be achieved by the use of dichlofluanid, but powdery mildew is more difficult to control with conventional fungicides where infection is severe, whilst control of the two soil-borne diseases has been impracticable. Now, however, as shown in Table 10.IV.6, these and other minor diseases have been controlled by benomyl, with the expected exception of red core. Control of *Botrytis* fruit rot has generally been of the same order as that given by dichlofluanid, as found in 1970 by O'Riordain *et al.* (1971), but in 1971 these authors found benomyl and thiophanate methyl inferior to dichlofluanid. Maas (1970) reported that benomyl was more effective than thiabendazole against grey mould, and much more effective against mildew. Against this disease, pyrazophos and a substituted azepine were less effective than triarimol and benomyl (Jordan, 1971).

Full systemic action is likely to be more readily achieved in the herbaceous strawberry plant than with top fruits, and it is probably significant that Howard (1971a) found benomyl effective against *Colletotrichum fragariae* whereas twelve non-systemic fungicides were not. Even so, phloem translocation does not seem to occur to any extent, as Lockhart *et al.* (1969) found benomyl ineffective as a foliar spray against *Verticillium* wilt although giving good control as a soil treatment in the planting hole. Using such a benomyl treatment, Jordan (1972a) could detect the pathogen within the tissues of a visually healthy plant, suggesting that the effect was only fungistatic and exposing the limitations of the compound especially for use in runner beds.

SUB-TROPICAL FRUIT

Table 10.IV.7 summarises reports on the control of citrus and grape diseases by systemic fungicides. The most important diseases of citrus are the post-harvest rots due to *Penicillium digitatum* ('green mould') and *P. italicum* ('blue mould'), which for convenience are covered together in the table. The current control measures involve the use of diphenyl wraps, or post-harvest treatment with *o*-phenyl phenol or 2-aminobutane, but these may well be superseded by benomyl and thiabendazole. Of these

TABLE 10.IV.7 **Control of diseases of sub-tropical fruits**

Host	*Causal fungus*	*Disease*	*Compound*	*References*
Citrus (mainly orange)	*Cercospora* sp.	Pink pitting	Benomyl	Suit and Ducharme (1971)
	Deuterophoma tracheiphila	'Mal secco'	Benomyl	Elia (1969); Salerno and Somma (1971)
	Diplodia natalensis	Stem-end rot	Benomyl	Brown (1968); Brown and McCornack (1970)
	Elsinoe fawcettii	Scab	Benomyl	Hearn and Childs (1969); Hearn *et al.* (1971); Phelps (1969)
	Guignardia citricarpa	Black spot	Benomyl	McOnie *et al.* (1969)
	Mycosphaerella sp.	Greasy spot	Benomyl	Phelps (1969)
	Penicillium digitatum and *P. italicum*	Penicillium rots†	Benomyl	Baron and Cayuela (1969); Brown (1968); Brown and McCornack (1970); Delp and Klöpping (1968); Gutter (1969a‡, 1969b, 1970); Harding (1968); McCornack and Brown (1970); Tugwell and Wicks (1969)
			Thiabendazole	Baron and Cayuela (1968); Cuillé and Bur-Ravault (1969)‡, Gutter (1969a)‡; Harding (1967‡, 1968); McCornack and Brown (1970); Seberry and Baldwin (1968); Tugwell and Wicks (1969)
			Thiophanate, thiophanate methyl	Ishii (1970)
	Unknown	Fruit rot	Benomyl, thiabendazole	Smoot and Melvin (1970)
	Unknown	Die back	Benomyl	Hearn and Fenton (1970)

Grape	*Botrytis cinerea*	Grey mould	Benomyl	Dietrich and Brechbuhler (1970); Frazao (1970); Thiollière (1970)
			Thiophanate, thiophanate methyl	Formigoni *et al.* (1971)
	Guignardia bidwellii	Black rot	Benomyl	Delp and Klöpping (1968); Klos *et al.* (1968); Welch (1968); Zehr (1971b)
			Triarimol	Zehr (1971b)
	Melanconium fuligineum	Bitter rot	Benomyl	Delp and Klöpping (1968); Welch (1968)
	Uncinula necator	Powdery mildew	Benomyl	Bottalico (1969); Braun (1971)
			Thiophanate, thiophanate methyl	Formigoni *et al.* (1971)
			Triarimol	Braun (1971); Farrant *et al.* (1970); Gramlich *et al.* (1969); Kovacs and Caumo (1971); Rogers (1971)

† Generally treated by post-harvest dips
‡ Includes use in wax formulation

two, there are indications that the former is the more active (Baron and Cayuela, 1969; Brown, 1968; Gutter, 1970). Many diseases of less economic importance are also controlled by these compounds, and the thiophanate fungicides may be expected to have a similar spectrum of control. Thiabendazole has been found to be toxic *in vitro* to *Diplodia natalensis, Phomopsis citri* and *Penicillium digitatum* (Brown *et al.*, 1967) and Whiteside (1970) demonstrated the antisporulant activity of this compound and of benomyl against the perithecial stage of *Mycosphaerella* sp. which causes 'greasy spot' disease. The systemic properties of benomyl have proved useful in the control of 'mal secco' by soil application to orange seedlings (Elia, 1969; Salerno and Somma, 1971).

Of grape diseases the most serious are powdery mildew, *Botrytis* fruit rot and downy mildew (caused by *Plasmopara viticola*). Conventionally, a different fungicide is used for each disease – sulphur or dinocap for powdery mildew, dichlofluanid for *Botrytis*, and a copper or dithiocarbamate fungicide for downy mildew. There is now the prospect of controlling the two first-named with benomyl or thiophanate fungicides; however, as might be expected, none of the systemic fungicides in current use is active against downy mildew (e.g. Welch, 1968). Thiophanate methyl (Formigoni *et al.*, 1971) and benomyl can be used without affecting fermentation processes.

TROPICAL FRUIT

Disease control under conditions of high temperature and rainfall is difficult to achieve with conventional fungicides, and under such conditions systemic compounds seem likely to have particular advantages (Byrde, 1970) despite their high cost. The latter is probably a major factor in their current extensive use as dips rather than as sprays.

The most important banana pathogens are *Mycosphaerella musicola (Cercospora musarum)* which causes 'Sigatoka' leaf spot, *Gloeosporium musarum* which causes anthracnose (stem-end rot, black end), and *Fusarium oxysporum* f. *cubense* which causes the 'Panama' wilt disease. No report appears to exist of the attempted control of the latter disease with a systemic fungicide. Copper and/or oil sprays are conventionally used against Sigatoka disease, but anthracnose is difficult to control with fungicides (Wardlaw, 1961).

Table 10.IV.8 summarises the control to date of banana and pineapple diseases, and thiabendazole and benomyl have clearly undergone extensive testing. There is some evidence that, of the two, benomyl tends to be the more effective against banana anthracnose (e.g. Burden, 1969; Long,

TABLE 10.IV.8 **Control of diseases of tropical fruits**

Host	*Causal fungus*	*Disease*	*Compound*	*References*
Banana	*Fusarium roseum*	*Fusarium* fruit rot	Benomyl†, thiabendazole†	Shillingford (1970)
	Gloeosporium musarum	Anthracnose	Benomyl†	Allen (1970); Burden (1969); Frossard (1969, 1970a); Long (1970); Ogawa *et al.* (1968b, 1969); Shillingford (1970); Tsai *et al.* (1968)
			Parbendazole†, 2-phenyl-benzimidazole†	Allen (1970)
			Thiabendazole†	Allen (1970); Beaudoin *et al.* (1969); Burden (1969); Frossard (1969); Long (1970); Scott and Roberts (1967); Shillingford (1970); Swarts *et al.* (1969)
			Thiophanate methyl	Ishii (1970)
	Mycosphaerella fijiensis	Black leaf streak	Benomyl	Long (1971)
	M. musicola (Cercospora musarum)	Sigatoka leaf spot	Benomyl	Brun and Melin (1970); Melin (1970); Stover (1969)
	Nigrospora sphaerica	'Squirter' disease	Benomyl† and related compounds†	Allen (1970)
			Thiabendazole†	Allen (1970); Rippon *et al.* (1970)
	Five fungi	Stem-end rot complex	Thiabendazole	Weinke *et al.* (1969)
Pineapple	*Ceratocystis (Thielaviopsis) paradoxa*	Fruit rot	Benomyl†	Frossard (1970b); Ogawa *et al.* (1969)
			Thiabendazole†	Frossard (1968, 1970b)
		Butt rot	Benomyl	Rohrbach and Apt (1971)

† As post-harvest dip

1970). Thiabendazole was found by Beaudoin *et al.* (1969) to be effective against this disease provided it was applied within three days of inoculation. The performance of benomyl against both *M. musicola* (Brun and Melin, 1970) and *M. fijiensis* (Long, 1971) was greatly enhanced by the addition of oil. Stover (1969), in a detailed examination of the mode of action of benomyl against *M. musicola*, showed that it reduced germination, sporulation and spot formation, had translaminar activity and could be translocated from roots to leaf. There was also some evidence of an increased host resistance following its use.

CONCLUSIONS

Even though not fully systemic in most of the fruit crops described above, these fungicides represent a marked advance in chemical disease control. Benomyl and the thiophanate fungicides control a wide range of fruit diseases, while triarimol and triforine show particular promise against powdery mildews. Thiabendazole, although not always quite as effective as benomyl, is particularly favoured for post-harvest use, at least partly because of long experience of its use as an anthelminthic in clinical and veterinary practice.

Resistance to the new fungicides by fruit pathogens has just been reported, and may well develop with increasing usage. There remains a need for compounds capable of full systemic activity in woody plants: the ultimate feasibility of this is encouraged by the demonstration by Helton and Williams (1970) of the translocation in apple trees of the antibiotic derivative cycloheximide (actidione) semicarbazone.

The Phycomycetes, represented among fruit pathogens by *Phytophthora cactorum, P. syringae, P. fragariae, Plasmopara viticola* and *Rhizopus* spp., also present a particular challenge. It is nevertheless clear that in the meantime many fruit diseases will be better controlled than hitherto by the use of the systemic fungicides now available.

References

Acosta, A. and Livingston, J. E. (1955) Effects of calcium sulfamate and sodium sulfanilate on small grains and on stem rust development. *Phytopathology* **45**, 503.

Adlung, K. G. and Drandarevski, C. A. (1971) The evaluation of 'CELA W524', a systemic fungicide, for the control of powdery mildew and apple scab. *Proc. 6th Br. Insectic. Fungic. Conf.* **2**, 577.

Aelbers, E. (1970) Thiophanate and thiophanate methyl, two new fungicides with systemic action. *Proc. 22nd int. Symposium for Phytopharmacy and Phytiatry, Ghent* (May 1970)

Aelbers, E. (1971) Thiophanate and thiophanate methyl, two new fungicides with systemic action. *Mededelingen Fakulteit Landb. wetensch. Gent* **36**, 126.

Aikman, D. P. and Anderson, W. P. (1971) A quantitative investigation of a peristaltic model for phloem translocation. *Ann. Bot.* **35**, 761.

Akai, S. (1955) Chemotherapeutic application of some compounds to rice plants and outbreak of *Helminthosporium* leaf spot. *Univ. Kyoto Forsch. PflKrankh.* **5**, 45.

Akai, S. and Kato, H. (1962) Internal application of pentachlorophenol to rice plants and the outbreak of *Helminthosporium* leaf spot. *Univ. Kyoto Forsch. PflKrankh.* **7**, 35.

Albersheim, P., Jones, T. M. and English, P. D. (1969) Biochemistry of the cell wall in relation to infective processes. *Ann. Rev. Phytopathol.* **7**, 171.

Albert, J. J. and Groves, A. B. (1966) Eradicant action of certain fungicides and fungicide combinations on *Venturia inaequalis*. *Phytopathology* **56**, 583.

Allam, A. I. and Sinclair, J. B. (1969) Degradation of DMOC (Vitavax) in cotton seedlings. *Phytopathology* **69**, 1548.

Allen, P. M. (1969) Ph.D. Thesis, University of Illinois, Urbana.

Allen, P. M. and Gottlieb, D. (1970) Mechanism of action of the fungicide thiabendazole, 2-(4'-thiazolyl) benzimidazole. *Appl. Microbiol.* **20**, 919.

Allen, R. N. (1970) Control of black-end and squirter diseases in bananas with benzimidazole and salicylanilide compounds. *Aust. J. exp. Agric. Anim. Husb.* **10**, 490. (*Chem. Abs.* **74**, 12092x, 1971.)

van Andel, O. M. (1958) Investigations on plant chemotherapy. II. Influence of amino-acids on the relation plant-pathogen. *Tijdschr. PlZiekt.* **64**, 307.

van Andel, O. M. (1962a) Fluorophenylalanine as a systemic fungicide. *Nature, Lond.* **194**, 790.

van Andel, O. M. (1962b) Growth regulating effects of amino acids and dithiocarbamic acid derivatives and their possible relation with chemotherapeutic activity. *Phytopath. Z.* **45**, 66.

van Andel, O. M. (1966a). Mode of action of L-threo-β-phenylserine as a chemotherapeutant of cucumber scab. *Nature, Lond.* **211**, 326.

van Andel, O. M. (1966b) Amino acids and plant diseases. *A. Rev. Phytopath.* **4**, 349.

van Andel, O. M. (1968) Shifts in disease resistance induced by growth regulators. *Neth. J. Pl. Pathol, Suppl. 1* **74**, 113.

Andersen, A. S. and Rowell, J. B. (1960) Half-life effectiveness against stem rust of systemic chemicals in wheat seedlings. *Phytopathology* **50**, 627.

Andersen, A. S. and Rowell, J. B. (1962) Duration of protective activity in wheat seedlings of various compounds against stem rust. *Phytopathology* **52**, 909.

Anderson, H. W. and Gottlieb, D. (1952) Plant disease control with antibiotics. *Econ. Bot.* **6**, 294.

Anderson, H. W. and Nienow, I. (1947) Effect of streptomycin on higher plants. *Phytopathology* **37**, 1.

Anderson, W. P. and Reilley, E. J. (1968) A study of the exudation of excised maize roots after removal of the epidermis and outer cortex. *J. exp. Bot.* **19**, 19.

Anon. (1970) Systemic fungicides against blast. *The IRRI Reporter* **6**, 1.

Anon. (1971a) *Technical manual concerning the activity of tridemorph.* Badische Anilin und Soda-fabrik.

Anon. (1971b) Benomyl drench effective against celery leaf spot. *Grower* **76**, 13.

Anon. (1971c) Ring the changes with chemicals. *Grower* **75**, 1333.

Ark, P. A. and Dekker, J. (1958) Uptake of antibiotic GS1 by seeds and plants. *Phytopathology* **48**, 391.

Ark, P. A. and Thompson, J. P. (1958) Prevention of antibiotic injury with Na–K chlorophyllin. *Pl. Dis. Reptr* **42**, 1203.

Arnold, W. R., Lade, D. H. and Christensen, C. D. (1971) Efficacy of triarimol (EL-273) against *Venturia inaequalis* and *Podosphaera leucotricha. Phytopathology* **61**, 884 (Abs.).

Arny, D. C. and Leben, C. (1954) Vapor action of certain mercury seed treatment materials. *Phytopathology* **44**, 380.

Asakawa, M., Misato, T. and Fukunaga, K. (1963) Studies on the prevention of the phytotoxicity of blasticidin S. *Pesticide and Technique (Tokyo)* **8**, 24.

Ashida, J. (1965) Adaptation of fungi to metal toxicants. *Phytopathology* **3**, 153.

van Assche, C. and Vanachter, A. (1970) Systemic fungicides to control fungal diseases in vegetables. *Parasitica* **26**, 117.

Ayers, M. R. L. W. and Gunary, D. (1971) Thiabendazole – broad spectrum fungicide – field experience of its performance on a range of crops in the United Kingdom. *Proc. 6th Br. Insectic. Fungic. Conf.* **2**, 341.

Aytoun, R. S. C. (1956) The effects of griseofulvin on certain phytopathogenic fungi. *Ann. Bot., Lond. (N.S.)* **20**, 297.

Baker, E. A. (1971) Chemical and physical characteristics of cuticular membranes.

Ecology of leaf-surface micro-organisms (eds. T. F. Preece and C. H. Dickinson), p. 55. Academic Press.

Baldwin, R. E. (1970) Control of snapbean rust with systemic and non-systemic fungicides. *Phytopathology* **60**, 1013.

Banfield, W. M. (1968) Dutch elm disease recurrence and recovery in American Elm. *Phytopathol. Z.* **62**, 21.

Bant, J. H. and Storey, I. F. (1952) Hot-water treatment of celery seed in Lancashire. *Pl. Path.* **1**, 81.

Baron, J. M. Q. and Cayuela, J. C. (1969) Trials on the effect of fungicide 1.991 on the preservation of Valencia oranges and comparative studies on the effect of other fungicides. (In Spanish.) *Ann. Inst. nac. Invest. agron.* **18**, 285. (*Rev. Pl. Path.* **49**, 2861, 1970.)

Bartels-Schooley, J. and MacNeill, B. H. (1971) A comparison of the modes of action of three benzimidazoles. *Phytopathology* **61**, 816.

Bateman, D. F. and Millar, R. L. (1966) Pectic enzymes and tissue degradation. *Ann. Rev. Phytopathol.* **4**, 119.

Bates, A. N., Spencer, D. M. and Wain, R. L. (1963) Investigations on fungicides. VII. The antifungal activity of certain hydroxy nitroalkanes and related compounds. *Ann. appl. Biol.* **51**, 153.

Batts, C. C. V. and Elliott, C. S. (1952) Indications of effect of yellow rust on yield of wheat. *Pl. Path.* **1**, 130.

Beattie, B. B. and Outbred, N. L. (1970) Benzimidazole derivatives as post-harvest fungicides to control rotting of pears, cherries and apricots. *Aust. J. exp. Agric. Anim. Husb.* **10**, 652. (*Rev. Pl. Path.* **50**, 1301, 1971.)

Beaudoin, C., Champion, J. and Mallessard, R. (1969) Essais de traitements des Bananes au thiabendazole. *Fruits d'outre-mer* **24**, 89. (*Rev. appl. Mycol.* **48**, 2497, 1969.)

Bebbington, R. M. Brooks, D. H., Geoghegan, M. J. and Snell, B. K. (1969) Ethirimol: a new systemic fungicide for the control of cereal powdery mildews. *Chem. Ind.* 1512.

Beckman, C. H. (1958) Growth inhibition as a mechanism in Dutch elm disease therapy. *Phytopathology* **48**, 172.

Beckman, C. H. (1959) Dutch elm disease control with polychlorobenzoic acid. *Phytopathology* **49**, 227.

Beckman, C. H. (1964) Host responses to vascular infection. *Ann. Rev. Phytopathol.* **2**, 231.

Beckman, C. H. (1969a) The mechanics of gel formation by swelling of simulated plant cell wall membranes and perforation plates of banana root vessels. *Phytopathology* **59**, 837.

Beckman, C. H. (1969b) Plasticizing of walls and gel induction in banana root vessels. infected with *Fusarium oxysporum*. *Phytopathology* **59**, 1477.

Beckman, C. H., Halmos, S. and Mace, M. E. (1962) The interaction of host, pathogen and soil temperature in relation to susceptibility to Fusarium wilt of banana. *Phytopathology* **52**, 134.

Beckman, C. H. and Howard, F. L. (1957) Chemical inhibition of springwood development in relation to infection and symptoms of Dutch elm disease. *Phytopathology* **47**, 3.

Beckman, C. H., Kuntz, J. E., Riker, A. J. and Berbee, J. G. (1953) Host responses associated with the development of oak wilt. *Phytopathology* **43**, 448.

Beckman, C. H., Mace, M. E., Halmos, S. and McGahan, M. W. (1961) Physical barriers associated with resistance in *Fusarium* wilt of bananas. *Phytopathology* **51**, 507.

Beckman, C. H., Stessel, G. J. and Howard, F. L. (1969) *Verticillium* spp. and associated fungi from certified potato seed tubers. *Pl. Dis. Reptr* **53**, 771.

Beckman, C. H. and Zaroogian, G. E. (1967) Origin and composition of vascular gel in infected banana roots. *Phytopathology* **57**, 11.

Bennett, M. (1962) An approach to the chemotherapy of silver leaf disease (*Stereum purpureum* (Fr.) Fr.) of plum trees. *Ann. appl. Biol.* **50**, 515.

Bennett, M. (1971) Effect of fungicides on inoculum potential of apple canker disease (*Nectria galligena*). *Proc. 6th Br. Insectic. Fungic. Conf.* **1**, 98.

Bent, K. J. (1967) Vapour action of fungicides against powdery mildews. *Ann. appl. Biol.* **60**, 251.

Bent, K. J. (1970) Fungitoxic action of dimethirimol and ethirimol. *Ann. appl. Biol.* **66**, 103.

Bent, K. J., Cole, A. M., Turner, J. A. W. and Woolner, M. (1971) Resistance of cucumber powdery mildew to dimethirimol. *Proc. 6th Br. Insectic. Fungic. Conf.* **1**, 274.

Berg, G. A. van den and Bollen, G. J. (1971) Effect of benomyl on incidence of wilting of *Callistephus chinensis* caused by *Phytophthora cryptogea*. *Acta bot. neerl.* **20**, 256.

Betgem, J. (1960) Veldproeven met het 3-phenyl derivaat van 5-amino-1-bis(dimethylamido)-fosforyl-triazole-1,2,4. *Meded. Landb. Hogesch. Opzoekstns. Gent* **25**, 1227.

Beute, M. K. and Milholland, R. D. (1971) Eradication of *Botryosphaeria corticis* from blueberry propagation wood. *Pl. Dis. Reptr* **54**, 122.

Beyries, A. (1969) Effectiveness of some systemic fungicides against *Corticium solani* by seed treatment of bean and radish. *Phytiat.–Phytopharm.* **18**, 107.

Bhate, D. S., Lavate, W. V. and Acharya, S. P. (1964) Isolation, crystallisation and physicochemical properties of neopentaene, a new antifungal antibiotic. *Hindustan Antibiot. Bull.* **6**, 153.

Biehn, W. L. (1969) Evaluation of seed and soil treatment for control of *Rhizoctonia* scurf and *Verticillium* wilt of potato. *Pl. Dis. Reptr* **53**, 425.

Biehn, W. L. (1970) Evaluation of seed treatments for control of seed-borne *Verticillium* wilt of potato. *Pl. Dis. Reptr* **54**, 254.

Biehn, W. L. (1972) Specificity of *p*-fluorophenylalanine for inhibiting sporulation of *Ceratocystis ulmi*. *Phytopathology* **62**, 493.

Biehn, W. L. and Dimond, A. E. (1969) Reduction of tomato *Fusarium* wilt symptoms by 1-(butylcarbamoyl)-2-benzimidazole carbamic acid, methyl ester. *Phytopathology* **59**, 397.

Biehn, W. L. and Dimond, A. E. (1970) Protective action of benomyl against Dutch elm disease. *Phytopathology* **60**, 571 (abs.).

Blanpied, G. D. and Purnasiri, A. (1968) Thiabendazole control of *Penicillium* rot of McIntosh apples. *Pl. Dis. Reptr* **52**, 867.

Blanpied, G. D. and Purnasiri, A. (1970) Thiabendazole control of post-harvest decay. *HortScience* **5**, 476. (*Rev. Pl. Path.* **50**, 3033, 1971.)

Bollen, G. J. (1971) Resistance to benomyl and some chemically related compounds in strains of *Penicillium* species. *Neth. J. Pl. Pathol.* **77**, 187.

Bollen, G. J. (1972) A comparison of the *in vitro* antifungal spectrum of thiophanates and benomyl. *Neth. J. Pl. Pathol.* (in press).

Bollen, G. J. and Fuchs, A. (1970) On the specificity of the *in vitro* and *in vivo* antifungal activity of benomyl. *Neth. J. Pl. Path.* **76**, 299.

Bollen, G. J. and Scholten, G. (1971) Acquired resistance to benomyl and some other systemic fungicides in a strain of *Botrytis cinerea* in cyclamen. *Neth. J. Pl. Path.* **77**, 83.

Bolley, H. L. (1891) Potato scab and possibilities of prevention. *Bull. N. Dakota Agric. Exp. Sta. no. 4,* p. 1.

Bolley, H. L. (1906) Tree feeding. *Rep. N. Dakota Agric. Exp. Sta. no. 17,* p. 104.

Bompeix, G. and Morgat, F. (1969) Lutte chimique contre les pourritures des pommes en conservation: éfficacité du benomyl et du thiabendazole. *C. r. hebd. Séanc. Acad. Agric. Fr.* **55**, 776. (*Rev. Pl. Path.* **49**, 515, 1970.)

Bondoux, P. (1967) L'eau chaude et le thiabendazole en traitement curatif contre les parasites lenticellaires des Pommes. *C. r. hebd. Séanc. Acad. Agric. Fr.* **53**, 1314. (*Rev. appl. Mycol.* **47**, 2205, 1968.)

Bonner, J. (1961) On the mechanics of auxin-induced growth. *Plant Growth Regulation,* p. 307. Ames: Iowa State Univ. Press.

Booth, A., Moorby, J., Davies, C. R., Jones, H. and Wareing, P. F. (1962) Effects of indolyl-3-acetic acid on the movement of nutrients within plants. *Nature, Lond.* **194**, 204.

Borecka, H., Borecki, Z. and Millikan, D. F. (1971) Control of apple scab, bitter rot and sawfly in Poland with the use of some newer fungicides. *Pl. Dis. Reptr* **55**, 828.

Borecki, Z., Masternak, H., Puchala, Z. and Millikan, D. F. (1970) Preliminary observations on Pezicula canker control in Poland. *Pl. Dis. Reptr* **54**, 640.

Borecki, Z., Millikan, D. F., Puchala, Z. and Bystydzienska, K. (1971) Effectiveness of benomyl and other chemicals for the control of leaf diseases on black currant in Poland. *Pl. Dis. Reptr* **55**, 932.

Borecki, Z. and Nowacka, H. (1969) Control of apple scab *(Venturia inaequalis)* with modern fungicides. *Biul. Inst. Ochr. Rósl, Poznán* **44**, 195. (*Rev. Pl. Path.* **49**, 1537j, 1970.)

Borum, D. E. and Sinclair, J. B. (1967) Systemic activity of 2,3-dihydro-5-carboxanilide-6-methyl-1,4-oxathiin against *Rhizoctonia solani* in cotton seedlings. *Phytopathology* **57**, 805.

van den Bos, B. G. (1960) Investigations on pesticidal phosphorus compounds. III. The structure of the reaction product of 3-amino-5-phenyl-1,2,4-triazole and bis(dimethylamido) phosphoryl chloride (WP155). *Rec. Trav. chim. Pays-Bas* **79**, 1129.

van den Bos, B. G., Koopmans, M. J. and Huisman, H. O. (1960) Investigations on pesticidal phosphorus compounds. I. Fungicides, insecticides and acaricides derived from 3-amino-1,2,4-triazole. *Rec. Trav. chim. Pays-Bas* **79**, 807.

van den Bos, B. G., Schipperheyn, A. and van Deuren, F. W. (1966) Investigations on pesticidal phosphorus compounds. V. The structure of the products of the reaction of 3-alkyl-5-anilino-1,2,4-triazoles with bis(dimethylamido) phosphoryl chloride. *Rec. Trav. chim. Pays-Bas* **85**, 429.

van den Bos, B. G., Shoot, C. J., Koopmans, M. J. and Meltzer, J. (1961) Investigations on pesticidal phosphorus compounds. IV. N-bis(dimethylamido) phosphoryl heterocycli. *Rec. Trav. chim. Pays-Bas* **80**, 1040.

Bottalico, A. (1969) Preliminary trials with benomyl against powdery mildew on grapevine in Apulia. (In Italian.) *Atti Giorn. fitopatol., 1969,* 433. (*Rev. Pl. Path.* **49**, 2144, 1970.)

Bouchereau, P. and Atkins, J. G. Jr. (1950) Studies on the action of copper seed treatments for water-planted rice. *Phytopathology* **40**, 3.

Boudru, M. (1935) La maladie de l'orme en Belgique. *Bull. Soc. chim. Belg.* **42**, 508.

Bourdin, J., Douchet, J. P. and Thollière, J. (1969) Efficacité du benomyl sur tavelures et oidium des arbres fruitiers à pepins. *Phytiat.-Phytopharm.* **18**, 15. (*Rev. Pl. Path.* **49**, 1422, 1970.)

Bowen, M. R. and Wareing, P. F. (1971) Further investigations into hormone-directed transport systems. *Planta* **99**, 120.

Braun, A. J. (1971) Grape: powdery mildew (*Uncinula necator*). *Fungicide–Nematicide Tests* **26**, 56.

Braun, A. J. and Pringle, R. B. (1959) Pathogen factors in the physiology of disease. Toxins and other metabolites. *Plant Pathology: Problems and Progress 1908–1958* (ed. C. S. Holton) p. 88. Madison: Univ. Wisconsin Press.

Brener, W. D. and Beckman, C. H. (1968) A mechanism of enhanced resistance to *Ceratocystis ulmi* in American elms treated with sodium trichlorophenylacetate. *Phytopathology* **58**, 555.

Brian, P. W. (1952a) Systemic fungicides. *Pl. Prot. Overseas Rev.* **3**, 5.

Brian, P. W. (1952b) Antibiotics as systemic fungicides and bactericides. *Ann. appl. Biol.* **39**, 434.

Brian, P. W. (1954) The use of antibiotics for control of plant diseases caused by bacteria and fungi. *J. appl. Bact.* **17**, 142.

Brian, P. W. (1960) Griseofulvin. *Trans. Br. mycol. Soc.* **43**, 1.

Brian, P. W., Wright, J. M., Stubbs, J. and Way, A. M. (1951) Uptake of antibiotic metabolites of soil micro-organisms by plants. *Nature, Lond.* **167**, 347

Briggs, G. E. and Robertson, R. N. (1957) Apparent free space. *A. Rev. Pl. Physiol.* **8**, 11.

Brooks, D. H. (1971a) Powdery mildew of barley and its control. *Outlook on Agriculture* **6**, 122.

Brooks, D. H. (1971b) Observation on the effects of mildew on growth of spring and winter barley. In press.

Brooks, F. T. and Bailey, M. A. (1919) Silver leaf disease (including observations upon the injection of trees with antiseptics). *J. Pomol.* **1**, 81.

Brooks, F. T. and Brenchley, G. H. (1931) Further injection experiments in relation to *Stereum purpureum*. *New Phytol.* **30**, 128.

Brooks, F. T. and Storey, H. H. (1923) Silver leaf disease. *J. Pomol.* **3**, 117.

Brown, G. E. (1968) Experimental fungicides applied pre-harvest for control of post-harvest decay in Florida citrus fruit. *Pl. Dis. Reptr* **52**, 844.

Brown, G. E. and McCornack, A. A. (1970) Benlate, an experimental pre-harvest fungicide for control of post-harvest citrus fruit decay. *Proc. Fla St. hort. Soc.* **82**, 39. (*Rev. Pl. Path.* **50**, 2277, 1971.)

Brown, G. E., McCornack, A. A. and Smoot, J. J. (1967) Thiabendazole as a post-harvest fungicide for Florida citrus fruit. *Pl. Dis. Reptr* **51**, 95.

Brown, I. F. (1970) Biological activity of a new broad spectrum systemic fungicide. *Proc. 7th Int. Congr. Plant Protection* p. 206 (abs.).

Brown, I. F. and Hall, H. R. (1971) Studies on the activity of triarimol (EL-273) against certain powdery mildew fungi. *Phytopathology* **61**, 886 (abs.).

Brown, I. F., Hall, H. R. and Miller, J. R. (1970) EL-273, a curative fungicide for the control of *Venturia inaequalis*. *Phytopathology* **60**, 1013 (abs.).

Brown, I. F. (Jr), Whaley, J. W., Taylor, H. M. and van Heynigen, E. M. (1967) The structure–activity relationships of a new group of pyridine alkane and carbinol fungicides. *Phytopathology* **57**, 805 (abs.).

Brown, J. G. and Boyle, A. M. (1944) Effect of penicillin on a plant pathogen. *Phytopathology* **34**, 760.

Brun, J. and Melin, P. (1970) Utilization of benomyl mixed with oil for the treatment against cercosporiose of bananas. *Summaries of Papers. VIIth Int. Cong. Plant Prot., Paris,* p. 231.

Bryant, J. H. and Hitchman, D. J. (1971) The control of apple mildew with triarimol (EL 273). *Proc. 6th Br. Insectic. Fungic. Conf.* **1**, 126.

Bucheneau, G. W. (1970) *Fungicide–Nematicide Tests* **26**, 104.

Buchenauer, H. and Erwin, D. C. (1971) Control of *Verticillium* wilt of cotton by spraying foliage with benomyl and thiabendazole solubilized with hydrochloric acid. *Phytopathology* **61**, 433.

Buchenauer, H. and Grossmann, F. (1969) Einfluss von Morphaktinen auf verschiedene Tomatenkrankheiten. *Dtsch. Bot. Ges., Neue Folge* **3**, 149.

Burchill, R. T. and Hutton, K. E. (1965) The suppression of ascospore production to facilitate the control of apple scab *(Venturia inaequalis)* (Cke.) (Wint.) *Ann. appl. Biol.* **56**, 285.

Burchill, R. T. and Swait, A. A. J. (1971) Apple scab *(Venturia inaequalis)*. Epidemiology: suppression of ascospores by post-harvest sprays. *Rep. E. Malling Res. Stn, 1970,* p. 106.

Burchill, R. T. and Williamson, C. J. (1971) Comparison of some new fungicides for the control of scab and powdery mildew of apple. *Pl. Path.* **20**, 173.

Burden, O. J. (1969) Control of ripe fruit rots of bananas by the use of post-harvest fungicide dips. *Aust. J. exp. Agric. Anim. Husb.* **9**, 655. (*Rev. Pl. Path.* **49**, 1736, 1970.)

Butler, G. W. (1953) Ion uptake by young wheat plants. II. *Physiologia Pl.* **6**, 617.

Butt, D. J. (1971) Apple mildew *(Podosphaera leucotricha)*: fungicide trials. *Rep. E. Malling Res. Stn, 1970,* p. 108.

Byrde, R. J. W. (1956) The varietal resistance of fruits to brown rot. I. Infection experiments with *Sclerotinia fructigena* Aderh. and Ruhl. on certain dessert, culinary and cider varieties of apples. *J. hort. Sci.* **31**, 188.

Byrde, R. J. W. (1957) The varietal resistance of fruits to brown rot. II. The nature of resistance in some varieties of cider apples. *J. hort. Sci.* **32**, 227.

Byrde, R. J. W. (1959) A note on the control of brown rot of apples by griseofulvin. *Pl. Path.* **8**, 90.

Byrde, R. J. W. (1963) Natural inhibitors of fungal enzymes and toxins in disease resistance. *Conn. Agric. Exp. Sta. Bull.* **663**, 31.

Byrde, R. J. W. (1969) Systemic fungicides, including recent developments in the agricultural sphere. *Proc. 5th Br. Insectic. Fungic. Conf.* **3**, 675.

Byrde, R. J. W. (1970) The new systemic fungicides and their potential use in the tropics. *Trop. Sci.* **12**, 105.

Byrde, R. J. W. (1972) Apple diseases (with particular reference to mildew and canker control). *Proc. 6th Br. Insectic. Fungic. Conf.* **3**, 648.

Byrde, R. J. W. and Ainsworth, G. C. (1958) The chemotherapy of fungal diseases. *Symp. Soc. gen. Microbiol.* **8**, 309.

Byrde, R. J. W., Crowdy, S. H. and Woodcock, D. (1953) Studies on systemic fungicides. III. The activity of certain chlorine-substituted β-naphthols and naphthyloxy-*n*-aliphatic carboxylic acids as systemic fungicides. *Ann. appl. Biol.* **40**, 152.

Byrde, R. J. W., Harper, C. W. and Holgate, M. E. (1969) Spraying trials against apple mildew and apple scab at Long Ashton, 1968. *Rep. agric. hort. Res. Stn Univ. Bristol, 1968,* p. 146.

Byrde, R. J. W., Harper, C. W. and Holgate, M. E. (1970) Spraying trials against apple mildew and apple scab at Long Ashton, 1969. *Rep. agric. hort. Res. Stn Univ. Bristol, 1969,* p. 189.

Byrde, R. J. W., Harper, C. W., Holgate, M. E. and Hunter, T. (1972) Residual effects of fungicidal spray programmes on apple trees (in preparation).

Byrde, R. J. W. and Melville, S. C. (1971) Chemical control of blossom wilt of apple. *Pl. Path.* **20**, 48.

Campos, A. and Borlaug, N. E. (1956) Wetting and penetrating agents combined with fungicides as protectants and eradicants for wheat stem rust. *Phytopathology* **46**, 8.

Canny, M. J. (1971) Translocation: mechanisms and kinetics. *A. Rev. Pl. Physiol.* **22**, 237.

Cargo, C. A. and Dewey, D. H. (1970) Thiabendazole and benomyl for the control of postharvest decay of apples. *HortScience,* **5**, 259.

Caroselli, N. E. and Howard, F. L. (1942) Response of diseased maple trees to chemotherapy and fertilization. *Phytopathology* **32**, 21.

Carter, G. A. and Wain, R. L. (1964) Investigations on fungicides. IX. The fungitoxicity, phytotoxicity and systemic fungicidal activity of some inorganic salts. *Ann. appl. Biol.* **53**, 291.

Catling, W. S. (1971a). Private communication.

Catling, W. S. (1971b) The control of powdery mildew, canker and other diseases of apple with benomyl. *Proc. 6th Br. Insectic. Fungic. Conf.* **1**, 103.

Cavell, B. D., Hemingway, R. J. and Teal, G. (1971) Some aspects of the metabolism and translocation of the pyrimidine fungicides. *Proc. 6th Br. Insectic. Fungic. Conf.* **2**, 431.

Chandler, W. A. (1968) Preharvest fungicides for peach brown rot control. *Pl. Dis. Reptr* **52**, 695.

Chapman, R. A. (1951) Relation of specific chemotherapeutants to the infection court. *Phytopathology* **41**, 6.

Charles, A. (1953) Uptake of dyes into cut leaves. *Nature, Lond.* **171**, 435.

Chin, W. T., Stone, G. M. and Smith, A. E. (1969) The fate of carboxin in soil, plants and animals. *Proc. 5th Br. Insectic. Fungic. Conf.* p. 322.

Chin, W. T., Stone, G. M. and Smith, A. E. (1970a) Metabolism of carboxin (Vitavax) by barley and wheat plants. *J. Agric. Food Chem.* **18**, 709.

Chin, W. T., Stone, G. M. and Smith, A. E. (1970b). Degradation of carboxin (Vitavax) in water and soil. *J. Agric. Food Chem.* **18**, 731.

Cimanowski, J., Masternak, A. and Millikan, D. F. (1970) Effectiveness of benomyl for controlling apple powdery mildew and cherry leaf spot in Poland. *Pl. Dis. Reptr* **54**, 81.

Cleland, R. (1963a) Independence of effects of auxin on cell wall methylation and elongation. *Pl. Physiol.* **38**, 12.

Cleland, R. (1963b) The occurrence of auxin-induced pectin methylation in plant tissue. *Pl. Physiol.* **38**, 738.

Clemons, G. P. and Sisler, H. D. (1969) Formation of a fungitoxic derivative from benlate. *Phytopathology* **59**, 705.

Clemons, G. P. and Sisler, H. D. (1971) Localization of the site of action of a fungitoxic benomyl derivative. *Pesticide Biochem. and Physiol.* **1**, 32.

Clifford, D. R., Fieldgate, D. M., Watkins, D. A. M. and Woodcock, D. (1971) The fungicidal activities of nitrophenols. *Proc. 2nd Int. Congress of Pestic. Chem. Tel Aviv* **5**, 281.

Clifford, D. R. and Hislop, E. C. (1971) The activities of some systemic fungicides against powdery mildew fungi. *Proc. 6th Br. Insectic. Fungic. Conf.* **2**, 438.

Clifford, D. R., Hislop, E. C. and Holgate, M. (1970) Some factors affecting the activities of dinitrophenol fungicides. *Pestic. Sci.* **1**, 18.

Cole, H., Boyle, J. S. and Smith, C. B. (1970) Effect of benomyl and certain cucumber viruses on growth, powdery mildew and element accumulation by cucumber plants in the greenhouse. *Pl. Dis. Reptr* **54**, 141.

Cole, H., Taylor, B. and Duich, J. (1968) Evidence of differing tolerances to fungicides among isolates of *Sclerotinia homoeocarpa. Phytopathology* **58**, 683.

Cole, M. (1958) Oxidation products of leuco-anthocyanins as inhibitors of fungal polygalacturonase in rotting apple fruit. *Nature, Lond.* **181**, 1596.

Cole, R. J., Gilchrist, A. J. and Soper, D. (1971) The control of diseases of apples and pears in the United Kingdom with thiophanate methyl. *Proc. 6th Br. Insectic. Fungic. Conf.* **1**, 118.

Colquhoun, T. T. (1940) Effect of manganese on powdery mildew of wheat. *J. Aust. Inst. agric. Sci.* **6**, 54.

Connor, S. R. and Heuberger, J. W. (1968) Apple scab. V. Effect of late season application of fungicides on prevention of perithecial development by *Venturia inaequalis. Pl. Dis. Reptr* **52**, 654.

Cooper, D., Banthorpe, D. V. and Wilkie, D. (1967) Modified ribosomes conferring resistance to cycloheximide in mutants of *Saccharomyces cerevisiae. J. Mol. Biol.* **26**, 347.

Corbett, D. C. M. and Hide, G. A. (1971) Chemical control of *Verticillium dahliae* and *Heterodera rostochiensis* on potatoes. *Proc. 6th Br. Insectic. Fungic. Conf.*, p. 258.

Corbin, J. B. (1971a) Benomyl beneficial for European canker control. *Orchard N.Z.* **44**, 55, 57. (*Rev. Pl. Path.* **50**, 3944, 1971).

Corbin, J. B. (1971b) Peach: brown rot *(Monilinia fructicola). Fungicide–Nematicide Tests* **26**, 47.

Corbin, J. B. and Kitchener, R. A. (1970) Peach: brown rot *(Monilinia fructicola);* scab *(Venturia carpophila);* stigmina *(Stigmina carpophila). Fungicide–Nematicide Tests* **25**, 43.

Corden, M. E. and Dimond, A. E. (1957) Relation of plant-growth regulating properties of naphthalene aliphatic acids to their activity in inducing disease resistance. *Phytopathology* **47**, 518.

Corden, M. E. and Dimond, A. E. (1959) The effect of growth-regulating substances on disease resistance and plant growth. *Phytopathology* **49**, 68.

Corden, M. E. and Edgington, L. V. (1960) A calcium requirement for growth-regulator-induced resistance to *Fusarium* wilt of tomato. *Phytopathology* **50**, 625.

Corke, A. T. K., Goldfinch, S. and Hunter, T. (1971) Bitter rot (caused by *Gloeosporium* spp.). *Rep. agric. hort. Res. Stn Bristol, 1970,* p. 115.

Corke, A. T. K., Hunter, T. and Goldfinch, S. (1972) The presence of benomyl (and/or breakdown products) in Cox flowers. *Rep. agric. hort. Res. Stn Bristol, 1971,* p. 147.

Covey, R. P., Jr. (1971) The effect of methionine on the development of apple powdery mildew. *Phytopathology* **61**, 346.

Covey, R. P. (1971) Orchard evaluation of two new fungicides for the control of apple powdery mildew. *Pl. Dis. Reptr* **55**, 514.

Coyier, D. L. (1970) Control of storage decay in 'd'Anjou' pear fruit by preharvest applications of benomyl. *Pl. Dis. Reptr* **54**, 647.

Coyier, D. L. (1971) Control of powdery mildew on apples with various fungicides as influenced by seasonal temperature. *Pl. Dis. Reptr* **55**, 263.

Crafts, A. S. (1933) The use of arsenical compounds in the control of deep-rooted perennial weeds. *Hilgardia* **7**, 361.

Crafts, A. S. (1961) *Translocation in plants.* Holt, Rinehart and Winston.

Crafts, A. S. and Crisp, C. E. (1971) *Phloem transport in plants.* Freeman and Co.

Crane, G. L. and Jackson, C. (1969) Fungicidal control of *Ascochyta chrysanthemi. Proc. Fla. St. Hort. Soc.* **82**, 379.

Cremlyn, R. J. W. (1961) Systemic fungicides. *J. Sci. Fd Agric.* **12**, 805.

Crosier, W. F., Nash, G. T. and Crosier, D. C. (1970) Differential reaction of *Pythium* sp. and five isolates of *Rhizoctonia solani* to fungicides on pea seeds. *Pl. Dis. Reptr* **54**, 349.

Cross, M. J. and Jacobs, L. (1968) Some observations on the biology of spores of *Verticillium malthousei. Mushr. Sci.* **7**, 239.

Crosse, R., McWilliams, R. and Rhodes, A. (1964) Some relations between chemical structure and antifungal effects of griseofulvin analogues. *J. gen. Microbiol.* **34**, 51.

Crosse, R., McWilliams, R., Rhodes, A. and Dunn, A. T. (1960) Antifungal action of streptomycin–copper chelate against *Phytophthora infestans* on tomato *(Lycopersicon esculentum). Ann. appl. Biol.* **48**, 270.

Crowdy, S. H. (1959) Uptake and translocation of organic chemicals by higher plants. *Plant Pathology, Problems and Progress, 1908–1958,* (ed. C. S. Holton), p. 231. Madison: University of Wisconsin Press.

Crowdy, S. H. (1970) Factors influencing the activity of systemic fungicides. *Proc. 7th Int. Congr. Pl. Prot* (Paris), 217.

Crowdy, S. H. (1972) Chapter 5 in *Systemic Fungicides* (ed. R. W. Marsh). London: Longman.

Crowdy, S. H. and Davies, M. E. (1952) Studies on systemic fungicides. II. Behaviour of groups of reported chemotherapeutants. *Phytopathology* **42**, 127.

Crowdy, S. H., Gardner, D., Grove, J. F. and Pramer, D. (1955) The translocation of antibiotics in higher plants. I. Isolation of griseofulvin and chloramphenicol from plant tissue. *J. exp. Bot.* **6**, 371.

Crowdy, S. H., Green, A. P., Grove, J. F., McCloskey, P. and Morrison, A. (1959a) The translocation of antibiotics in higher plants. 3. The estimation of griseofulvin relatives in plant tissues. *Biochem. J.* **72**, 230.

Crowdy, S. H., Grove, J. F., Hemming, H. G. and Robinson, K. C. (1956) The translocation of antibiotics in higher plants. II. The movement of griseofulvin in broad bean and tomato. *J. exp. Bot.* **7**, 42.

Crowdy, S. H., Grove, J. F. and McCloskey, P. (1959b) The translocation of antibiotics in higher plants. 4 Systemic fungicidal activity and chemical structure in griseofulvin relatives. *Biochem. J.* **72**, 241.

Crowdy, S. H. and Pramer, D. (1955) Movement of antibiotics in higher plants. *Chemy. Ind.* 160.

Crowdy, S. H. and Rudd Jones, D. (1956) Partition of sulphonamides in plant roots: a factor in their translocation. *Nature, Lond.* **178**, 1165.

Crowdy, S. H., Rudd Jones, D. and Witt, A. V. (1958) The translocation of sulphonamides in higher plants. II. *J. exp. Bot.* **9**, 206.

Crowdy, S. H. and Tanton, T. W. (1970) Water pathways in higher plants. I. *J. exp. Bot.* **21**, 102.

Crowdy, S. H. and Wain, R. L. (1950) Aryloxyaliphatic acids as systemic fungicides. *Nature, Lond.* **165**, 937.

Crowdy, S. H. and Wain, R. L. (1951) Studies on systemic fungicides. I. Fungicidal properties of the aryloxyalkylcarboxylic acids. *Ann. appl. Biol.* **38**, 318.

Cruger, G. (1969) Anwendung systemischer fungizide zur gurkenmehltaube kämpfung in giessverfahren. *Mitt. biol. Bund Anst. Ld-U. Forstw.* **132**, 216.

Cruickshank, I. A. M. (1963) Phytoalexins. *A. Rev. Phytopath.* **1**, 351.

Cruickshank, I. A. M. (1966) Defence mechanisms in plants. *Wld. Rev. Pest Control* **5**, 161.

Cruickshank, I. A. M. and Jacks, H. (1953) Seed disinfection. IX. Control of onion smut *Urocystis cepulae* Frost. *N.Z. J. Sci. and Tech. A* **35**, 390.

Cuillé, J. and Bur-Ravault, L. (1969) Traitements des oranges contre les *Penicillium* avec des formules a base de thiabendazole. *Fruits d'outre-mer* **24**, 421. (*Rev. Pl. Path.* **49**, 759, 1970.)

de Rooy – see under R.

de Waarde – see under W.

Daines, R. H. (1970) Effects of fungicide dip treatments and dip temperatures on postharvest decay of peaches. *Pl. Dis. Reptr* **54**, 764.

Daines, R. H. and Snee, R. D. (1969) Control of blue mould of apples in storage. *Phytopathology* **59**, 792.

D'Armini, M. (1953) Esperienze di endoprevenzione condotte con carbonato di litio contro l'oidio del tabacco. *Tabacco, Roma* **57**, 319.

Darpoux, H. (1970) Actions du triarimol sur les tavelures du poirier et du pommier *Societé de Phytiatrie-Phytopharmacie Séance de 20.1.70.*

Darpoux, H., Catelot, Mme and Gorse, Mlle (1958a) Étude preliminaire sur l'action systèmique et endothérapique du sel de manganese de la 2-pyridinethione-1-oxyde. *Phytiat.-Phytopharm.* **7**, 107.

Darpoux, H., Halmos, E. and Leblanc, R. (1958b) Étude de l'action systémique de diverses substances la plupart antibiotiques. *Ann. Épiphyties* **9**, 387.

Darpoux, H., Sharon, T., Ventura, E. and Bourdin, J. (1966) Essais d'efficacité du thiabendazole sur la tavelure du poirier *(Venturia pirina). Phytiat.-Phytopharm.* **15**, 121. (*Rev. appl. Mycol.* **46**, 530, 1967.)

Darpoux, H., Ventura, E. and Cassini, R. (1968) Activité fongicide du thiabendazole vis-à-vis des principaux champignons parasites des semences de céréales. *Phytiat.-Phytopharm.* **17**, 219.

Davies, J. M. L. (1970) Studies on variation of *Cercosporella herpotrichoides* Fron. Ph.D. Thesis, University of Wales.

Davies, M. E., Weeden, S. M. and Martin, T. J. (1971) Studies of powdery mildew control in spring barley using chloraniformethan. *Proc. 6th. Br. Insect. Fungic. Conf.* p. 42.

Davis, D. (1952) Chemotherapeutic activity may be independent of fungitoxicity. *Phytopathology* **42**, 6.

Davis, D., Becker, H. J. and Rogers, E. F. (1959) The chemotherapy of wheat and bean rust diseases with sydnones. *Phytopathology* **49**, 821.

Davis, D., Chaiet, L., Rothrock, J. W., Deak, J., Halmos, S. and Barger, J. D. (1960) Chemotherapy of cereal rusts with a new antibiotic. *Phytopathology* **50**, 841.

Davis, D. and Dimond, A. E. (1952) Altering resistance to disease with synthetic organic chemicals. *Phytopathology* **42**, 563.

Davis, D. and Dimond, A. E. (1953) Inducing disease resistance with plant growth-regulators. *Phytopathology* **43**, 137.

Davis, D. and Dimond, A. E. (1956) Site of disease resistance induced by plant growth regulators in tomato. *Phytopathology* **46**, 551.

Davis, D., Lo, C. P. and Dimond, A. E. (1954) Chemotherapeutic activity of unsubstituted heterocyclic compounds. *Phytopathology* **44**, 680.

Davis, J. R. and Callihan, R. H. (1970) Evaluation of several fungicides in Idaho for control of *Rhizoctonia* on potato. *Phytopathology* **60**, 1533.

Davis, J. R., Groskopp, M. D. and Callihan, R. H. (1971a) Seed and soil treatments for control of *Rhizoctonia* on stems and stolons of potato. *Pl. Dis. Reptr* **55**, 550.

Dawes, E. A. and Large, P. J. (1968) Class II reactions: synthesis of small molecules, *Biochemistry of bacterial growth* (eds. J. Mandelstam and K. McQuillen), p. 203. New York: John Wiley and Sons.

Day, L. H. (1928). Pear blight control in California. *California Agric. Ext. Service Circ.* **20**, 1.

Deese, D. and Stahmann, M. E. (1960) Role of pectic enzymes in susceptibility and resistance to *Fusarium* and *Verticillium* wilts of plants. *Phytopathology* **50**, 633.

Defosse, L. (1970) Essais de lutte chimique contre le piétin-verse, *Cercosporella herpotrichoides* Fron, avec un fongicide systemique. *Parasitica* **26**, 4.

Dekhuijzen, H. M. (1961) A paper chromatographic method for the demonstration of fungitoxic transformation products of sodium dimethyldithiocarbamate in plants. *Meded. Landb-Hoogesch. OpzoekStns Gent* **26**, 1542.

Dekhuijzen, H. M. (1964) The systemic action of dimethyldithiocarbamates on cucumber scab caused by *Cladosporium cucumerinum* and the conversion of these compounds by plants. *Neth. J. Pl. Path.* **70**, Suppl. 1.

Dekhuijzen, H. M. and Dekker, J. (1971) Mechanism of resistance of *Cladosporium cucumerinum* against 6-azauracil and 6-azauridine monophosphate. *Pesticide Biochem. and Physiol.* **1**, 11.

Dekker, J. (1955) Internal seed disinfection by an antibiotic from *Streptomyces rimosus. Nature, Lond.* **175**, 689.

Dekker, J. (1957) Internal seed disinfection of peas infected by *Ascochyta pisi* by means of the antibiotics rimocidin and pimaricin and some aspects of the parasitism of this fungus. *Tijdschr. PlZiekt.* **63**, 65.

Dekker, J. (1961a) Systemic activity of procaine hydrochloride on powdery mildew. *Tiljdschr. PlZiekt.* **67**, 25.

Dekker, J. (1961b) Mode of action of procaine hydrochloride and some derivatives on powdery mildew. *Meded. LandbHoogesch. OpzoekStns. Gent* **26**, 1378.

Dekker, J. (1962) Systemic control of powdery mildew by 6-azauracil and some other purine and pyrimidine derivatives. *Meded. LandbHoogesch. Opzoek-Stns. Gent* **27**, 1214.

Dekker, J. (1963a) Antibiotics in the control of plant diseases. *A. Rev. Microbiol.* **17**, 243.

Dekker, J. (1963b) Effect of kinetin on powdery mildew. *Nature, Lond.* **197**, 1027.

Dekker, J. (1967) Conversion of 6-azauracil in sensitive and resistant strains of

Cladosporium cucumerinum. Mechanism of action of fungicides and antibiotics (ed. W. Girbardt), p. 333. *Akademie-Verlag* Berlin.

Dekker, J. (1968a) Acquired resistance to fungicides. *World Rev. Pest Control* **8**, 79.

Dekker, J. (1968b) The development of resistance in *Cladosporium cucumerinum* against 6-azauracil, a chemotherapeutant of cucumber scab, and its relation to biosynthesis of RNA-precursors. *Neth. J. Pl. Path.* **74**, Suppl. 1, 127.

Dekker, J. (1969a) L-Methionine-induced inhibition of powdery mildew and its reversal by folic acid. *Neth. J. Pl. Path.* **75**, 182.

Dekker, J. (1969b) Antibiotics. *Fungicides. Vol. II* (ed. D. C. Torgeson), p. 579. New York: Academic Press.

Dekker, J. (1971a) Selective action of systemic fungicides and development of resistance. *Proceedings 2nd International Congress of Pesticide Chemistry IUPAC, Tel-Aviv, 1971*, p. 305.

Dekker, J. (1971b) Selective action of fungicides and development of resistance in fungi to fungicides. *Proc. 6th Br. Insectic. Fungic. Conf.* **3**, 715.

Dekker, J., van Andel, O. M. and Kaars Sijpesteijn, A. (1958) Internal seed disinfection with pyridine-2-thiol-N-oxide and a derivative. *Nature, Lond.* **181**, 1017.

Dekker, J. and van der Hoek-Scheuer, R. G. (1964) A microscopic study of the wheat-powdery mildew relationship after application of the systemic compounds procaine, griseofulvin and 6-azauracil. *Neth. J. Pl. Path.* **70**, 142.

Dekker, J. and Oort, A. J. P. (1964) Mode of action of 6-azauracil against powdery mildew. *Phytopathology* **54**, 815.

Dekker, J. and Roosje, G. S. (1968) Effect of 6-azauracil against apple powdery mildew and apple scab. *Neth. J. Pl. Path.* **74**, 219.

Delp, C. J. and Klöpping, H. L. (1968) Performance attributes of a new fungicide and mite ovicide candidate. *Pl. Dis. Reptr* **52**, 95.

Derbyshire, D. M. and Crisp, A. F. (1971) Vegetable storage diseases in East Anglia. *Proc. 6th Br. Insectic. Fungic. Conf.*, p. 167.

Deverall, B. J. (1964) Substances produced by pathogenic organisms that induce symptoms of disease in higher plants. *14th Symp. Soc. gen. Microbiol.*, p. 165.

Deverall, B. J. and Wood, R. K. S. (1961) Chocolate spot of beans *(Vicia fabae* L.*)* – interaction between phenolase of host and pectic enzymes of the pathogen. *Ann. appl. Biol.* **49**, 473.

Dickens, J. W. and Sharp, M. K. (1970) Mercury-tolerant *Pyrenophora avenae* in seed oat samples from England and Wales. *Pl. Pathol.* **19**, 93.

Dickins, L. E. and Oshima, N. (1971) Chemotherapeutants evaluated for control of dwarf bunt of wheat. *Pl. Dis. Reptr* **55**(7), 613.

Dickson, J. G. (1959) Chemical control of cereal rusts. *Bot. Rev.* **25**, 486.

Dietrich, J. V. and Brechbuhler, C. (1970) Control of *Botrytis cinerea* Pers. in vineyards. *Summaries of Papers, VIIth Int. Cong. Plant Prot., Paris*, p. 253.

Dimond, A. E. (1947) Oxyquinoline benzoate aids suppression of symptoms of Dutch elm disease. *Phytopathology* **37**, 848.

Dimond, A. E. (1953) Progress in plant chemotherapy. *Agric. Chem.* **8**, 34.

Dimond, A. E. (1959) Plant chemotherapy. *Plant Pathology – Problems and Progress, 1908–1958*, (ed. C. S. Holton), p. 221. Wisconsin: University Press.

Dimond, A. E. (1962) Objectives in plant chemotherapy. *Phytopathology* **52**, 1115.

Dimond, A. E. (1963a) The modes of action of chemotherapeutic agents in plants. *Conn. Agric. Exp. Sta. Bull. no. 663*, p. 62.

Dimond, A. E. (1963b) The selective control of plant pathogens. *Wld Rev. Pest Control* **2**, 7.

Dimond, A. E. (1965) Natural models for plant chemotherapy. *Advances in Pest Control Research* (ed. R. L. Metcalf) **6**, 128. New York: Wiley Interscience Publishers Inc.

Dimond, A. E. (1970) Biophysics and biochemistry of the vascular wilt syndrome. *Ann. Rev. Phytopathol.* **8**, 301.

Dimond, A. E. and Chapman, R. A. (1951) Chemotherapeutic properties of two compounds against *Fusarium* wilts. *Phytopathology* **41**, 11.

Dimond, A. E. and Davis, D. (1952) 2-carboxymethylmercaptobenzothiazole salts as chemotherapeutants for plant diseases. *Phytopathology* **42**, 7.

Dimond, A. E. and Davis, D. (1953) The chemotherapeutic activity of benzothiazole and related compounds for *Fusarium* wilt of tomato. *Phytopathology* **43**, 43.

Dimond, A. E., Davis, D., Chapman, R. A. and Stoddard, E. M. (1952) Plant chemotherapy as evaluated by the *Fusarium* wilt assay on tomatoes. *Conn. Agric. Exp. Sta. Bull.* no. 557, 1.

Dimond, A. E. and Horsfall, J. G. (1959) Plant chemotherapy. *A. Rev. Pl. Physiol.* **10**, 257.

Dimond, A. E., Plumb, G. H., Stoddard, E. M. and Horsfall, J. G. (1949) An evaluation of chemotherapy and vector control by insecticides for combating Dutch elm disease. *Conn. Agric. Exp. Sta. Bull.* no. 531, 1.

Dimond, A. E. and Waggoner, P. E. (1953) On the nature and role of vivotoxins in plant disease. *Phytopathology* **43**, 229.

Doma, S., Clifford, D. R. and Byrde, R. J. W. (1971) Experiments with some of the newer systemic fungicides on apple and marrow. *Pestic. Sci.* **2**, 197.

Doodson, J. K. and Saunders, P. J. W. (1969) Observations on the effects of some systemic fungicides applied to cereals in trials at the NIAB. *Proc. 5th Br. Insect. Fungic. Conf.*, p. 1.

Douchet, J. P., Lhoste, J. and Quere, G. (1971) Thiophanate et methylthiophanate, nouveaux fongicides systèmiques actifs sur la tavelure et l'oidium des arbres fruitiers à pepins. *Meded. Fac. Landbouw. RijksUniversiteit Gent* **36**, 135.

Drakes, D. and Fletcher, J. T. (1971) Experiments on the control of wet bubble *(Mycogone perniciosa)* at Fairfield E.H.S. *M.G.A. Bull.* **259**, 300.

Eaton, F. M. (1930) The effect of boron on powdery mildew and spot blotch of barley. *Phytopathology* **20**, 967.

Ebben, M. H. (1970a) Tomato wilt caused by *Verticillium* spp. *A. Rep. Glasshouse Crops Res. Inst.* (1969), p. 119.

Ebben, M. H. (1970b) Cucumber diseases. *A. Rep. Glasshouse Crops Res. Inst.* (1969), p. 120.

Ebben, M. H. (1971) Cucumber stem rot caused by *Mycosphaerella melonis. A. Rep. Glasshouse Crops Res. Inst.* (1970), p. 137.

Ebben, M. H. and Last, F. T. (1969) Benomyl applied to soil for the control of some pathogens of tomato and cucumber. *Proc. 5th Br. Insectic. Fungic. Conf.*, p. 315.

Ebenebe, C., Fehrmann, H. and Grossman, F. (1971) Effect of a new systemic fungicide, piperazin-1, 4-diyl-bis-(1-(2,2,2,-trichloroethyl) formamide) against wheat leaf rust. *Pl Dis. Reptr* **55**, 691.

Eble, T. E., Bergy, M. E., Large, C. M., Herr, R. R. and Jackson, W. G. (1959) Isolation, purification and properties of streptovicins A and B. *Antibiot. Ann.* p. 555.

Eckert, J. W. (1967) Application and use of postharvest fungicides. *Fungicides* (ed. D. C. Torgeson) **1**, 287. Academic Press, Inc., New York.

Eckert, J. W. (1969) Chemical treatments for control of post-harvest diseases. *Wld Rev. Pest Control* **8**, 138.

Edgington, L. V. (1963) A chemical that retards development of Dutch elm disease. *Phytopathology* **53**, 349.

Edgington, L. V. and Barron, G. L. (1967) Fungitoxic spectrum of oxathiin compounds. *Phytopathology* **57**, 1256.

Edgington, L. V. and Busch, L. V. (1967) Control of *Rhizoctonia* stem canker in potato. *Can. Pl. Dis. Surv.* **47**, 28.

Edgington, L. V., Corden, M. E. and Dimond, A. E. (1961) The role of pectic substances in chemically induced resistance to *Fusarium* wilt of tomato. *Phytopathology* **51**, 179.

Edgington, L. V. and Dimond, A. E. (1959) Correlation between the wilt resistance of plants and the nature of their pectic substances. *Phytopathology* **49**, 538.

Edgington, L. V. and Dimond, A. E. (1960) Effect of charge of ionized chemotherapeutants on their translocation through xylem. *Phytopathology* **50**, 239.

Edgington, L. V. and Dimond, A. E. (1964) The effect of adsorption of organic cations to plant tissue on their use as systemic fungicides. *Phytopathology* **54**, 1193.

Edgington, L. V., Khew, K. L. and Barron, G. L. (1971) Fungitoxic spectrum of benzimidazole compounds. *Phytopathology* **61**, 32.

Edgington, L. V. and Walker, J. C. (1958) Influence of calcium and boron nutrition on development of *Fusarium* wilt of tomato. *Phytopathology* **48**, 324.

Edgington, L. V., Walton, G. S. and Miller, P. M. (1966) Fungicide selective for Basidiomycetes. *Science* **153**, 307.

Edney, K. L. (1970) Some experiments with thiabendazole and benomyl as post-harvest treatments for the control of storage rots of apples. *Pl. Path.* **19**, 189.

Elgersma, D. M. (1967) Factors determining resistance of elms to *Ceratocystis ulmi*. *Phytopathology* **57**, 641.

Elgersma, D. M. (1970) Length and diameter of xylem vessels as factors in resistance of elms to *Ceratocystis ulmi*. *Neth. J. Pl. Pathol.* **76**, 179.

Elia, M. (1969) Brief preliminary notes on endotherapeutic trials against citrus 'mal secco'. *Inftore fitopatol.* **19**, 403. (*Rev. Pl. Path.* **49**, 2482, 1970.)

Elias, R. S., Shephard, M. C., Snell, B. K. and Stubbs, J. (1968) 5-*n*-Butyl-2-dimethylamino-4-hydroxy-6-methylpyrimidine: a systemic fungicide. *Nature, Lond.* **219**, 1160.

El-Nakeeb, M. A. and Lampen, J. O. (1964a) Complexing griseofulvin by nucleic acids of fungi and its relation to griseofulvin sensitivity. *Biochem J.* **92**, 59P.

El-Nakeeb, M. A. and Lampen, J. O. (1964b) Binding of tritiated griseofulvin by ribosomes of *Microsporum gypseum*. *Bacteriol. Proc.* **15**, 7.

El-Nakeeb, M. A., McLellan, W. L., Jr. and Lampen, J. O. (1965a) Antibiotic action of griseofulvin on dermatophytes. *J. Bacteriol.* **89**, 557.

El-Nakeeb, M. A., Moustafa, A. and Lampen, J. O. (1965b) Uptake of griseofulvin by the sensitive dermatophyte, *Microsporum gypseum*. *J. Bacteriol.* **89**, 564.

El-Nakeeb, M. A., Moustafa, A. and Lampen, J. O. (1965c) Distribution of griseofulvin taken up by *Microsporum gypseum*: complexes of the antibiotic with cell constituents. *J. Bacteriol.* **89**, 1075.

El-Zayat, M. M., Lukens, R. J., Dimond, A. E. and Horsfall, J. G. (1968) Systemic action of nitrophenols against powdery mildew. *Phytopathology* **58**, 434.

Endo, A., Kakiki, K. and Misato, T. (1970) Mechanism of action of the antifungal agent polyoxin D. *J. Bacteriol.* **104**, 189.

Endo, A. and Misato, T. (1969) Polyoxin D, a competitive inhibitor of UDP-N-acetylglucosamine: chitin N-acetylglucosaminyltransferase in *Neurospora crassa*. *Biochem. Biophys. Res. Comm.* **37**, 718.

Engelhard, A. W. (1969) Preventive and residual fungicidal activity of three benzimidazole compounds and zinc ion + maneb against *Diplocarpon rosae* on two rose cultivars. *Pl. Dis. Reptr* **53**, 537.

Ennis, H. L. and Lubin, M. (1964) Cycloheximide: aspects of inhibition of protein synthesis in mammalian cells. *Science* **146**, 1474.

Epstein, A. H. (1969) Enhancing efficacy of benomyl to control *Venturia inaequalis*. *Phytopathology* **60**, 1291 (abs.).

Epstein, E. (1955) Passive permeation and active transport of ions into plant roots. *Pl. Physiol. Lancaster* **30**, 529.

Erwin, D. C. (1970) Progress in the development of systemic fungitoxic chemicals for control of plant diseases. *Pl. Prot. Bull. F.A.O.* **18**, 73.

Erwin, D. C., Hee, H. and Sims, J. J. (1968a) The systemic effect of 2-benzimidazolecarbamic acid, methyl ester, on *Verticillium* wilt of cotton. *Phytopathology* **58**, 528.

Erwin, D. C., Sims, J. J., Borum, D. E. and Childers, J. R. (1971) Detection of the systemic fungicide, thiabendazole, in cotton plants and soil by chemical analysis and bioassay. *Phytopathology* **61**, 964.

Erwin, D. C., Sims, J. J. and Partridge, J. (1968b) Evidence for the systemic fungitoxic activity of 2-(4′-thiazoly)-benzimidazole in the control of *Verticillium* wilt of cotton. *Phytopathology* **58**, 860.

Esau, K. (1965) *Plant anatomy*. 2nd ed. Wiley: New York.

Esau, K., Engelman, E. M. and Bisalputra, T. (1963) What are translocation strands? *Planta* **50**, 617.

Eschrich, W. (1967) Bidirektionelle Translocation in Siebröhren. *Planta* **73**, 37.

Eschrich, W. (1970) Biochemistry and fine structure of phloem in relation to transport. *A. Rev. Pl. Physiol.* **21**, 193.

Esuruoso, O. F. and Wood, R. K. S. (1971) The resistance of spores of resistant strains of *Botrytis cinerea* to quintozene, tecnazene and dicloran. *Ann. appl. Biol.* **68**, 271.

Evans, E. (1968) *Plant diseases and their chemical control*. Blackwells Scientific Publications, Oxford.

Evans, E. (1971) Systemic fungicides in practice. *Pest. Sci.* **2**, 192.

Evans, E., Couzens, B. J. and Griffiths, W. (1965) Timing experiments on the control of potato blight with copper fungicides in the United Kingdom. *Wld Rev. Pest Control* **4**, 84.

Evans, E. and Hawkins, J. H. (1971) Timing of fungicidal sprays for control of mildew on spring barley. *Proc. 6th Br. Insectic. Fungic. Conf.*, p. 33.

Evans, E., Marshall, J., Couzens, B. J. and Runham, R. L. (1970) The curative activity of non-ionic surface active agents against some powdery mildew diseases. *Ann. appl. Biol.* **65**, 473.

Evans, E., Marshall, J. and Hammond, C. (1971) An evaluation of benomyl as a scab/mildew fungicide in the United Kingdom. *Pl. Path.* **20**, 1.

Evans, E., Richard, M. and Whitehead, R. (1969) Effect of benomyl on some diseases of spring barley. *Proc. 5th Br. Insectic. Fungic. Conf.*, p. 610.

Evans, E. and Saggers, D. T. (1962) A new systemic rust therapeutant. *Nature, Lond.* **195**, 619.

Fargher, R. G., Galloway, L. D. and Probert, M. E. (1930) The inhibitory action of certain substances on the growth of mould fungi. *Mem. Shirley Inst.* **9**, 37.

Farkas, G. L. and Király, Z. (1962) Role of phenolic compounds in the physiology of plant diseases and disease resistance. *Phytopath. Z.* **44**, 105.

Farrant, D. M., Gramlich, J. V. and Schwer, J. F. (1970) EL-273 (Triarimol). A summary of field results in Europe, the Middle East and South Africa for the control of important diseases of fruits and cereals. *Summaries of Papers, VIIth Int. Cong. Plant Prot., Paris,* p. 208.

Fawcett, C. H. and Spencer, D. M. (1969) Natural antifungal compounds. *Fungicides Vol. 2,* (ed. D. C. Torgeson), p. 637. New York and London: Academic Press.

Fawcett, C. H. and Spencer, D. M. (1970) Plant chemotherapy with natural products. *A. Rev. Phytopath.* **8**, 403.

Fawcett, C. H., Spencer, D. M. and Wain, R. L. (1955) Investigations on fungicides. I. Fungicidal and systemic fungicidal activity in certain aryloxyalkanecarboxylic acids. *Ann. appl. Biol.* **43**, 553.

Fawcett, C. H., Spencer, D. M. and Wain, R. L. (1957) Investigations on fungicides. II. Aryloxy- and arylthio-alkanecarboxylic acids and their activity as fungicides and systemic fungicides. *Ann. appl. Biol.* **45**, 158.

Fawcett, C. H., Spencer, D. M. and Wain, R. L. (1958) Investigations on fungicides. IV. (Aryloxythio)trichloromethanes. *Ann. appl. Biol.* **46**, 651.

Fawcett, C. H., Spencer, D. M. and Wain, R. L. (1969) The isolation and properties of a fungicidal compound present in the seedlings of *Vicia faba. Neth. J. Pl. Path.* **75**, 72.

Fawcett, C. H., Spencer, D. M., Wain, R. L., Fallis, A. G., Jones, Sir E. R. H., LeQuan, M., Page, C. B., Thaller, V., Shubrook, D. C. and Whitham, P. M. (1968) Natural acetylenes. Part XXVIII. An antifungal acetylenic furanoid keto-ester (wyerone) from shoots of broad bean *(Vicia faba* L.) Fam. *Papilionaceae J. chem. Soc.* 2455.

Fawcett, C. H., Spencer, D. M., Wain, R. L., Jones, Sir E. R. H., LeQuan, M., Page, C. B. and Thaller, V. (1965) An antifungal acetylenic keto-ester from a plant of the *Papilionaceae* family. *Chem. Commun.* **1**, 422.

Fehrmann, H. F. (1970) Bekämpfung der Halmbruchkrankheit des Weizers mit Benomyl. *Nachrichtenbl. Deutsch. Pflanzenschutz* (Stuttgart) **22**, 136.

Feldman, A. W. and Caroselli, N. E. (1951) Soil and tree acidity in relation to susceptibility to Dutch elm disease. *Phytopathology* **41**, 12.

Feldman, A. W., Caroselli, N. E. and Howard, F. L. (1950) Physiology of toxin production by *Ceratostomella ulmi. Phytopathology* **40**, 341.

Fensom, D. S. (1957) The bioelectrical potentials of plants and their functional significance. I. *Can. J. Bot.* **35**, 573.

Fensom, D. S. (1959) The bioelectrical potentials of plants and their functional significance. III. *Can. J. Bot.* **37**, 1003.

Fensom, D. S., Clattenburg, R., Chung, T., Lee, D. R. and Arnold, D. C. (1968) Moving particles in intact sieve tubes of *Heracleum mantegazzianum. Nature, Lond.* **219**, 531.

Findlayson, D. G. and Campbell, C. J. (1971) Fungicides for preventing clubroot of cauliflower in loam and peat soils. *Can. Pl. Dis. Surv.* **51**, 122.

Finney, J. R. and Hall, D. W. (1971) The effect of an autumn attack of mildew *(Erysiphe graminis* DC.) on the growth and development of winter barley. (In press.)

Fletcher, J. T. (1971) Systemic fungicides – Progress and promise. *Proc. 13th East. Reg. Grow. Conf. Oaklands 1971,* p. 40.

Ford, J. H., Klomparens, W. and Hamner, C. L. (1958) Cycloheximide (Acti-dione) and its agricultural uses. *Pl. Dis. Reptr* **42**, 680.

Formigoni, A., Castagna, G. and Ciocca, P. (1971) Two years of trials in Italy on 1,2-bis-(3-ethoxy-carbonyl-2-thioureido) benzene, a new systemic fungicide. *Meded. Fac. Landbouw. RijksUniversiteit Gent* **36**, 96.

Forsyth, F. R. (1957) Effect of ions of certain metals on the development of stem rust in the wheat plant. *Nature, Lond.* **179**, 217.

Forsyth, F. R. and Peturson, B. (1959a) Chemical control of cereal rusts. IV. The influence of nickel compounds on wheat, oat and sunflower rusts in the greenhouse. *Phytopathology* **49**, 1.

Forsyth, F. R. and Peturson, B. (1959b) Control of leaf rust of wheat with inorganic nickel. *Pl. Dis. Reptr* **43**, 5.

Forsyth, F. R. and Peturson, B. (1960) Control of leaf and stem rust of wheat by zineb and inorganic nickel salts. *Pl. Dis. Reptr* **44**, 208.

Foy, C. L. (1961) Absorption, distribution and metabolism of 2,2-dichloropropionic acid II. *Pl. Physiol. Lancaster* **36**, 698.

Frank, A. (1971) Studies on the metabolism of 2-(2-furyl)benzimidazole in certain mammals. *Acta Pharmacol. Toxicol.* **29**, suppl. 2, 1.

Franke, W. (1967) Mechanisms of foliar penetration of solutions. *A. Rev. Pl. Physiol.* **18**, 281.

Franklin, T. J. and Snow, G. A. (1971) *Biochemistry of antimicrobial action.* Chapman and Hall Ltd., London.

Frazao, A. (1970) Chemical control of *Botrytis cinerea* in vineyards. *Summaries of Papers, VIIth Int. Cong. Plant Prot., Paris*, p. 258.

Freeman, J. A. and Pepin, H. S. (1967) A systemic fungicide (Fungicide 1991) for the control of gray mould and powdery mildew in strawberries and raspberries. *Can. Pl. Dis. Surv.* **47**, 104.

Freeman, J. A. and Pepin, H. S. (1968) A comparison of two systemic fungicides with non-systemic for control of fruit rot and powdery mildew of strawberries. *Can. Pl. Dis. Surv.* **48**, 120.

Frick, E. L. (1971) Fungicides. *Rep. E. Malling Res. Stn, 1970*, p. 132.

Fridrich, J. H. and Gilpatrick, J. D. (1971a) Apple: scab *(Venturia inaequalis). Fungicide–Nematicide Tests* **26**, 10.

Fridrich, J. H. and Gilpatrick, J. D. (1971b) Apple: rust *(Gymnosporangium juniperi-virginianae)*; mites *(Panonychus ulmi)*; fruit finish. *Fungicide–Nematicide Tests* **26**, 11.

Fripp, Y. J. and Dettman, E. B. (1969) Thiabendazole as a post-harvest treatment against *Sclerotinia fructicola* in dessert peaches. *Aust. J. exp. Agric. Anim. Husb.* **9**, 9. *(Rev. appl. Mycol.* **48**, 2479, 1969.)

Fron, G. (1936) La maladie de l'orme. *C. r. hebd. Séanc. Acad. Agric. Fr.* **22**, 1081.

Frossard, P. (1968) Essais de désinfection des pedoncules d'Ananas contre le *Thielaviopsis paradoxa. Fruits d'outre-mer* **23**, 207.

Frossard, P. (1969) Action du thiabendazole et du benlate sur l'anthracnose des Bananes et son champignon pathogène: *Colletotrichum musae. Fruits d'outre-mer* **24**, 365. (*Rev. Pl. Path.* **49**, 523, 1970.)

Frossard, P. (1970a) Précisions sur les propriétés du benomyl (benlate) vis-à-vis de l'anthracnose de blessures des Bananes. *Fruits d'outre-mer* **25**, 265. (*Rev. Pl. Path.* **50**, 2391, 1971.)

Frossard, P. (1970b) Désinfection des Ananas contre *Thielaviopsis paradoxa. Fruits d'outre-mer* **25**, 785. (*Rev. Pl. Path.* **50**, 3069, 1971.)

Frost, A. J. P. B. and Pattisson, N. (1971) Some results obtained in the United Kingdom using dodemorph for the control of powdery mildew in roses and other ornamentals. *Proc. 6th Brit. Insectic. Fungic. Conf.*, p. 349.

Froyd, J. D. and Johnson, H. G. (1966) Sugar beets. *Cercospora* leaf spot. *Cercospora beticola. Fungicide–nematicide tests, results of 1966*, **22**, 98. American Phytopathological Society.

Fuchs, A. (1964) Effect of treatment with anthraquinone derivatives on *Fusarium* infected tomato plants. *Host–parasite relations in plant pathology* (eds. Z. Király and G. Ubrizsy), p. 175. Budapest: Research Institute for Plant Protection.

Fuchs, A., Doma, S. and Vörös, J. (1971) Laboratory and greenhouse evaluation of a new systemic fungicide, N,N'-bis-(1-formamido-2,2,2-trichloroethyl)piperazine (CELA W 524). *Neth. J. Pl. Path.* **77**, 42.

Fuchs, A., Homans, A. L. and de Vries, F. W. (1970) Systemic activity of benomyl against *Fusarium* wilt of pea and tomato plants. *Phytopath. Z.* **69**, 330.

Fuchs, A., Viets-Verwey, M. and de Vries, F. W. (1972) Metabolic conversion in plants of the systemic fungicide triforine (N,N'-bis-(1-formamido-2,2,2-trichloroethyl)piperazine (CELA W 524). *Phytopathol. Z.* **75**, 111.

Fuchs, W. H. and Bauermeister, R. (1958) Uber die Wirkung von Thiosemicarbazid auf die Entwicklung von *Puccinia graminis tritici* auf Weizen. *Naturwissenschaften* **45**, 343.

Fukunaga, K., Misato, T. and Asakawa, M. (1955) Blasticidin, a new anti-phytopathogenic fungal susbstance. *Bull. Agr. Chem. Soc. Japan* **19**, 181.

Futrell, M. C. and Berry, J. W. (1964) The action of propionate on wheat stem rust and sorghum head smut. *Phytopathology* **54**, 893.

Gaff, D. F., Chambers, T. C. and Markus, K. (1964) Studies of extrafascicular movement of water in the leaf. *Aust. J. Biol. Sci.* **17**, 581.

Gage, R. S. and Aronoff, S. (1960) Translocation. III. *Pl. Physiol. Lancaster* **35**, 53.

Gandy, D. G. (1971a) Experiments on the use of benomyl (Benlate) against *Verticillium. M.G.A. Bull.* **257**, 184.

Gandy, D. G. (1971b) Unpublished results.

Gandy, D. G. (1972) Observation on the development of *Verticillium malthousei* in mushroom crops and the role of cultural practices in its control. *Mushr. Sci.* **8**. (In press.)

Garibaldi, A. (1968) Sull 'efficacia del tiabendazolo contro la fialoforosi del garofano: Nota preliminaire *Inftore Fitopatol.* **18**, 52.

Gassner, G. and Hassebrauk, K. (1936) Untersuchungen zur Frage der Getreiderostbekämpfung mit chemischen Mitteln. *Phytopath. Z.* **9**, 427.

Gaunt, R. E. and Manners, J. G. (1971) Host–parasite relations in loose smut of wheat. II. *Ann. Bot.* **35**, 1141.

Geary, T. F. and Kuntz, J. E. (1962) Effect of growth regulators on oak wilt development. *Phytopathology* **52**, 733.

Georgopoulos, S. G. (1963) Tolerance to chlorinated nitrobenzenes in *Hypomyces solani* f. *cucurbitae* and its mode of inheritance. *Phytopathology* **53**, 1086.

Georgopoulos, S. G. and Sisler, H. D. (1970) Gene mutation eliminating antimycin A-tolerant electron transport in *Ustilago maydis. J. Bacteriol.* **103**, 745.

Georgopoulos, S. G. and Zaracovitis, C. (1967) Tolerance of fungi to organic fungicides. *Ann. Rev. Phytopathol.* **5**, 109.

Gerecke, M., Kyburz, E., Planta, C. von and Brossi, A. (1962) Griseofulvin IV. Synthesis of griseofulvin analogues. *Helv. Chim. Acta.* **45**, 2241.

Germar, B. (1934) Über einige Wirkungen der Kieselsäure in Getreidepflanzen insbesondere auf deren Resistenz gegenüber Mehltau. *Z. Pflernähr. Düngung BodenKde A* **35**, 102.

Geuther, T. (1895) Über die Einwirkung von Formaldehydelösungen auf Getreidebrand. *Ber. dt. pharmacol. Ges.* **5**, 325.

Gigante, R. (1935) Ricerche sopra l'influenza del boro sulla resistenza della piante agli attachi parassitari. *Boll. Staz. Patol. veg. Roma* **15**, 471.

Gilchrist, A. J. and Cole, R. J. (1971) Control of diseases of soft fruit and other crops in the United Kingdom with thiophanate methyl. *Proc. 6th Br. Insectic. Fungic. Conf.* **2**, 332.

Gilmour, J. (1971) Fungicidal control of mildew on spring barley in South-East Scotland. *Proc. 6th Br. Insect. Fungic. Conf.*, p. 63.

Gilpatrick, J. D. (1969) Systemic activity of benzimidazoles as soil drenches against powdery mildew of apples and cherries. *Pl. Dis. Reptr* **53**, 721.

Gilpatrick, J. D. (1970) Apple: scab *(Venturia inaequalis). Fungicide–Nematicide Tests* **25**, 12.

Gilpatrick, J. D. and Szkolnik, M. (1970) Unusual after-infection control of apple scab with α(2,4-dichlorophenyl)α-phenyl-5-pyrimidine methanol against *Venturia inaequalis. Phytopathology* **60**, 1287.

Goodman, R. N. (1959) The influence of antibiotics on plants and plant disease control. *Antibiotics – Their Chemistry and Non-Medical Uses* (ed. H. S. Goldberg), p. 322. Princeton: van Nostrand.

Goodman, R. N. (1962a) The impact of antibiotics upon plant disease control. *Advances in Pest Control Research Vol. V* (ed. R. L. Metcalf), p. 1. New York: Wiley Interscience Publishers Inc.

Goodman, R. N. (1962b) Systemic effects of antibiotics. *Antibiotics in Agriculture* (ed. M. Woodbine), p. 165. London: Butterworths.

Goodman, R. N. (1967) Uses of antibiotics in plant pathology and production. *Antimicrob. Ag. Chemother. 1966*, 747.

Goodman, R. N. and Dowler, W. M. (1958) The absorption of streptomycin as influenced by growth regulators and humectants. *Pl. Dis. Reptr* **42**, 122.

Goodman, R. N., Király, Z. and Zaitlin, M. (1967) *The Biochemistry and Physiology of Infectious Plant Disease.* Princeton: van Nostrand.

Gottlieb, D. (1957) The effect of metabolites on antimicrobial agents. *Phytopathology* **47**, 59.

Gottlieb, D. and Kumar, K. (1970) The effect of thiabendazole on spore germination. *Phytopathology* **60**, 1451.

Gould, C. J. and Miller, V. L. (1970) Effectiveness of benzimidazole fungicides in controlling *Fusarium* basal rot of narcissus. *Pl. Dis. Reptr* **54**, 377.

Gould, C. J. and Miller, V. L. (1971) Improved control of *Fusarium* basal rot with benomyl and thiabendazole. *Pl. Dis. Reptr* **55**, 425.

Graham-Bryce, I. J. and Coutts, J. (1971) Interactions of pyrimidine fungicides with soil and their influence on uptake by plants. *Proc. 6th Br. Insectic. Fungic. Conf.*, p. 419.

Gramlich, J. V., Schwer, J. F. and Brown, I. F. Jr. (1969) Characteristics and field performance of a new, broad-spectrum systemic fungicide. *Proc. 5th Br. Insectic. Fungic. Conf.* **2**, 576.

Gray, R. A. (1956) Increasing the absorption of streptomycin by leaves and flowers with glycerol. *Phytopathology* **46**, 105.

Gray, R. A. (1958) The downward translocation of antibiotics in plants. *Phytopathology* **48**, 71.

Gray, L. E. and Sinclair, J. B. (1971) Systemic uptake of ^{14}C-labelled 2-(4′-thiazolyl) benzimidazole in soybean. *Phytopathology* **61**, 523.

Greenaway, W. (1972) (private communication).

Greenaway, W. and Cowan, J. W. (1970) The stability of mercury resistance in *Pyrenophora avenae*. *Trans. Brit. mycol. Soc.* **54**, 127.

Grosjean, J. (1951) Onderzoekingen over de mogelijkheid van een bestrijding van de loodglansziekte volgens de boorgat-methode. *Tijdschr. PlZiekt.* **57**, 103.

Grosse-Brauckmann, E. (1957) Über den Einfluss der Kieselsäure auf den Mehltaubefall von Getreide bei unterschiedlicher Stickstoffdüngung. *Phytopath. Z.* **30**, 112.

Grosse-Brauckmann, E. (1958) Massnahmen zur Mehltaubekämpfung bei verschiedener Gerstensorten, ihre Erfolgsaussichten und ihr Einfluss auf den Mineralstoffgehalt. *Z. PflKrankh. PflPath. PflSchutz.* **65**, 689.

Grossmann, F. (1957) Untersuchungen über die innertherapeutische Wirkung organischer Fungizide. I. Thiocarbamate und Thiurame. *Z. PflKrankh. PflPath. PflSchutz.* **64**, 718.

Grossmann, F. (1958a) Untersuchungen über die innertherapeutische Wirkung organischer Fungizide. II. Chlornitrobenzole. *Z. PflKrankh. PflPath. PflSchutz.* **65**, 594.

Grossmann, F. (1958b) Über die Hemmung pektolytischer Enzyme von *Fusarium oxysporum* f. *lycopersici* durch Gerbstoffe. *Naturwiss.* **45**, 113.

Grossmann, F. (1961) Die systemische Wirkung von Sulfonamiden gegen Pflanzenkrankheiten, hervorgerufen durch nicht obligate Erreger. *Z. PflKrankh. PflPath. PflSchutz.* **68**, 633.

Grossmann, F. (1962a) Untersuchungen über die Hemmung pektolytischer Enzyme von *Fusarium oxysporum* f. *lycopersici*. I. Hemmung durch definierte Substanzen *in vitro*. *Phytopath. Z.* **44**, 361.

Grossmann, F. (1962b) Therapeutische Wirkung von Pektinase-Hemmstoffen gegen die *Fusarium*-Welke der Tomate. *Naturwissenschaften* **49**, 138.

Grossmann, F. (1962c) Untersuchungen über die Hemmung pektolytischer Enzyme von *Fusarium oxysporum* f. *lycopersici*. II. Hemmung durch Pflanzenextrakte *in vitro*. *Phytopath. Z.* **45**, 1.

Grossmann, F. (1962d) Untersuchungen über die Hemmung pektolytischer Enzyme von *Fusarium oxysporum* f. *lycopersici*. III. Wirkung einiger Hemmstoffe *in vivo*. *Phytopath. Z.* **45**, 139.

Grossmann, F. (1963) Fungitoxische, enzymehemmende und therapeutische Wirkungen der Rufiansäure (1,4-Dioxy-Anthrachinon-Sulfonsäure-(2)). *Meded. LandbHoogesch. OpzoekStns. Gent* **28**, 604.

Grossmann, F. (1968a) Studies on the therapeutic effects of pectolytic enzyme inhibitors. *Neth. J. Pl. Path.* **74**, **Suppl. 1**, 91.

Grossmann, F. (1968b) Conferred resistance in the host. *Wld Rev. Pest Control* **7**, 176.

Grove, J. F. (1963) Griseofulvin. *Quart. Rev.* **17**, 1.

Grover, R. K. and Chopra, B. L. (1970) Adaptation of *Rhizoctonia* species to two oxathiin compounds and manifestation of the adapted isolates. *Acta Phytopathological (Hung)* **5**, 113.

Grümmer, G. and Beyer, H. (1959) The influence exerted by species of *Camelina* on flax by means of toxic substances. *Biology of Weeds* (ed. J. L. Harper), p. 153. Oxford: Blackwell.

Gunning, B. E. S. and Pate, J. S. (1969) 'Transfer cells': plant cells with wall ingrowths, specialised in relation to short distance transport of solutes. Their occurrence, structure and development. *Protoplasma* **68**, 107.

Gutter, Y. (1969a) Comparative effectiveness of benomyl, thiabendazole and other antifungal compounds for post-harvest control of Penicillium decay in Shamouti and Valencia oranges. *Pl. Dis. Reptr* **53**, 474.

Gutter, Y. (1969b) Effectiveness of pre-inoculation and post-inoculation treatments with sodium orthophenylphenate, thiabendazole and benomyl for green mold control in artificially inoculated Eureka lemons. *Pl. Dis. Reptr* **53**, 479.

Gutter, Y. (1970) Influence of application time on effectiveness of fungicides for green mold control in artificially inoculated oranges. *Pl. Dis. Reptr* **54**, 325.

Hacker, R. G. and Vaughn, J. R. (1957a) Cycloheximide analogues cause pre-infection resistance to *Puccinia graminis* var. *tritici* in spring wheat. *Phytopathology* **47**, 14.

Hacker, R. G. and Vaughn, J. R. (1957b) Chemically induced resistance to stem rust of wheat by derivatives of Acti-dione. *Pl. Dis. Reptr* **41**, 442.

Hagborg, W. A. F. (1970) Carboxanilido systemic chemical in the control of leaf and stem rusts of wheat. *Can. J. Pl. Sci.* **50**, 631.

Hagborg, W. A. F. (1971) 'Plantavax' emulsifiable concentrate in the control of leaf and stem rusts in wheat. *Can. J. Pl. Sci.* **51**, 239.

Hagborg, W. A. F., Spencer, J. F. T. and Chelack, W. S. (1961) Antibiotic P-9 in the control of cereal rusts. *Can. J. Bot.* **39**, 1725.

ten Haken, P. and Dunn, C. L. (1971) Structure-activity relationships in a group of carboxanilides systemically active against broad bean rust *(Uromyces fabae)* and wheat rust *(Puccinia recondita)*. *Proc. 6th Br. Insectic. Fungic. Conf.*, **2**, 453.

Hall, D. W. (1971) Control of mildew in winter barley with ethirimol. *Proc. 6th Br. Insectic. Fungic. Conf.*, p. 26.

Hallam, N. D. and Juniper, B. E. (1971) The anatomy of the leaf surface. *Ecology of leaf surface micro-organisms* (ed. T. F. Preece and C. H. Dickinson), p. 3. Academic Press.

Hamilton, J. M. (1931) Studies of the fungicidal action of certain dusts and sprays in the control of apple scab. *Phytopathology* **21**, 445.

Hamilton, J. M. (1948) Evaluation of organic fungicides for the control of *Venturia inaequalis* (Cke.) Wint. *Phytopathology* **38**, 313.

Hamilton, J. M. and Szkolnik, M. (1953) Factors involved in the performance of cycloheximide (acti-dione) against *Coccomyces hiemalis*. *Phytopathology* **43**, 109.

Hamilton, J. M. and Szkolnik, M. (1958) Movement of dodecylguanidine derivatives through the leaf in the control of apple scab and cedar-apple rust fungi. *Phytopathology* **48**, 262.

Hamilton, J. M., Szkolnik, M. and Sondheimer, E. (1956) Systemic control of cherry leaf spot fungus by foliar sprays of Acti-dione derivatives. *Science, N.Y.* **123**, 1175.

Hammett, K. R. W. (1968) Root application of a systemic fungicide for control of powdery mildews. *Pl. Dis. Reptr* **52**, 754.

Hansing, E. D. (1967) Systemic oxathiin fungicide for control of loose smut *(Ustilago tritici)* of winter wheat. *Phytopathology* **57**, 814 (abs.).

Harding, P. R. (1967) Wax emulsion additives for control of storage decay in lemons. *Pl. Dis. Reptr* **51**, 781.

Harding, P. R. (1968) Comparison of Fungicide 1991, thiabendazole and sodium orthophenylphenate for control of Penicillium molds of postharvest citrus fruit. *Pl. Dis. Reptr* **52**, 623.

Hardison, J. R. (1967) Chemotherapeutic control of stripe smut *(Ustilago striiformis)* in grasses by two derivatives of 1,4-oxathiin. *Phytopathology* **57**, 242.

Harper, C. W. and Byrde, R. J. W. (1972) Apple: scab *(Venturia inaequalis)*; powdery mildew *(Podosphaera leucotricha)*. *Fungicide–Nematicide Tests* **27**, 15.

Harper, F. R. (1968) Control of root rot in garden peas with a soil fungicide. *Pl. Dis. Reptr* **52**, 565.

Harrow, K. M. (1970) Onion smut: an examination of a potential hazard to commercial growers in New Zealand. *N.Z. Comml Grow. for March 1970*, p. 15.

Hart, H. and Allison, J. L. (1939) Toluene compounds to control plant disease. *Phytopathology* **29**, 978.

Harvey, A. E. and Helton, A. W. (1962) *In vitro* effects of seventeen chemical agents on *Cytospora cincta*. *Pl. Dis. Reptr* **46**, 593.

Hassebrauk, K. (1938) Weitere Untersuchungen über Getreiderostbekämpfung mit chemischen Mitteln. *Phytopath. Z.* **11**, 14.

Hassebrauk, K. (1940) Abschliessende Untersuchungen über die feldmässige Verwendungsmöglichkeit von *p*-Toluolsulfonamid als innertherapeutisch wirkendes Getreiderostbekämpfungsmittel. *Phytopath. Z.* **12**, 509.

Hassebrauk, K. (1951) Untersuchungen über die Einwirkung von Sulfonamiden und Sulfonen auf Getreideroste. I. Die Beeinflussing des Fruktifikationsvermögens. *Phytopath. Z.* **17**, 384.

Hassebrauk, K. (1952) Untersuchungen über die Einwirkung von Sulfonamiden und Sulfonen auf Getreideroste. II. Weitere Untersuchungen über die rosthemmende Wirkung. *Phytopath. Z.* **18**, 453.

Hastie, A. C. (1970) Benlate-induced instability of *Aspergillus* diploids. *Nature, Lond.* **226**, 771.

Hastie, A. C. (1971) Benomyl-induced instability of *Aspergillus* diploids. *Proc. 6th Br. Insectic. Fungic. Conf. at Brighton, 1971* **1**, 283.

Hastie, A. C. and Georgopoulos, S. G. (1971) Mutational resistance to fungitoxic benzimidazole derivatives in *Aspergillus nidulans*. *J. Gen. Microb.* **67**, 371.

Hay, S. J. B. (1971) The control of apple mildew with HOE2873. *Proc. 6th Br. Insectic. Fungic. Conf.* **1**, 134.

Hay, J. R. and Thimann, K. V. (1956) The fate of 2,4-dichlorophenoxyacetic acid in bean seedlings. II. *Pl. Physiol. Lancaster* **31**, 446.

Headford, D. W. R. and Douglas, D. (1967) Tuber necrosis following the desiccation of potato foliage with diquat. *Weed Res.* **7**, 131.

Hearn, C. J. and Childs, J. F. L. (1969) A systemic fungicide effective against sour orange scab disease. *Pl. Dis. Reptr* **53**, 203.

Hearn, C. J., Childs, J. F. L. and Fenton, R. (1971) Comparison of benomyl and copper sprays for control of sour orange scab of citrus. *Pl. Dis. Reptr* **55**, 241.

Hearn, C. J. and Fenton, R. (1970) Benomyl sprays for control of twig dieback of 'Robinson' tangerine. *Pl. Dis. Reptr* **54**, 869

Heggeness, H. G. (1942) Effect of borax applications on the incidence of rust on flax. *Pl. Physiol.* **17**, 143.

Helser, T. L., Davies, J. E. and Dahlberg, J. E. (1972) Mechanism of kasugamycin resistance in *Escherichia coli*. *Nature, Lond.* **235**, 6.

Helton, A. W. and Kochan, W. J. (1967) Chemotherapeutic effects of cycloheximide thiosemicarbazone, phytoactin and 8-quinolinol benzoate, with and without dimethyl sulfoxide and pruning on the Cytospora disease of prune trees. *Pl. Dis. Reptr* **51**, 340.

Helton, A. W. and Rohrbach, K. G. (1967) Chemotherapy of Cytospora canker disease in peach trees. *Phytopathology* **57**, 442.

Helton, A. W. and Williams, R. E. (1970) Systemic distribution of three compounds in sprayed 'Jonathan' apple trees. *Pl. Dis. Reptr* **54**, 996.

Heyns, A. J., Carter, G. A., Rothwell, K. and Wain, R. L. (1966) Investigations on fungicides. XIII. The systemic fungicidal activity of certain N-carboxymethyl-dithiocarbamic acid derivatives. *Ann. appl. Biol.* **57**, 33.

Hickey, K. D. (1971a) (ed.) Pome fruit disease reports. *Fungicide–Nematicide Tests* **26**, 4.

Hickey, K. D. (1971b) (ed.) Stone fruit disease reports. *Fungicide–Nematicide Tests* **26**, 42.

Hide, G. A. and Hirst, J. M. (1971) Chemical control of tuber diseases. *Rep. Rothamsted exp. Stn for 1970* **1**, 140.

Hide, G. A., Hirst, J. M. and Griffith, R. L. (1969) Control of potato tuber diseases with systemic fungicides. *Proc. 5th Br. Insectic. Fungic. Conf.*, p. 310.

Hignett, R. C. and Kirkham, D. S. (1967) The role of extracellular melanoproteins of *V. inaequalis* in host susceptibility. *J. gen. Microbiol.* **48**, 269.

Hilborn, M. T. (1953) Effect of various chemicals on infection by *Rhizoctonia solani* and *Verticillium albo-atrum*. *Phytopathology* **43**, 475.

Hilborn, M. T. (1969) Blueberry: powdery mildew *(Microsphaera penicillata* var. *vaccinii)*. *Fungicide–Nematicide Tests* **24**, 34.

Hill-Cottingham, D. G. and Lloyd-Jones, C. P. (1968) Relative mobility of some nitrogenous compounds in the xylem of apple shoots. *Nature, Lond.* **220**, 389.

Hiltner, L. (1915) Über die Erfolge der im Herbst 1914 in Bayern durch geführten Beizung des Saatgutes von Winterroggen und Winterweizen mit sublimathaltigen Mittleln. *Prakt. Bl. Pflbau. PflSchutz.* **18**, 65.

Hine, R. B., Johnson, D. L. and Wenger, C. J. (1969) The persistence of two benzimidazole fungicides in soil and their fungistatic activity against *Phymatotrichum omnivorum*. *Phytopathology* **59**, 798.

Hinosan Information Bulletin, 1971.

Hirst, J. M. (1971) Mechanical dusting of tubers with fungicide. *Rep. Rothamsted exp. Stn for 1970* **1**, 141.

Hislop, E. C. (1967) Observations on the vapour phase activity of some foliage fungicides. *Ann. appl. Biol.* **60**, 265.

Hock, W. K. and Sisler, H. D. (1969a) Specificity and mechanism of antifungal action of chloroneb. *Phytopathology* **59**, 627.

Hock, W. K. and Sisler, H. D. (1969b) Metabolism of chloroneb by *Rhizoctonia solani* and other fungi. *J. Agric. Food Chem.* **17**, 123.

Hock, W. K., Schreiber, L. R. and Roberts, B. R. (1970) Factors influencing uptake, concentration and persistence of benomyl in American elm seedlings. *Phytopathology* **60**, 1619.

Hocking, D. and White, P. J. (1965) Fungicides for arabica coffee. III. Determination of translaminar, curative activity against leaf rust. *PANS* **B11**, 273.

Hoeven, E. P. van der and Bollen, G. J. (1972) The effect of benomyl on antagonism towards fungi causing foot rot in rye. *Acta bot. neerl.* **21** (in press).

Hoffmann, J. A. (1969) *Fungicide–Nematicide Tests* **25**, 129.

Hoffmann, J. A. (1970) *Fungicide–Nematicide Tests* **26**, 146.

Hoffmann, J. A. (1971) Control of common and dwarf bunt of wheat by seed treatment with thiabendazole. *Phytopathology* **61**, 1071.

Hoffmann, P. F. (1951) Screening chemotherapeutants for control of oak wilt. *Proc. Iowa Acad. Sci.* **58**, 139.

Hoffmann, P. F. (1952) Early trials in oak wilt chemotherapy. *Phytopathology* **42**, 11.

Hoffmann, P. F. (1953) Chemicals for therapy of oak wilt. *Phytopathology* **43**, 475.

Holloway, P. J. (1971) The chemical and physical characteristics of leaf surfaces. *Ecology of leaf surface micro-organisms* (ed. T. F. Preece and C. H. Dickinson), p. 39. Academic Press.

Holly, K. (1956) Penetration of chlorinated phenoxyacetic acids into leaves. *Ann. appl. Biol.* **44**, 195.

Holowczak, J., Kuć, J. and Williams, E. B. (1962) Metabolism of DL- and L-phenylalanine in *Malus* related to susceptibility and resistance to *Venturia inaequalis*. *Phytopathology* **52**, 699.

Hope, A. B. and Stevens, P. G. (1952) Electrical potential differences in bean roots and their relation to salt uptake. *Aust. J. sci. Res. Ser. B* **5**, 445.

Horner, C. E. (1963) Chemotherapeutic effects of streptomycin on establishment and progression of systemic downy mildew infection in hops. *Phytopathology* **53**, 472.

Horner, C. E. and Maier, C. R. (1957) Antibiotics eliminate systemic downy mildew from hops. *Phytopathology* **47**, 525.

Horsfall, J. G. (1945) *Fungicides and their Action*. Waltham, Mass.: Chronica Botanica.

Horsfall, J. G. (1956) *Principles of Fungicidal Action*. Waltham, Mass.: Chronica Botanica.

Horsfall, J. G. (1961) Chemotherapy of plant disease. *Scientia, Sixième Série, Aug. 1961*, 1.

Horsfall, J. G. and Dimond, A. E. (1951a) Plant chemotherapy. *Trans. N.Y. Acad. Sci.* **13**, 338.

Horsfall, J. G. and Dimond, A. E. (1951b) Plant chemotherapy. *A. Rev. Microbiol.* **5**, 209.

Horsfall, J. G. and Dimond, A. E. (1957) Interaction of tissue sugar, growth substances and disease susceptibility. *Z. PflKrankh. PflPath. PflSchutz.* **64**, 415.

Horsfall, J. G., Dimond, A. E., Stoddard, E. M. and Chapman, R. A. (1951) Chemotherapy for controlling plant diseases. *Proc. 2nd int. Congr. Crop Prot. London 1949*, 1.

Horsfall, J. G. and Lukens, R. J. (1968) Aldehyde traps as antisporulants for fungi. *Conn. Agr. Exp. Station Bull.* **694**, 28 pp. New Haven.

Horsfall, J. G. and Zentmyer, G. A. (1942) Antidoting the toxins of plant diseases. *Phytopathology* **32**, 23.

Hotson, H. H. (1952) The inhibition of wheat stem rust by sulfadiazine and the reversal of this effect by para-aminobenzoic acid and folic acid. *Phytopathology* **42**, 11.

Hotson, H. H. (1953) Some chemotherapeutic agents for wheat stem rust. *Phytopathology* **43**, 659.

Houseworth, L. D., Brunton, E. W. and Tweedy, B. G. (1971) The effect of triarimol on oxidation of glucose and acetate to carbon dioxide and on spore germination of *Aspergillus niger* and *Cladosporium cucumerinum*. *Phytopathology* **61**, 896 (abs.).

Howard, C. M. (1971a) Control of strawberry anthracnose with benomyl. *Pl. Dis. Reptr* **55**, 139.

Howard, C. M. (1971b) Strawberry: common leaf spot *(Mycosphaerella fragariae)*; leaf blight *(Dendrophoma obscurans)*. *Fungicide–Nematicide Tests* **26**, 60.

Howard, F. L. (1941) Antidoting of toxins of *Phytophthora cactorum* as a means of plant disease control. *Science, N.Y.* **94**, 345.

Howard, F. L. and Caroselli, N. E. (1941) The role of tree injection in the control of bleeding canker of hardwoods. *Phytopathology* **31**, 12.

Howard, F. L. and Horsfall, J. G. (1959) Therapy. *Plant Pathology Vol. I* (eds. J. G. Horsfall and A. E. Dimond), p. 563. New York and London: Academic Press.

Howard, F. L. and Machardy, W. E. (1969) The basis of elm twig wilt by *Ceratocystis ulmi* and its bearing on chemotherapy. *XI Internat. Botan. Congr. Abstr.*, p. 95. Seattle.

Howard, F. L. and Sorrell, M. B. (1943) Cationic phenyl mercury compounds as specific apple-scab eradicants on foliage. *Phytopathology* **33**, 1114.

Huang, K. T., Misato, T. and Asuyama, H. (1964) Selective toxicity of blasticidin S to *Piricularia oryzae* and *Pellicularia sasakii*. *J. Antibiotics* (Tokyo) A, **17**, 71.

Hudson, J. C. and Jaffer, A. A. (1970) Effect of systemic fungicides on bean rust *(Uromyces appendiculatus* (Pers.) Lev.*)*. *Misc. Rep. trop. Pestic.* no. 726, 7 pp.

Ibrahim, I. A. (1951) Effect of 2,4-D on stem rust development in oats. *Phytopathology* **41**, 951.

Ingram, J. (1971) Black currant: American gooseberry mildew/leaf spot 1970. *Rep. Luddington exp. Hort. Stn for 1970*, 40.

Ishii, K. (1970) New fungicides, thiophanate and its derivatives. *Summaries of Papers, VIIth Int. Cong. Plant Prot., Paris,* 200.

Ishiyama, T., Okamoto, H., Sato, K., Nakamura, K. and Nakamura, M. (1965a) Translocation of kasugamycin in the leaf tissue. *Ann. phytopath. Soc. Japan* **30**, 111 (abs.).

Ishiyama, T., Okamoto, H., Sato, K. and Nakamura, M. (1965b) Control of rice blast by application of kasugamycin through roots. *Ann. Phytopath. Soc. Japan* **30**, 111 (abs.).

Ishiyama, T., Sato, K., Nakamura, T., Takecuhi, T. and Umezawa, H. (1967) Absorption and translocation by rice plants of ^{14}C-kasugamycin. *J. Antibiotics Ser. B.* **20**, 357.

Isono, K., Nagatsu, J., Kobinata, K., Sasaki, K. and Suzuki, S. (1967) Studies on polyoxins, antifungal antibiotics. II. Isolation and characterisation of polyoxins C, D, E, F, G, H and I. *Agr. Biol. Chem.* **31**, 190.

Jacobs, H. L. (1928) Injection of shade trees for the control of insects and diseases. *18th A. Conv. Davey Tree Expert Co.*, Kent, Ohio.

Jacobsen, B. J. and Williams, P. H. (1970) Control of cabbage clubroot using benomyl fungicide. *Pl. Dis. Reptr* **54**, 456.

Jacobsen, B. J. and Williams, P. H. (1971) Histology and control of *Brassica oleracea* seed infection by *Phoma lingam*. *Pl. Dis. Reptr* **55**, 934.

Jain, S. S. (1956) A preliminary note on the inactivation of *Fusarium oxysporum* f. *psidii* in guava plants by chemotherapeutic treatment. *Indian. J. Hort.* **13**, 102.

James, C. S. (1971) Private communication.

James, C. S., Gramlich, J. V. and Darge, G. D. (1971). Control of barley powdery mildew *(Erysiphe graminis)* with triarimol. British Association of Grain, Seed, Feed and Agricultural Merchants Ltd. Unpublished Report.

Jank, B. and Grossmann, F. (1971) 2-Methyl-5,6-dihydro-4-H-pyran-3-carboxylic acid anilide: a new systemic fungicide against smut diseases. *Pestic. Sci.* **2**, 43.

Jarvis, P. and Thaine, R. (1971) Strands in sections of sieve elements cut in a cryostat. *Nature, Lond.* **232**, 236.

Jaworski, E. G. and Hoffman, P. F. (1963) Chemotherapeutic control of wheat leaf rust with phenylhydrazones. *Phytopathology* **53**, 639.

Jefferys, E. G. (1952) The stability of antibiotics in soils. *J. gen. Microbiol.* **7**, 295.

Jhooty, J. S. and Behar, D. S. (1970) Evaluation of different benomyl treatments for control of *Rhizoctonia* damping-off of peas. *Pl. Dis. Reptr* **54**, 1049.

Johnson, M. P. and Bonner, J. (1956) The uptake of auxin by plant tissue. *Physiologia Pl.* **9**, 102.

Johnston, H. W. (1970) Control of powdery mildew of wheat by soil-applied benomyl. *Pl. Dis. Reptr* **54**, (2) 91.

Jones, B. M. and Swartwout, H. G. (1961a) Systemic control of powdery mildew, *Sphaerotheca pannosa*, of roses with the semicarbazone derivative of actidione (cycloheximide), *Pl. Dis. Reptr* **45**, 366.

Jones, B. M. and Swartwout, H. G. (1961b) Systemic control of powdery mildew of roses, caused by the fungus *Sphaerotheca pannosa*, with derivatives of cycloheximide. *Pl. Dis. Reptr* **45**, 794.

Jones, J. P. and Barnett, R. D. (1968) Control of loose smut of wheat and barley in Arkansas with fungicides. U.S. Dept. of Agr. CR42–68, 3.

Jones, J. P., Overman, A. J. and Geraldson, C. M. (1971) Fumigants for the control of *Verticillium* wilt of tomato. *Pl. Dis. Reptr* **55**, 26.

Jones, J. P. and Woltz, S. S. (1969) Effect of ethionine and methionine on the growth, sporulation and virulence of *Fusarium oxysporum* f. sp. *lycopersici*, race 2. *Phytopathology* **59**, 1464.

Jooste, W. J. (1969) Control of ear and tassel smut of maize with 1,4-oxathiin derivatives. *Phytophylactica* **1**, 235.

Jordan, V. W. L. (1971) Strawberry: mildew (*Sphaerotheca macularis*). *Fungicide–Nematicide Tests* **26**, 61.

Jordan, V. W. L. (1972a) Evaluation of fungicides for the control of *Verticillium* wilt (*V. dahliae* Kleb.) of strawberry. *Ann. appl. Biol.* **70**, 163.

Jordan, V. W. L. (1972b) Diseases of strawberries: *Verticillium* wilt. *Rep. agric. hort. Res. Stn Bristol, 1971*, p. 140.

Jung, K. U. and Bedford, J. L. (1971) The evaluation of application time of tridemorph for the control of mildew in barley and oats in the United Kingdom. *Proc. 6th Br. Insect. Fungic. Conf.*, p. 75.

Kaars Sijpesteijn, A. (1961a) Chemotherapy and synthetic systemic fungicides. *Proc. 9th int. Bot. Congr. Montreal, 1959* **1**, 457.

Kaars Sijpesteijn, A. (1961b) New developments in the systemic combat of fungal diseases of plants. *Tijdschr. PlZiekt.* **67**, 11.

Kaars Sijpesteijn, A. (1969) Enzymic transformations of fungicides in the host plant. *Wld Rev. Pest Control* **8**, 138.

Kaars Sijpesteijn, A. (1970) Biochemical modes of action of agricultural fungicides. *Wld Rev. Pest Control* **9**, 85.

Kaars Sijpesteijn, A. and van der Kerk, G. J. M. (1965) Fate of fungicides in plants. *A. Rev. Phytopath.* **3**, 127.

Kaars Sijpesteijn, A. and Pluijgers, C. W. (1962) On the action of phenylthioureas as systemic compounds against fungal diseases of plants. *Meded. Landb-Hoogesch. OpzoekStns. Gent* **27**, 1199.

Kaars Sijpesteijn, A., Pluijgers, C. W., Verloop, A. and Tempel, A. (1968) Phenylazothioformamide, the active principle in the fungitoxic action of 1-phenylthiosemicarbazide. *Ann. appl. Biol.* **61**, 473.

Kaars Sijpesteijn, A., Rombouts, J. E., van Andel, O. M. and Dekker, J. (1958) Investigations on the activity of pyridine-2-thiol-N-oxide as a systemic fungicide. *Meded LandbHoogesch. OpzoekStns. Gent* **23**, 824.

Kaars Sijpesteijn, A. and Sisler, H. D. (1968) Studies on the mode of action of phenylthiourea, a chemotherapeutant for cucumber scab. *Neth. J. Pl. Path.* **74**, 121.

Kable, P. F. (1970) Eradicant action of fungicides applied to dormant peach trees for control of brown rot *(Monilinia fructicola)*. *J. hort. Sci.* **45**, 143.

Kado, M., Maeda, T., Yoshinaga, E., Iwakura, T. and Uchida, T. (1965) Studies on the rice blast control. I. Chemical structure and fungicidal activities of benzyl phosphoric esters. *Ann. Phytopathol. Soc. Japan* **30**, 109 (abs.).

Kado, M. and Yoshinaga, E. (1969) Fungicidal action of organophosphorus compounds. *Residue Rev.* **25**, 133.

Kakiki, K., Maeda, T., Abe, T. and Misato, T. (1969) Studies on the mode of action of organophosphorus fungicide Kitazin. Part I: Effect on respiration, protein synthesis, nucleic acid synthesis, cell wall synthesis and leakage of cellular substances from mycelia of *Pyricularia oryzae*. *Nippon Nogeikagaku Kaishi* **43**, 37.

Kantzes, J. G. and Weaver, L. O. (1971) Pear: leaf spot *(Entomosporium maculatum)*. *Fungicide–Nematicide Tests* **26**, 40.

Kappas, A. and Georgopoulos, S. G. (1970) Genetic analysis of dodine resistance in *Nectria haematococca* (syn. *Hypomyces solani*). *Genetics* **66**, 617.

Kappas, A. and Georgopoulos, S. G. (1971) Independent inheritance of avirulence and dodine resistance in *Nectria haematococca* var. *cucurbitae*. *Phytopathology* **61**, 1093.

Kavanagh, T. (1970) Black spot *(Diplocarpon rosae)* of rose. *Res. Rep. An Foras Taluntais, Hort. Div.* 1970, p. 87.

Kavanagh, T. and O'Callaghan, T. F. (1969) A comparison of fungicides for the control of Botrytis of outdoor and protected strawberries. *Proc. 5th Br. Insectic. Fungic. Conf.* **1**, 139.

Kavanagh, T. and O'Kennedy, N. D. (1971) Evaluation of apple scab *(Venturia inaequalis)* and canker *(Nectria galligena)* control programmes in Ireland 1961–1971. *Proc. 6th Br. Insectic. Fungic. Conf.* **1**, 110.

Keen, N. T., Long, M. and Malca, I. (1970) Induction and repression of β-galactosidase synthesis by *Verticillium albo-atrum*. *Physiolog. Plantarum* **23**, 691.

Keen, N. T., Wang, M. C., Long, M. and Erwin, D. C. (1971) Dimorphism in *Verticillium albo-atrum* as affected by initial spore concentration and antisporulant chemicals. *Phytopathology* **61**, 1266.

Keiding, J. (1967) Persistence of resistant populations after the relaxation of the selection pressure. *Wld Rev. Pest Control* **6**, 115.

Keil, H. L. and Frohlich, H. P. (1961) U.S. Patent 2971880.

Keil, H. L., Frohlich, H. P. and van Hook, J. O. (1958) Chemical control of cereal rusts. I. Protective and eradicative control of rye leaf rust in the greenhouse with various chemical compounds. *Phytopathology* **48**, 652.

Kennedy, J. (1961) Fumigation of agricultural products. XVII. Control of *Ascochyta* blight of peas by fumigation. *J. Sci. Fd. Agric.* **12**, 96.

Kent, N. L. (1941a) Absorption, translocation and ultimate fate of lithium in the wheat plant. *New Phytol.* **40**, 291.

Kent, N. L. (1941b) The influence of lithium salts on certain cultivated plants and their parasitic diseases. *Ann. appl. Biol.* **28**, 189.

van der Kerk, G. J. M. (1956) The present state of fungicidal research. *Meded. LandbHoogesch. OpzoekStns. Gent* **21**, 305.

van der Kerk, G. J. M. (1961) New developments in organic fungicides. *Soc. chem. Ind. Monograph* **15**, 67.

Kerridge, D. (1958) The effect of actidione and other antifungal agents on nucleic acid and protein synthesis in *Saccharomyces carlsbergensis. J. Gen. Microbiol.* **19**, 497.

Keyworth, W. G. and Dimond, A. E. (1952) Root injury as a factor in the assessment of chemotherapeutants. *Phytopathology* **42**, 311.

King, J. M. (1971) The control of *Botrytis* pod rot in dwarf beans. *Proc. 6th Br. Insectic. Fungic. Conf.*, p. 173.

Kingsland, G. (1970) Barley leaf stripe control by 'Vitavax'. *Phytopathology* **60**, 584.

Király, Z., Bócsa, E. and Vörös, J. (1962) Effect of treatment of wheat seed with gibberellic acid on bunt infection. *Phytopathology* **52**, 171.

Kirby, A. H. M. (1971) Phytotoxic effects of some new fungicides for apple mildew control. *Meded. Fac. Landbouw. RijksUniversiteit Gent* **36**, 355.

Kirby, A. H. M., Butt, D. J. and Williamson, C. J. (1970) Trials in 1968 of new fungicides on apple and pear. *Rep. E. Malling Res. Stn, 1969*, p. 171.

Kirkham, D. S. (1957) The significance of polyphenolic metabolites of apple and pear on the host relations of *Venturia inaequalis* and *Venturia pirina. J. gen. Microbiol.* **17**, 491.

Kirkham, D. S. and Flood, A. E. (1963) Some effects of respiration inhibitors and *o*-coumaric acid on the inhibition of sporulation in *Venturia inaequalis. J. gen. Microbiol.* **32**, 123.

Kirkham, D. S. and Hunter, L. D. (1964) Systemic antifungal activity of *iso*butyl *o*-coumarate in apple. *Nature, Lond.* **201**, 638.

Kirkham, D. S. and Hunter, L. D. (1965) Studies of the *in vivo* activity of esters of *o*-coumaric and cinnamic acids against apple scab. *Ann. appl. Biol.* **55**, 359.

Kishi, K. (1971) History of chemical control for the main diseases and pests of Japanese pear (Nijusseiki). *Japan Pesticide Inf.* **6**, 11. (*Rev. Pl. Path.* **50**, 3965, 1971.)

Klöpping, H. L. (1960) U.S. patent 2933502; U.S. patent 2933504.

Klos, E. J. and Fronek, F. R. (1964) Chemical eradication of cherry leaf spot fungus. *Quart. Bull. Mich. State Coll. agric. Exp. Sta.* **47**, 65.

Klos, E. J., Robb, J. and Miller, R. (1968) Grape: black rot and powdery mildew; *Guignardia bidwellii* and *Uncinula necator. Fungicide–Nematicide Tests* **23**, 58.

Klos, E. J., Yoder, K., Nowacka, H. and Riggle, J. (1971) Apple: scab (*Venturia inaequalis*). *Fungicide–Nematicide Tests* **26**, 26.

Knight, B. C. (1971) Private communication.

Koch, W. (1971) Behaviour of commercial systemic fungicides in conventional (non-systemic) tests. *Pestic. Sci.* **2**, 207.

König, K. H., Pommer, E. H. and Saune, W. (1965) N-substituted tetrahydro-1,4-oxazines: a new class of fungicidal compounds. *Angew. Chem. (Int. Ed.)* **4**, 336.

Koopmans, M. J. (1960) Systemic fungicidal action of some 5-amino-1-(bis (dimethylamido)phosphoryl)-1,2,4-triazoles. *Meded. Fac. Landbouwwetensch. Rijksuniv. Gent* **25**, 1221.

Kosuge, T. (1969) The role of phenolics in host response to infection. *Ann. Rev. Phytopathol.* **7**, 195.

Kovacs, A. I. and Caumo, B. (1971) Grape: powdery mildew *(Uncinula necator). Fungicide–Nematicide Tests* **26**, 57.

Kozaka, T. (1969) Chemical control of rice blast in Japan. *Rev. Pl. Protec. Res.* **2**, 53.

Kradel, J., Effland, H. and Pommer, E. H. (1969) Erfchrungen bei der Bekämpfung des Getreidemehlfaus (*Erysiphe graminis). Gesund Pflanzen* **21**, 121.

Kradel, J. and Pommer, E. H. (1967) Some remarks and results on the control of powdery mildew in cereals. *Proc. 4th Br. Insectic. Fungic. Conf.,* **1**, 170.

Kradel, J., Pommer, E. H. and Effland, H. (1969) Response of barley varieties to the control of powdery mildew with cyclomorph and tridemorph. *Proc. 5th Br. Insectic. Fungic. Conf.,* p. 16.

Kramer, P. J. (1957) Outer space in plants. *Science, N.Y.* **125**, 633.

Krüger, W. (1959) Antibiotics as seed protectants. *S. Afr. J. Agr. Sci.* **2**, 207.

Krüger, W. (1960) Antibiotics as seed protectants (1958/59). *S. Afr. J. agric. Sci.* **3**, 409.

Krüger, W. (1961) The activity of antibiotics in soils. II. Movement, stability and biological activity of antibiotics in soils and their uptake by tomato plants. *S. Afr. J. agric. Sci.* **4**, 301.

Kuć, J. (1963) The role of phenolic compounds in disease resistance. *Conn. Agric. Exp. Sta. Bull.* no. 663, p. 20.

Kuć, J. (1968) Biochemical control of disease resistance in plants. *Wld Rev. Pest Control* **7**, 42.

Kuć, J., Williams, E. B. and Shay, J. R. (1957) Increase in resistance to apple scab following injection of host with phenylthiourea and D-phenylalanine. *Phytopathology* **47**, 21.

Kuiper, J. (1965) Failure of hexachlorobenzene to control common bunt of wheat. *Nature, Lond.* **206**, 1219.

Kuiper, J. (1968) Field evaluation of a derivative of 1,4-oxathiin for the control of cereal smuts. *Aust. J. exp. Agric. Anim. Husb.* **8**, 756.

Kurtzmann, R. H. (1966) Xylem sap flow as affected by metabolic inhibitors and girdling. *Pl. Physiol. Lancaster* **41**, 641.

Kylin, A. and Hylmö, B. (1957) Uptake and transport of sulphate in wheat. Active and passive components. *Physiologia Pl.* **10**, 467.

Lade, D. H. (1970) Apple: scab *(Venturia inaequalis)*; frogeye leaf spot *(Physalospora obtusa). Fungicide–Nematicide Tests* **25**, 25.

Lade, D. H. and Christensen, C. D. (1971) Triarimol, (EL-273), a new fungicide for tart cherry disease control. *Phytopathology* **61**, 899 (abs.).

Lafon, R. and Bugaret, Y. (1968) Les traitements des semences d'oignon contre la pourriture blanche. *Phytiat.-Phytopharm.* **18**, 23.

Land, B. G. van't, and Dekker, J. (1972) Effect of antimetabolites and fungicides on elongation of germination hyphae of powdery mildew *in vitro. Neth. J. Pl. Path.* **78** (in press).

Leach, S. S. (1970) Evaluation of post-harvest-prestorage fungicidal treatments for the control of *Fusarium* tuber rot of potatoes. *Phytopathology* **60**, 1299.

Leben, C. and Arny, D. C. (1952) Seed treatment experiments with helixin B. *Phytopathology* **42**, 469.

Leben, C., Arny, D. C. and Keitt, G. W. (1953) Small grain seed treatment with the antibiotic, Helixin B. *Phytopathology* **43**, 391.

Leben, C., Arny, D. C. and Keitt, G. W. (1954) Effectiveness of certain antibiotics for the control of seed-borne diseases of small grains. *Phytopathology* **44**, 704.

Leben, C., Boone, D. M. and Keitt, G. W. (1955) *Venturia inaequalis* (Cke) Wint. IX. Search for mutants resistant to fungicides. *Phytopathology* **45**, 467.

Leben, C. and Keitt, G. W. (1952) Studies on helixin in relation to plant disease control. *Phytopathology* **42**, 168.

Leben, C. and Keitt, G. W. (1954) Antibiotics and plant disease. Effects of antibiotics in control of plant diseases. *J. agric. Fd Chem.* **2**, 234.

Lee, B. K. and Wilkie, D. (1965) Sensitivity and resistance of yeast strains to actidione and actidione derivatives. *Nature, Lond.* **206**, 90.

Lemaire, J. M., Doussinault, G. and Jouan, B. (1970) Nouvelle perspective de lutte contre le piétin-verse des céréales par traitement des semences. *C.R. Acad. Agri. France* **56**, 643.

Lemin, A. J. and Ford, J. H. (1960) Isocycloheximide. *J. Org. Chem.* **25**, 244.

Lemin, A. J. and Magee, W. E. (1957) Degradation of cycloheximide derivatives in plants. *Pl. Dis. Reptr* **41**, 447.

Lester, E. (1971) Where do we stand with mildew. *Arable Farmer, March 1971*, p. 32.

Levitt, J. (1957) The significance of 'apparent free space' in ion absorption. *Physiologia Pl.* **10**, 882.

Lewis, D. (1963) Structural gene for the methionine activating enzyme and its mutation as a cause of resistance to ethionine. *Nature, Lond.* **200**, 151.

Lewis, F. H. (1968) Apple: scab, mildew, mites. *Venturia inaequalis, Podosphaera leucotricha, Panonychus ulmi* and *Tetranychus telarius. Fungicide–Nematicide Tests* **23**, 43.

Lewis, F. H. (1971) Apple: scab *(Venturia inaequalis)*; powdery mildew *(Podosphaera leucotricha)*; rust *(Gymnosporangium juniperi-virginianae). Fungicide–Nematicide Tests* **26**, 31.

Lindberg, G. D. (1969) *Fungicide–Nematicide Tests* **25**, 92.

Linder, P. J., Craig, J. C., Cooper, F. E. and Mitchell, J. W. (1958) Movement of 2,3,6-trichlorobenzoic acid from one plant to another through their root systems. *Ag. Food Chem.* **6**, 356.

Line, R. F. (1970) *Fungicide–Nematicide Tests* **26**, 147.

Little, R. and Doodson, J. K. (1971) The comparison of yields of some spring barley varieties in the presence of mildew and when treated with a fungicide. *Proc. 6th Br. Insectic. Fungic. Conf.*, p. 91.

Livingston, J. E. (1953) The control of leaf and stem rust of wheat with chemotherapeutants. *Phytopathology* **43**, 496.

Livingston, J. E. and Hilborn, M. T. (1959) The control of plant diseases by chemotherapy. *Econ. Bot.* **13**, 3.

Locke, S. B. (1969) Botran tolerance of *Sclerotium cepivorum* isolates from fields with different Botran treatment histories. *Phytopathology* **59**, 13.

Lockhart, C. L. (1971) Blueberry (low bush): red leaf *(Exobasidium vaccinii). Fungicide–Nematicide Tests* **26**, 55.

Lockhart, C. L., MacNab, A. A. and Bolwyn, B. (1969) A systemic fungicide for control of *Verticillium* wilt in strawberries. *Can. Pl. Dis. Surv.* **49**, 46.

Lockwood, J. L., Leben, C. and Keitt, G. W. (1954) Production and properties of antimycin A from a new *Streptomyces* isolate. *Phytopathology* **44**, 438.

Long, P. G. (1970) Commercial control of stem-end rot disease *(Gloeosporium musarum)* of banana fruit. *Pl. Dis. Reptr* **54**, 93.

Long, P. G. (1971) Control of black leaf streak disease of bananas with benomyl. *Pl. Dis. Reptr* **55**, 50.

Loux, H. M. (1961) Process for manufacture of certain alkyl esters of benzimidazole carbamic acids. U.S. patent 3 010 968.

Lovrekovich, L. and Farkas, G. L. (1963) Kinetin as an antagonist of the toxic effect of *Pseudomonas tabaci. Nature, Lond.* **198**, 710.

Ludwig, R. A. (1960) Toxins. *Plant Pathology Vol. II* (eds. J. G. Horsfall and A. E. Dimond) p. 315. New York and London: Academic Press.

Lukens, R. J. (1970) Melting-out of Kentucky bluegrass, a low-sugar disease. *Phytopathology* **60**, 1276.

Lyda, S. D. and Burnett, E. (1970) Influence of benzimidazole fungicides on *Phymatotrichum omnivorum* and root rot of cotton. *Phytopathology* **60**, 726.

Maas, J. L. (1970) Fungicidal control of Botrytis fruit rot and powdery mildew on leaves of strawberries. *Pl. Dis. Reptr* **54**, 883.

Maas, J. L. (1971) Timing of benomyl application in relation to control of Botrytis strawberry rot. *Pl. Dis. Reptr* **55**, 883.

Maas, J. L. and MacSwan, I. C. (1970) Post-harvest fungicide treatments for reduction of Penicillium decay of Anjou pear. *Pl. Dis. Reptr* **54**, 887.

Maas, J. L. and Scott, D. H. (1971) Control of storage decay of strawberry plants with benomyl preharvest treatment. *Pl. Dis. Reptr* **55**, 1009.

McCallan, S. E. A. and Wellman, R. H. (1943) A greenhouse method of evaluating fungicides by means of tomato foliage diseases. *Contr. Boyce Thompson Inst. Pl. Res.* **13**, 93.

McClintock, J. A. (1931) The relation of canker treatment to fireblight control. *Phytopathology* **21**, 901.

McClure, T. T. and Cation, D. (1951) Comparison of Acti-dione with some other spray chemicals for control of cherry leaf spot in Michigan. *Pl. Dis. Reptr* **35**, 393.

McCornack, A. A. and Brown, G. E. (1970) Benlate, an experimental postharvest citrus fungicide. *Proc. Fla St. hort. Soc.* **82**, 235. (*Rev. Pl. Path.* **50**, 2278, 1971.)

McDonnell, P. F. (1970) Control of Nectria decay of apple fruits by benomyl. *Pl. Dis. Reptr* **54**, 83.

McDonnell, P. F. (1971) Control of Nectria decay of apple fruits by post-harvest chemical dip treatments. *Pl. Dis. Reptr* **55**, 771.

Mace, M. E. and Solit, E. (1966) Interactions of 3-indoleacetic acid and 3-hydroxytyramine in *Fusarium* wilt of banana. *Phytopathology* **56**, 245.

Macer, R. C. F., Cook, I. K. and Dash, C. R. (1969) Laboratory and field trials in the United Kingdom with the systemic fungicide carboxin. *Proc. 5th Br. Insectic. Fungic. Conf.*, p. 55.

McIntosh, D. L. (1969) A low-volume post-harvest spray of benomyl prevents ascospore production in apple leaves infected by *Venturia inaequalis. Pl. Dis. Reptr* **53**, 816.

MacKenzie, D. R., Cole, H. and Nelson, R. R. (1971) Qualitative inheritance of fungicide tolerance in a natural population of *Cochliobolus carbonum. Phytopathology* **61**, 458.

MacLennan, D. H., Kuć, J. and Williams, E. B. (1963) Chemotherapy of the apple scab disease with butyric acid derivatives. *Phytopathology* **53**, 1261.

McMillan, R. T. (1970) Post-harvest control of *Sclerotinia sclerotiorum* of pole beans. *Proc. Fla. St. hort. Soc.* **82**, 139.

McNabb, H. S., Heybroek, H. M. and MacDonald, W. L. (1970) Anatomical factors in resistance to Dutch elm disease. *Neth. J. Pl. Pathol.* **76**, 196.

McNall, E. C. (1960) Metabolic studies on griseofulvin and its mechanism of action. *Antibiot. Ann.*, p. 674.

McNew, G. L. and Sundholm, N. K. (1949) The fungicidal activity of substituted pyrazoles and related compounds. *Phytopathology* **39**, 721.

McOnie, K. C., Kellerman, C. and Kruger, D. J. (1969) Benlate, a highly promising new fungicide for the control of black spot. *S. Afr. Citrus J., 1969*, no. 423: p. 7. (*Rev. Pl. Path.* **49**, 2075, 1970.)

MacSwan, I. C. (1971a) Prune (Early Italian): brown rot fruit rot (*Monilinia fructicola*). *Fungicide–Nematicide Tests* **26**, 54.

MacSwan, I. C. (1971b) Blackberry (evergreen): storage rots (*Botrytis* sp., *Rhizopus* sp., *Penicillium* sp. and *Alternaria* sp.). *Fungicide–Nematicide Tests* **26**, 55.

Maeda, T., Abe, H., Kakiki, K. and Misato, T. (1970) Studies on the mode of action of organophosphorus fungicide, Kitazin. *Agr. Biol. Chem.* **34**, 700.

Magendans, J. F. C. and Dekker, J. (1966) A microscopic study of powdery mildew on barley after application of the systemic compound wepsyn. *Neth. J. Pl. Path.* **72**, 274.

Magie, R. O. (1970) Control methods for post-harvest diseases of flowers and bulbs. *Rep. 1st International Symp. on Flower Bulbs, Lisse, Netherlands (1970) Sect. 2*, p. 4.

Maier, C. R. (1960) Streptomycin absorption, translocation and retention in hops. *Phytopathology* **50**, 351.

Malone, J. P. (1968) Mercury-resistant *Pyrenophora avenae* in Northern Ireland seed oats. *Pl. Path.* **17**, 41.

Malz, H. *et al.* (1968) Belg. Pat. 701322.

Mankin, C. J. and Wood, L. S. (1969) *Fungicide–Nematicide Tests* **25**, 106.

Mankin, C. J. and Wood, L. S. (1970) *Fungicide–Nematicide Tests* **26**, 128.

Manners, K. B. and Corke, A. T. K. (1971) Bitter rot (caused by *Gloeosporium* spp.). *Rep. agric. hort. Res. Stn Bristol, 1970*, p. 116.

Mariouw Smit, F. (1969) Diaethyl-methylethoxycarbonyl-pyrazolo-pyrimidino-yl fosforothiolaat, een nieuw systemisch werkzaam meeldauw bestrijdingsmiddel. *Meded. Fac. Landbouwwetensch. Rijksuniv. Gent.* **34**, 763.

Martin, J. T. and Juniper, B. E. (1970) *The cuticles of plants.* London: Edward Arnold Ltd.

Mason, T. G. and Maskell, E. J. (1928a) Studies on the transport of carbohydrates in the cotton plant. I. *Ann. Bot. Lond.* **42**, 189.

Mason, T. G. and Maskell, E. J. (1928b) Studies on the transport of carbohydrates in the cotton plant. II. *Ann. Bot. Lond.* **42**, 571.

Mason, T. G. and Phillis, E. (1933) On the simultaneous movement of solutes in opposite directions through the phloem. *Ann. Bot. Lond.* **50**, 161.

Massee, G. (1903) On a method of rendering cucumber and tomato plants immune against fungus parasites. *J. R. hort. Soc.* **28**, 142.

Mathre, D. E. (1968) Uptake and binding of oxathiin systemic fungicides by resistant and sensitive fungi. *Phytopathology* **58**, 1464.

Mathre, D. E. (1970) Mode of action of oxathiin systemic fungicides. I. Effect of carboxin and oxycarboxin on the general metabolism of several Basidiomycetes. *Phytopathology* **60**, 671.

Mathre, D. E. (1971a) Mode of action of oxathiin systemic fungicides. III. Effect on mitochondrial activities. *Pest. Biochem. Physiol.* **1**, 216.

Mathre, D. E. (1971b) Mode of action of oxathiin systemic fungicides, structure-activity relationships. *J. Agr. Fd Chem.* **19**, 872.

Matta, A. (1963) Attivita polifenolossidasica resistenza indotta alla fusariosi nel Pomodoro. *Riv. Pat. veg., Pavia, Ser. 3*, **3**, 99.

Matta, A. and Garibaldi, A. (1970) Attivita di fungicidi sistemici contro le Verticillosi del Pomodoro e della melanzana. *Agricoltore Ital.* **70, (25 NS)**, 331.

Matta, A. and Gentile, A. T. (1971) Activation of the thiophanate systemic fungicides by plant tissues. *Meded. Fac. Làndbouwwetensch. Rijksuniv. Gent.* **36**, 1151.

Matta, A., Gentile, I. A. and Giai, I. (1967) Variazioni del contenuto in feloli solubili indotte del *Fusarium oxysporum* f. *lycopersici* in piante di pomodoro suscettibili e resistenti. *Ann. Fac. Sci. Agr. della Univ. degli Studi di Torino* **4**, 17.

Maude, R. B. (1966) Pea seed infection by *Mycosphaerella pinodes* and *Ascochyta pisi* and its control by seed soaks in thiram and captan suspensions. *Ann. appl. Biol.* **57**, 193.

Maude, R. B. and Kyle, A. M. (1970a) Seed treatments with benomyl and other fungicides for the control of *Ascochyta pisi* of peas. *Ann. appl. Biol.* **66**, 37.

Maude, R. B. and Kyle, A. M. (1970b) Laboratory screening tests of systemic compounds as potential dry seed dressings. *Rep. natn. Veg. Res. Stn for 1969*, p. 105.

Maude, R. B. and Kyle, A. M. (1971) Seed dressings with systemic fungicides for the control of seed-borne fungal pathogens. *Rep. natn. Veg. Res. Stn for 1970*, p. 106.

Maude, R. B. and Moule, C. G. (1971a) Leaf spot *(Septoria apiicola)* of celery. *Rep. natn. Veg. Res. Stn for 1970*, p. 107.

Maude, R. B. and Moule, C. G. (1971b) Dark leaf spot *(Alternaria brassicicola)* of cabbage. *Rep. natn. Veg. Res. Stn for 1970*, p. 107.

Maude, R. B. and Shuring, C. G. (1969) Seed treatments with 'Vitavax' for the control of loose smut of wheat and barley. *Ann. appl. Biol.* **64**, 259.

Maude, R. B. and Shuring, C. G. (1970) *Septoria* blight of celery. *Rep. natn. Veg. Res. Stn for 1969*, p. 103.

Maude, R. B., Vizor, A. S. and Shuring, C. G. (1969) The control of fungal seed-borne diseases by means of a thiram seed soak. *Ann. appl. Biol.* **64**, 245.

Melin, P. (1970) Nouvelles perspectives de lutte contre la cercosporiose du Bananier – premières indications sur les résultats obtenus avec le Benlate dans les bananeraies du Cameroun. *Fruits d'outre-mer* **25**, 141. (*Rev. Pl. Path.* **50**, 2389, 1971.)

Melnikov, N. N. (1971) Chemistry of pesticides. *Residue Rev.* **36**, 1.

Melville, S. C. and Jemmett, J. L. (1971) The effect of glume blotch on the yield of winter wheat. *Pl. Path.* **20**, 14.

Mercer, R. T. (1971) Some studies on the systemic activity of the thiophanate fungicides in plants. *Pestic. Sci.* **2**, 214.

Metcalfe, P. B. and Brown, J. F. (1969) Evaluation of nine fungicides in controlling flag smut of wheat. *Pl. Dis. Reptr.* **53**, 631.

Meyer, C. von, Greenfield, S. A. and Seidel, M. (1970) Wheat leaf rust: control by 4-*n*-butyl-1,2,4-triazole a systemic fungicide. *Science N.Y.* **169**, 997.

Meyer, H. (1950) Über den Einfluss von Cadmium auf die Krankheitsbereitschaft der Weizens für *Erysiphe graminis tritici* Marchal. *Phytopath. Z.* **17**, 63.

Meyer, R. W. and Parmeter, J. R. (1968) Changes in chemical tolerance associated with heterocaryosis in *Thanatephorus cucumeris*. *Phytopathology* **58**, 472.

Meynhardt, J. T. and Malan, A. H. (1963) Translocation of sugars in double stem grape vines. *S. Afr. J. agr. Sci.* **6**, 337.

Miller, P. M. (1970) Reducing discharge of ascospores of *Venturia inaequalis* with a spring application of benomyl, thiabendazole or urea. *Pl. Dis. Reptr* **54**, 27.

Milthorpe, F. L. and Moorby, J. (1969) Vascular transport and its significance in plant growth. *A. Rev. Pl. Physiol.* **20**, 117.

Minabe, M. (1951) On the relation between the disease susceptibility of naked barley and the silicification of epidermal cells of leaves. *Proc. Crop. Sci. Soc. Japan* **19**, 255.

Ministry of Agriculture, Fisheries and Food (1971) 1971 List of approved products and their uses for farmers and growers.

Misato, T. (1967) Blasticidin S. *Antibiotics.* Vol. I. *Mechanisms of action* (eds. D. Gottlieb and P. D. Shaw) p. 434. Springer, New York.

Misato, T. (1969) Mode of action of agricultural antibiotics developed in Japan. *Residue Rev.* **25**, 93.

Misato, T. (1970) The development of agricultural antibiotics in Japan. *Jap. Pesticide Information* No. 1, p. 15.

Misato, T., Ishii, M., Asakawa, M., Okimoto, Y. and Fukunaga, K. (1959) Antibiotics as protectant fungicides against rice blast. II. The therapeutic action of blasticidin-S. *Ann. phytopath. Soc. Japan* **24**, 302.

Mitchell, J. E., Hotson, H. H. and Bell, F. H. (1950) The use of sulfanilamide derivatives as eradicant fungicides for wheat stem rust. *Phytopathology* **40**, 873.

Mitchell, J. W., Smale, B. C., Daly, E. J., Preston, W. H. Jr., Pridham, T. G. and Sharpe, E. S. (1959) Absorption and translocation of the F-17 complex by bean plants and subsequent effect on the rust fungus, *Uromyces phaseoli typica. Pl Dis. Reptr* **43**, 431.

Mitchell, J. W., Smale, B. C. and Metcalf, J. W. (1960) Absorption and translocation of regulators and compounds used to control plant diseases and insects. *Adv. Pest Control Res.* **3**, 359.

Mitchell, J. W., Zaumeyer, W. J. and Preston, W. H. (1954) Absorption and translocation of streptomycin by bean plants and its effect on the halo and common blight organisms. *Phytopathology* **44**, 25.

Mizukami, T. (1970) 1969 evaluation of candidate pesticides (B) fungicides. *Jap. Pesticide Information* **4**, 12.

Moje, W., Kendrick, J. B. Jr. and Zentmyer, G. A. (1963) Systemic and fungicidal activity of D-, L- and DL-ethionine, S-alkyl-DL-homocysteine derivatives and methionine antagonists. *Phytopathology* **53**, 883.

Mokrzecki, S. (1903) Über die innere Therapie der Pflanzen. *Z. PflKrankh. PflPath. PflSchutz.* **13**, 257.

Moller, W. J. and Carter, M. V. (1969) A preliminary observation on apricot dieback prevention with chemicals. *Pl. Dis. Reptr* **53**, 828.

Moller, W. J., Wicks, T. J. and Steed, J. N. (1969) Field evaluation of a new fungicide (benomyl) for apple scab. *Aust. J. exp. Agric. Anim. Husb.* **9**, 461. (*Rev. Pl. Path.* **49**, 198, 1970.)

Moore, M. H. (1966) Glasshouse experiments on apple scab. II. Direct and translocated fungicidal activity. *Ann. appl. Biol.* **57**, 451.

Moore, M. H. (1967) Glasshouse experiments on apple scab. III. Fungicidal mixtures, curative translocation and the influence of mildew. *Ann. appl. Biol.* **59**, 239.

Moreshet, S. (1970) Effect of environmental factors on cuticular transpiration resistance. *Pl. Physiol. Lancaster* **46**, 815.

Morgan, N. G. (1952) Spray application problems. VI. A laboratory technique using *Botrytis fabae* on broad bean for the biological evaluation of fungicidal spray deposits. *Long Ashton 1952 Annual Report,* p. 99.

Morris, D. A. and Thomas, E. E. (1968) Distribution of ^{14}C-labelled sucrose in seedlings of *Pisum sativum* L. treated with indoleacetic acid and kinetin. *Planta* **83**, 276.

Moseman, J. G. (1968) Fungicidal control of smut diseases of cereals. U.S. Dept. of Agr. CR42–68.

Moss, V. D. (1961) Antibiotics for control of blister rust on western white pine. *For. Sci.* **7**, 380.

Moss, V. D., Viche, H. J. and Klomparens, W. (1960) Antibiotic treatment of western white pines infected with blister rust. *J. For.* **58**, 691.

Müller, A. (1926) *Die innere Therapie der Pflanzen,* Monographien der angewandten Entomologie, N. 8. Berlin: Paul Parey.

Müller, H. W. K. (1970) Results with benomyl against strawberry grey mould *(Botrytis cinerea). NachrBl. dt. PflSchutzdienst., Stuttg.* **22**, 81. (*Rev. Pl. Path.* **50**, 812, 1970.)

Müller, K. O., Klinkowski, M. and Meyer, G. (1939) Physiologisch-genetische Untersuchungen über die Resistenz der Kartoffel gegenüber *Phytophthora infestans. Naturwissenschaften* **27**, 765.

Müller, K. O., Mackay, J. H. E. and Friend, J. N. (1954) Effect of streptomycin on host–pathogen relationship of a fungal pathogen. *Nature, Lond.* **174**, 878.

Münch, E. (1927) Versuche über den Saftkreislauf. *Ber. Deutsch Bot. Ges.* **45**, 340.

Mussell, H. E. and Green, R. J., Jr. (1968) Production of polygalacturonase by *Verticillium albo-atrum* and *Fusarium oxysporum* f. sp. *lycopersici, in vitro* and in vascular tissue of susceptible and resistant hosts. *Phytopathology* **58**, 1061.

Mussell, H. W. and Green, R. J., Jr. (1970) Host colonization and polygalacturonase production by two tracheomycotic fungi. *Phytopathology* **60**, 192.

Nagatsu, J., Anzai, K. and Suzuki, S. (1962) Pathocidin, a new antifungal antibiotic. II. Taxonomic studies on the pathocidin producing organism *Streptomyces albus* var. *pathocidicus. J. Antibiotics (Tokyo)* **A15**, 103.

Nakamura, H. and Sakurai, L. (1962) Tolerance of *Piricularia oryzae* on blasticidin-S. *Ann. Phytopath. Soc. Japan* **27**, 84.

Nakanishi, T. and Oku, H. (1969) Mechanism of selective toxicity: absorption and detoxication of an antibiotic, ascochytine, by sensitive and insensitive fungi. *Phytopathology* **59**, 1563.

Nakanishi, T. and Oku, H. (1970) Mechanism of selective toxicity of fungicides: absorption, metabolism and accumulation of pentachloronitrobenzene by phytopathogenic fungi. *Ann. Phytopath. Soc. Japan* **36**, 67.

Napier, E. J., Rhodes, A., Turner, D. I., Toothill, J. P. R. and Dunn, A. (1957) Systemic action of captan against *Botrytis fabae* (chocolate spot of broad bean). *J. Sci. Fd Agric.* **8**, 467.

Napier, E. J., Turner, D. I. and Rhodes, A. (1956) The *in vitro* action of griseofulvin against pathogenic fungi of plants. *Ann. Bot. N.S.* **20**, 461.

Natti, J. J. (1967) Bean white mold control with foliar sprays. *Fm Res.* **33**, 10.

Natti, J. J. (1970) Bean seed orientation and performance of fungicide treatment of seed. *Phytopathology* **60**, 577.

Natti, J. J. and Crosier, D. C. (1971) Seed and soil treatments for control of bean root rots. *Pl. Dis. Reptr* **55**, 483.

Natti, J. J. and Szkolnik, M. (1954) Influence of growth regulators on resistance of beans to *Colletotrichum lindemuthianum. Phytopathology* **44**, 111.

Netzer, D. and Dishon, I. (1970) Effect of mode of application of benomyl on control of *Sclerotinia sclerotiorum* on musk melons. *Pl. Dis. Reptr* **54**, 909.

Netzer, D., Dishon, I. and Krikun, J. (1970) Control of some diseases on greenhouse grown vegetables with benomyl as related to studies of its movement. *Proc. 7th Int. Congr. Pl. Prot. (Paris),* p. 222.

Nevins, D. J., English, P. D. and Albersheim, P. (1968) Changes in cell wall polysaccharides associated with growth. *Pl. Physiol.* **43**, 914.

Nicholson, J. F., Meyer, W. A., Sinclair, J. B. and Buller, J. D. (1971) Turf isolates of *Sclerotinia homoeocarpa* tolerant to dyrene. *Phyt. Zeitschr.* **72**, 169.

Nickell, L. G., Gordon, P. N. and Goenaga, A. (1961) 2-*n*-Alkylmercapto-1,4,5,6-tetrahydropyrimidines, chemotherapeutic agents for plant rusts. *Pl. Dis. Reptr* **45**, 756.

Niemann, G. J. (1964) Investigations on the chemical control of powdery mildew. *Thesis, University of Utrecht.*

Niemann, G. J. and Dekker, J. (1966a) Evaluation of the activity of a number of procaine derivatives and analogous compounds against powdery mildew *(Sphaerotheca fuliginea)* on cucumber. *Ann. appl. Biol.* **57**, 53.

Niemann, G. J. and Dekker, J. (1966b) Activity against rust and powdery mildew of some *para*phenylenediamines and related compounds. *Neth. J. Pl. Path.* **72**, 213.

Noble, M., Maggarvie, Q. D., Hams, A. F. and Leafe, E. L. (1966) Resistance to mercury of *Pyrenophora avenae* in Scottish seed oats. *Pl. Path.* **15**, 23.

Noguchi, T. and Ohkuma, K. (1971a) Chemistry and metabolism of thiophanate fungicides. *Symp. Terminal Pesticide Residues, Tel Aviv,* p. 263.

Noguchi, T. and Ohkuma, K. (1971b) Relation of chemical structure to biological activity in thiophanate. *Proc. 2nd Int. Congress of Pesticide Chem.* **5**, 257.

North, C. P. (1962) Technique of injecting substances into woody plants. Chapter from *A decade of synthetic chelating agents in inorganic plant nutrition,* p. 138. Edited and published by A. Wallace, 2 278 Parnell Avenue, Los Angeles, U.S.A.

Norton, J. B. S. (1916) Internal action of chemicals on resistance of tomatoes to leaf diseases. *Md. Agric. Exp. Sta. Bull.* **192**, 17.

Noveroske, R. L., Kuć, J. and Williams, E. B. (1964a) Oxidation of phloridzin related to resistance of *Malus* to *Venturia inaequalis. Phytopathology* **54**, 92.

Noveroske, R. L., Williams, E. B. and Kuć, J. (1964b) Beta-glucosidase and phenoloxidase in apple leaves and their possible relation to resistance to *Venturia inaequalis. Phytopathology* **54**, 98.

O'Brien, T. P. and Carr, O. J. (1970) A suberised layer in the cell walls of the bundle sheath of grasses. *Aust. J. biol. Sci.* **23**, 275.

Ogawa, J. M., Manji, B. T. and Blair, E. (1968a) Efficacy of Fungicide 1991 in reducing fruit rot of stone fruits. *Pl. Dis. Reptr* **52**, 722.

Ogawa, J. M., Su, H. J., Tsai, Y. P., Chen, S. S. and Liang, C. H. (1968b) Protective and therapeutic action of 1-(butylcarbamoyl)-2-benzimidazole carbamic acid, methyl ester (F 1991) against the banana crown rot pathogens. *Pl. Prot. Bull. Taiwan* **10**, 1. (*Rev. Pl. Path.* **49**, 1100, 1970.)

Ogawa, J. M., Su, H. J., Tsai, Y. P. and Lee, I. M. (1969) Postharvest decay development as affected by systemic activity of benomyl in bananas, oranges and pineapples. *Phytopathology* **59**, 1043 (abs.).

Ohkuma, K. (1961) Chemical structure of toyocamycin. *J. Antibiotics (Tokyo)* **A19**, 343.

Ohmori, K. (1967) Studies on characters of *Piricularia oryzae* made resistant to kasugamycin. *J. Antibiotics (Tokyo),* Ser. A **20**, 109.

Okamoto, H. (1968) Fungicidal properties of kasugamycin. *Noyaku-Shunju.* **23**, 16.

Okimoto, Y. and Misato, T. (1963a) Antibiotics as protectant bactericide against bacterial leaf blight of rice plant. 2. Effect of cellocidin on growth, respiration and glycolysis of *Xanthomonas oryzae*. *Ann. Phytopath. Soc. Japan* **28**, 209.

Okimoto, Y. and Misato, T. (1963b) Antibiotics as protectant bactericide against bacterial leaf blight of rice plant. 3. Effect of cellocidin on TCA cycle, electron transport system and metabolism of protein in *Xanthomonas oryzae*. *Ann. Phytopath. Soc. Japan* **28**, 250.

Okuda, T., Suzuki, M., Egawa, Y. and Askino, K. (1958) Studies on cycloheximide and its new stereoisomeric antibiotics. *Chem. Pharm. Bull. (Tokyo)* **6**, 328.

Olsen, C. (1939) The employment for water culture experiments of distilled water containing traces of copper. *C.r. Trav. Lab. Carlsberg, Ser. chim.* **23**, 37.

Oort, A. J. P. and van Andel, O. M. (1960) Aspects of chemotherapy. *Meded. LandbHoogesch. OpzoekStns. Gent* **25**, 981.

Oort, A. J. P. and Dekker, J. (1960) Experiments with rimocidin and pimaricin, two fungicidal antibiotics with systemic action. *Proc. IVth int. Congr. Crop Prot., Hamburg,* 1957, **2**, 1565.

Oort, A. J. P. and Dekker, J. (1964) Systemic activity of 6-azauracil against powdery mildew. *Proc. Symp. Host–Parasite Relations, Budapest 1964*, p. 171.

Ó'Ŕiordáin, F., Kavanagh, T. and O'Callaghan, T. F. (1971) Comparisons of systemic and protectant fungicides for disease control on four soft fruits. *Proc. 6th Br. Insectic. Fungic. Conf.* **2**, 324.

Ost, W., von Bruchhausen, V. and Drandarevski, C. (1971) Transport of the systemic fungicide Cela W 524 in barley plants. (Part I). *Pestic. Sci.* **2**, 219.

Ost, W., Thomas, K., Jerchel, D. and Appel, K. R. (1969) Ger. offen. 1901421.

Otake, N., Takeuchi, S., Endo, T. and Yonehara, H. (1965) The structure of blasticidin S. *Tetrahedron Letters* No. 19, 1411.

Overbeek, J. van (1956) Absorption and translocation of plant regulators *A. Rev. Pl. Physiol.* **7**, 355.

Oxford, A. E., Raistrick, H. and Simonart, P. (1939) XXIX. Studies in the biochemistry of micro-organisms. LX. Griseofulvin $C_{17}H_{17}O_6Cl$, a metabolic product of *Penicillium griseofulvum* Dierckx. *Biochem. J.* **33**, 240.

Paisley, R. B. and Heuberger, J. W. (1971) Apple: scab *(Venturia inaequalis). Fungicide–Nematicide Tests* **26**, 34.

Palmiter, D. H. (1970) Apple: scab *(Venturia inaequalis). Fungicide–Nematicide Tests* **25**, 31.

Papavizas, G. C. and Davey, C. B. (1963) Effect of sulfur-containing amino compounds and related substances on Aphanomyces root rot of peas. *Phytopathology* **53**, 109.

Papavizas, G. C. and Lewis, J. A. (1970) Suppression of Aphanomyces root rot of peas by soil amendments, fungicides and fumigants. *Phytopathology* **60**, 1306.

Papavizas, G. C. and Lewis, J. A. (1971) Black root rot of bean and tobacco caused by *Thielaviopsis basicola* as affected by soil amendments and fungicides in the greenhouse. *Pl. Dis. Reptr* **55**, 352.

Parr, W. J. (1971) Private communication.

Parr, W. J. and Binns, E. S. (1970) Acaricidal activity of benomyl. *A. Rep. Glasshouse Crops Res. Inst. 1969*, p. 113.

Parr, W. J. and Binns, E. S. (1971) Cucumber pests. Integrated control of red spider mite. *A. Rep. Glasshouse Crops Res. Inst. 1970*, p. 119.

Partridge, A. D. and Rich, A. E. (1962) Induced tolerance to fungicides in three species of fungi. *Phytopathology* **52**, 1000.

Patel, P. V. and Johnson, J. R. (1968) Dominant mutation for nystatin resistance in yeast. *Appl. Microbiol.* **16**, 164.

Pathak, K. D. and Joshi, L. M. (1970) Seed treatment by systemic fungicides for the control of wheat rusts. *Indian Phytopath.* **13**, 655.

Patil, S. S. and Dimond, A. E. (1967a) Inhibition of *Verticillium* polygalacturonase by oxidation products of polyphenols. *Phytopathology* **57**, 492.

Patil, S. S. and Dimond, A. E. (1967b) Induction and repression of polygalacturonase synthesis in *Verticillium albo-atrum. Phytopathology* **57**, 825.

Patil, S. S. and Dimond, A. E. (1967c) Effect of sugars and sugar alcohols on production of polygalacturonase by *Fusarium oxysporum* f. sp. *lycopersici. Phytopathology* **57**, 825.

Patil, S. S. and Dimond, A. E. (1968) Repression of polygalacturonase synthesis in *Fusarium oxysporum* f. sp. *lycopersici* by sugars, and its effect on symptom reduction in infected tomato plants. *Phytopathology* **58**, 676.

Paul, R. and Tchelitcheff, S. (1955) Constitution chimique de l'inactone. Synthèse partielle de trois isomères de l'actidione. *Bull. Soc. Chim. France* **55**, 1316.

Paulus, A. O., Shibuya, F., Holland, A. H. and Nelson, J. (1970a) Timing interval for control of *Septoria apiicola* leaf spot of celery *(Apium graveolens). Pl. Dis. Reptr* **54**, 531.

Paulus, A. O., Shibuya, F., Nelson, J. and Harvey, D. A. (1970b) Systemic fungicide interval for control of sugarbeet *Cercospora* leaf spot. *Phytopathology* **60**, 1539.

Paulus, A. and Starr, G. H. (1951) Control of loose smut with antibiotics. *Agron. J.* **43**, 617.

Paulus, A. O., Voth, V., Shibuya, F., Bowen, H. and Holland, A. H. (1969) Fungicidal control of *Botrytis* fruit rot of strawberry. *Calif. Agric.* **23**, 15. (*Rev. appl. Mycol.* **48**, 1858, 1969.)

Peel, A. J. (1970) Further evidence for the relative immobility of water in the sieve tubes of willow. *Physiologia Pl.* **23**, 667.

Pellegrini, G., Bugiani, A. and Tenerini, I. (1965) Systemic properties of a new class of fungicides. *Phytopath. Z.* **52**, 37.

Pellissier, M., Lacasse, N. L. and Cole, H., Jr. (1971) Effect of benomyl on the response to ozone in Pinto bean. *Phytopathology* **61**, 131.

Pepin, H. S. and Ormrod, D. J. (1968) Control of mummy berry of highbush blueberry. *Can. Pl. Dis. Surv.* **48**, 132.

Person, C., Samborski, D. J. and Forsyth, F. R. (1957) Effect of benzimidazole on detached wheat leaves. *Nature, Lond.* **180**, 1294.

Pessanha, B. M. R., Iamamoto, T. and Scaranari, H. J. (1970) Benlate and cercobin – new fungicides in the control of leaf spots of strawberry *(Fragaria vesca* L.*)* (in Spanish). *Biológico* **36**, 121. (*Rev. Pl. Path.* **50**, 804, 1971.)

Petersen, D. and Cation, D. (1950) Exploratory experiments on the use of Actidione for the control of peach brown rot and cherry leaf spot. *Pl. Dis. Reptr* **34**, 5.

Peterson, C. A. and Edgington, L. V. (1969) Translocation of the fungicide benomyl in bean plants. *Phytopathology* **59**, 1044.

Peterson, C. A. and Edgington, L. V. (1970) Transport of the systemic fungicide benomyl in bean plants. *Phytopathology* **60**, 475.

Peterson, C. A. and Edgington, L. V. (1971) Transport of benomyl into various plant organs. *Phytopathology* **61**, 91.

Peturson, B., Forsyth, F. R. and Lyon, C. B. (1958) Chemical control of cereal rusts. II. Control of leaf rust of wheat with experimental chemicals under field conditions. *Phytopathology* **48**, 655.

Phelps, R. H. (1969) Control of two citrus diseases by low volume spraying. *Bull. Citrus Res., Univ. W. Indies no.* 16, 4 pp. (*Rev. Pl. Path.* **50**, 1234, 1971.)

Phelps, W. R., Kuntz, J. E. and Riker, A. J. (1957) Antibiotics delay oak wilt symptoms on inoculated northern pin oaks in central Wisconsin. *Phytopathology* **47**, 27.

Phelps, W. R., Kuntz, J. E. and Ross, A. (1966) A field evaluation of antibiotics and chemicals for control of oak wilt in northern pin oaks. *Pl. Dis. Reptr* **50**, 736.

Pierson, C. F. (1966) Fungicides for the control of blue mold rot of apples. *Pl. Dis. Reptr* **50**, 913.

Pitblado, R. E. and Edgington, L. V. (1971) Effect of various surfactants on the 'solubility' of benomyl. *Proc. Can. Phytopathological Soc. 37th Session,* p. 20.

van der Plank (1963) *Plant diseases: epidemics and control.* New York and London: Academic Press.

Pleisch, P., Lauber, J. and Preiser, F. A. (1971) Post-harvest disease control of pome fruits with thiabendazole. *Proc. 6th Br. Insectic. Fungic. Conf.* **1**, 181.

Pluijgers, C. W. (1959) Direct and systemic antifungal action of dithiocarbamic acid derivatives. *Thesis, University of Utrecht.*

Pluijgers, C. W. and Kaars Sijpesteijn, A. (1966) The antifungal activity of thiosemicarbazides *in vitro* and their systemic activity against cucumber scab *Cladosporium cucumerinum. Ann. appl. Biol.* **57**, 465.

Pomerleau, R. (1970) Pathological anatomy of the Dutch elm disease. Distribution and development of *Ceratocystis ulmi* in elm tissue. *Can. J. Bot.* **48**, 2043.

Pomerleau, R. and Mehran, A. R. (1966) Distribution of spores of *Ceratocystis ulmi* labelled with phosphorus-32 in green shoots and leaves of *Ulmus americana. Naturaliste can.* **93**, 577.

Pommer, E. H. (1970) The systemic efficacy of a new fungicide of the Furane carbon acid anilide group (BAS 3191 F). *VII Int. Congr. Pl. Prot. 1970,* (abs.).

Pommer, E. H. (1971) The systemic activity of a new fungicide of the furane carbonic acid anilide group (BAS 3191 F). *Proc. 2nd int. Congr. Pestic. Chem.* **5**, 397.

Pommer, E. H., Hampel, M. and Löcher, F. (1971) Results with 'BAS 3270F'. *Proc. 6th Br. Insectic. Fungic. Conf.* **2**, 587.

Pommer, E. H. and Kradel, J. (1967) Substitutierte dimethylmorpholin-derivate als neue fungizide zur bekämpfung echter mehltaupilze. *Meded. Rijksfaculteit Landbouw. Wetenschappen Gent* **32**, 735.

Pommer, E. H. and Kradel, J. (1969) Mebenil (BAS 3050F), a new compound with specific action against some Basidiomycetes. *Proc. 5th Br. Insectic. Fungic. Conf.* **2**, 563.

Pommer, E. H., Otto, S. and Kradel, J. (1969) Some results concerning the systemic action of tridemorph. *Proc. 5th Br. Insectic. and Fungic. Conf.* **2**, 347.

Ponchet, J. (1959) La maladie du piétin-verse des céréales: *Cercosporella herpotrichoides,* importance agronomique biologie, epiphytologie. *Ann. Épiphyt.* **10**, 45.

Porter, R. H. (1956) Seed treatment tests for control of barley loose smut. *Pl. Dis. Reptr* **40**, 106.

Potter, A. W. and Milburn, J. A. (1970) Sub-aqueous transpiration in new perspective. *New Phytol.* **69**, 961.

Powelson, R. L. and Cook, G. E. (1969) *Fungicide–Nematicide Tests* **25**, 97.

Powelson, R. L. and Beaver, R. G. (1970) *Fungicide–Nematicide Tests* **26**, 108.

Powelson, R. L. and Shaner, G. E. (1966) An effective chemical seed treatment for systemic control of seedling infection of wheat by stripe rust *(Puccinia striiformis). Pl. Dis. Reptr* **50**, 806.

Pramer, D. (1953) Observations on the uptake and translocation of five actinomycete antibiotics by cucumber seedlings. *Ann. appl. Biol.* **40**, 617.

Pramer, D. (1955) Absorption of antibiotics by plant cells. *Science N.Y.* **121**, 507.

Pramer, D. (1956) Absorption of antibiotics by plant cells. II. *Archs. Biochem. Biophys.* **62**, 265.

Pratella, C. and Tonini, G. (1968) Prevention and cure in the control of *Gloeosporium* rot (in Italian). *Frutticoltura* **30**, 235. (*Rev. Pl. Path.* **49**, 1708, 1970.)

Preston, W. H., Jr., Daly, E. G., Smale, B. C. and Mitchell, J. W. (1956) Effects of absorbed and translocated F-17 culture filtrate antibiotic factors on the bean rust organism. *Phytopathology* **46**, 469.

Preston, W. H., Mitchell, J. W. and Reeve, W. R. (1954) Movement of alphamethoxyphenylacetic acid from one plant to another through their root systems. *Science, N.Y.* **119**, 437.

Price, D. (1970) Tulip fire caused by *Botrytis tulipae. A. Rep. Glasshouse Crop Res. Inst. 1969*, p. 124.

Pridham, T. G. (1961) Plant disease antibiotics. *Developments in industrial microbiology, Vol. 2,* p. 141. New York: Plenum Press.

Pridham, T. G., Lindenfelser, L. A., Shotwell, O. L., Stodola, F. H., Benedict, R. G., Foley, C., Jackson, R. W., Zaumeyer, W. J., Preston, W. H., Jr. and Mitchell, J. W. (1956) Antibiotics against plant disease. I. Laboratory and greenhouse survey. *Phytopathology* **46**, 568.

Priest, D. and Wood, R. K. S. (1961) Strains of *Botrytis allii* resistant to chlorinated nitrobenzenes. *Ann. appl. Biol.* **49**, 445.

Pringle, R. B. and Scheffer, R. P. (1964) Host-specific plant toxins. *A. Rev. Phytopath.* **2**, 133.

Purdy, L. H. (1964) Inhibition of wheat stripe rust by the antibiotic phleomycin in the greenhouse. *Pl. Dis. Reptr* **48**, 159.

Raa, J. (1968) Natural resistance of apple plants to *Venturia inaequalis.* A biochemical study of its mechanism. *Thesis, Utrecht.* 100 pp.

Raa, J. and Kaars Sijpesteijn, A. (1968) A biochemical mechanism of natural resistance of apple to *Venturia inaequalis. Neth. J. Pl. Pathol.* **74**, 229.

Raabe, R. D. and Hurliman, J. H. (1971) Control of *Thielaviopsis basicola* root rot of Poinsettia with benomyl and thiabendazole. *Pl. Dis. Reptr* **55**, 238.

van Raalte, M. H. (1954) A test for the translocation of fungicide through plant tissues. *C.r. IIIe Congrès International de Phytopharmacie, Paris, 1952, Vol. 2,* p. 76.

van Raalte, M. H., Kaars Sijpesteijn, A., van der Kerk, G. J. M., Oort, A. J. P. and Pluijgers, C. W. (1955) Investigations on plant chemotherapy. *Meded. Landb-Hoogesch. OpzoekStns. Gent.* **20**, 543.

Rader, W. E., Monroe, C. M. and Whetstone, R. R. (1952) Tetrahydropyrimidine derivatives as potential foliage fungicides. *Science* **115**, 124.

Ragsdale, N. N. and Sisler, H. D. (1970) Metabolic effects related to fungitoxicity of carboxin. *Phytopathology* **60**, 1422.

Ramm, C. von, Lucas, G. B. and Marshall, H. V. (1962) The effect of maleic hydrazide on brown spot incidence of flue-cured tobacco. *Tobacco Sci.* **6**, 100.

Rangaswami, G. (1956) A preliminary report on the use of mycothricin complex in plants. *Pl. Dis. Reptr* **40**, 483.

Ranney, C. D. (1971) Studies with benomyl and thiabendazole on control of cotton diseases. *Phytopathology* **61**, 783.

Rao, K. V. (1960) E73: an anti-tumour substance II. structure. *J. Am. Chem. Soc.* **82**, 1130.

Rao, K. V. and Cullen, W. P. (1960) E73: an anti-tumour substance. *J. Am. Chem. Soc.* **82**, 1127.

Rao, S. S. and Grollman, A. P. (1967) Cycloheximide resistance in yeast: a property of the 60S ribosomal unit. *Biochem. Biophys. Res. Comm.* **29**, 696.

Rapilly, F. (1970) La détermination des dates de traitements fongicides appliqués par voie aérienne sur céréales en végétation-cas du blé d'hiver. *Phytiat.-Phytopharm.* **19**, 185.

Rawlinson, C. J. and Davies, J. M. L. (1971) Effect of eyespot *(Cercosporella herpotrichoides)* and seed treatments of benomyl, chlormequat and organomercury on rye. *Plant Pathology* **20**, 131.

Ray, J. (1901) Les maladies cryptogamiques des végétaux. *Revue gén. Bot.* **13**, 145.

Rea, B. L. and Brown, P. A. (1971) The control of barley powdery mildew using triarimol. *Proc. 6th Br. Insectic. Fungic. Conf.*, p. 52.

Reinbergs, E. and Edgington, L. V. (1968) Field control of loose smut in barley with the systemic fungicides 'Vitavax' and 'Plantavax'. *Can. J. Pl. Sci.* **48**, 31.

Renaud, R. (1968) Essais de traitements contre les Monilioses du Prunier sur fleurs et sur fruits. *Rev. Zool. agric. appl.* **3**, 81. (*Rev. appl. Mycol.* **48**, 1839, 1969.)

Renaud, R. (1970) Experiments with fungicides on Sclerotinia of French prune *(S. laxa* and *S. fructigena). Summaries of Papers, VIIth Int. Cong. Plant. Prot., Paris,* p. 267.

Rhodes, A. (1962) Status of griseofulvin in crop protection. *Antibiotics in Agriculture* (ed. M. Woodbine), p. 101. London: Butterworths.

Rhodes, A., Crosse, R., McWilliam, R., Toothill, J. P. R. and Dunn, A. T. (1957) Small-plot trials of griseofulvin as a fungicide. *Ann. appl. Biol.* **45**, 215.

Rhodes, A., Fantes, K. H., Boothroyd, B., McGonagle, M. P. and Crosse, R. (1961) Venturicidin: a new antifungal antibiotic of potential use in agriculture. *Nature, Lond.* **192**, 952.

Rhodes, R. C., Pease, H. L. and Brantley, R. K. (1971) Fate of ^{14}C-labelled chloroneb in plants and soils. *J. Agric. Food Chem.* **19**, 745.

Rice, E. L. (1948) Absorption and translocation of ammonium 2,4-dichlorophenoxyacetic acid by bean plants. *Bot. Gaz.* **107**, 301.

Rich, S. (1956) Foliage fungicides plus glycerine for the chemotherapy of cucumber scab. *Pl. Dis. Reptr* **40**, 620.

Rich, S. and Tomlinson, H. (1970) Private communication.

Richard, G. and Vallier, J. P. (1969) Treatment of cereal seed by the combination of carboxin with copper oxyquinolate. *Proc. 5th Br. Insectic. Fungic. Conf.*, p. 45.

Richardson, M. (1968) *Translocation in plants. Studies in Biology No. 10.* London: Edward Arnold Ltd.

Richmond, D. V. and Pring, R. J. (1971) The effect of benomyl on the fine structure of *Botrytis fabae. J. gen. Microbiol.* **66**, 79.

Riehm, E. (1914) Prüfung einiger Mittel zur Bekämpfung des Steinbrandes. *Zentbl. Bakt. ParasitKde (Abt. II)* **40**, 424.

Rippon, L. E., Glennie-Holmes, M. and Gilbert, S. (1970) Post-harvest treatment of bananas with thiabendazole. *Agric. Gaz. N.S.W.* **81**, 416. (*Rev. Pl. Path.* **50**, 816, 1970.)

Ristich, S. S. and Cohen, S. I. (1961) Control of disease systemically by root absorption of derivatives of 2-pyridinethiol-1-oxide. *Phytopathology* **51**, 578.

Roach, W. A. (1934) Injection for the diagnosis and cure of physiological diseases of fruit trees. *Ann. appl. Biol.* **21**, 319.

Roach, W. A. (1938) Plant injection for diagnostic and curative purposes. *Tech. Commun. Imp. Bur. Pl. Crops, no. 10.*

Roach, W. A. (1939) Plant injection as a physiological method. *Ann. Bot. N.S.* **3**, 155.

Roach, W. A. (1942) The use of plant injection in plant pathology. *Trans. Br. mycol. Soc.* **25**, 338.

Robards, A. W. (1968) A new interpretation of plasmodesmatal ultrastructure. *Planta* **82**, 200.

Robinson, H. J., Phares, H. F. and Graessle, O. E. (1964) Antimycotic properties of thiabendazole. *J. Invest. Dermatol.* **42**, 479.

Ŕodigin, M. N. and Krasnova, T. M. (1959) Seed treatment with microelements against brown rust of wheat. *Abstr. in Rev. appl. Mycol.* **39**, 405.

Rogers, W. E. (1971) Grape: powdery mildew *(Uncinula necator). Fungicide–Nematicide Tests* **26**, 58.

Rohrbach, K. G. and Apt, W. J. (1971) Field control of *Ceratocystis paradoxa* on pineapple asexual propagative parts. *Phytopathology* **61**, 1323 (abs.).

Rombouts, J. E. and Kaars Sijpesteijn, A. (1958) The chemotherapeutic effect of pyridine-2-thiol-N-oxide and some of its derivatives on plant diseases. *Ann. Appl. Biol.* **46**, 30.

Rooy, M. de (1969) Use of benzimidazole compounds for the control of some pathogenic fungi in flower bulb culture. *21st Int. Symp. Fytofarm., Fytiat., Ghent,* p. 931.

Rooy, M. de and Vink, J. M. (1969) Botrytis tulipae in tulpen. *Jversl. Lab. Bloemb Onderz. Lisse 1968–69,* 51.

Rosen, H. R. (1959) Preliminary report on the systemic activity of captan when applied to the soil for the control of black spot of roses. *Pl. Dis. Reptr* **43**, 1176.

Ross, R. G. (1971) Pear: scab *(Venturia pirina). Fungicide–Nematicide Tests* **26**, 41.

Ross, R. G. and Ludwig, R. A. (1957) A comparative study of fungitoxicity and phytotoxicity in an homologous series of N-*n*-alkylethylenethioureas. *Can. J. Bot.* **35**, 65.

Ross, R. G. and Sanford, K. H. (1971) Apple: spray injury. *Fungicide–Nematicide Tests* **26**, 35.

Rosser, W. R. (1969) Spray timing and the control of mildew in spring barley. *Proc. 5th Br. Insectic. Fungic. Conf.,* p. 20.

Rothwell, K. and Wright, S. T. C. (1967) Phytokinin activity in some new 6-substituted purines. *Proc. R. Soc. B.* **167**, 202.

Rowell, J. B. (1967) Control of leaf and stem rust of wheat by an 1,4-oxathiin derivative. *Pl. Dis. Reptr* **51**, 336.

Rowell, J. B. (1968a) Chemical control of the cereal rusts. *Ann. Rev. Phytopathol.* **6**, 243.

Rowell, J. B. (1968b) Control of leaf and stem rusts of wheat by combinations of soil application of an 1,4-oxathiin derivative with foliage sprays. *Pl. Dis. Reptr* **52**, 856.

Rowell, J. B. (1969) *Fungicide–Nematicide Tests* **25**, 98.

Rowell, J. B. (1970) *Fungicide–Nematicide Tests* **26**, 110.

Rudd Jones, D. (1956) The systemic action of sulphonamides against plant diseases. *Outlook on Agriculture* **1**, 111.

Rumbold, C. (1920a) Giving medicine to trees. *Am. For.* **26**, 359.

Rumbold, C. (1920b) The injection of chemicals into chestnut trees. *Am. J. Bot.* **7**, 1.

Ryan, E. W. and Kavanagh, T. (1971) Comparisons of fungicides for control of leaf spot *(Septoria apiicola)* of celery. *Ann. appl. Biol.* **67**, 121.

Ryan, E. W., Staunton, W. P. and Kavanagh, T. (1971) Control of diseases of five vegetable crops with systemic fungicides. *Proc. 6th Br. Insectic. Fungic. Conf.*, p. 355.

Salerno, M. and Edgington, L. V. (1963) Similarity of movement of N-alkyl-quaternary ammonium compounds in aqueous systems on paper chromatograms and in plant xylem. *Phytopathology* **53**, 605.

Salerno, M. and Somma, V. (1971) Observations on the systemic nature of benomyl in Sour Orange seedlings and results of trials against Citrus 'mal secco' (in Italian). *Phytopath. Mediterranea* **10**, 99. (*Rev. Pl. Path.* **50**, 3744, 1971.)

Samborski, D. J., Forsyth, F. R. and Person, C. (1958) Metabolic changes in detached wheat leaves floated on benzimidazole and the effect of these changes on rust reaction. *Can. J. Bot.* **36**, 591.

Samborski, D. J., Person, C. and Forsyth, F. R. (1960) Differential effects of maleic hydrazide on the growth of leaf and stem rusts of wheat. *Can. J. Bot.* **38**, 1.

Sampson, M. J. (1969) The mode of action of a new group of species-specific pyrimidine fungicides. *Proc. 5th Br. Insectic. Fungic. Conf.* **2**, 483.

Sands, D. C. and Zucker, M. L. (1971) Impaired amino acid biosynthesis in phytopathogenic pseudomonads. *Phytopathology* **61**, 909.

Sasaki, S., Ohta, N., Yamaguchi, I., Kureda, S. and Misato, T. (1968) Studies on polyoxin action. Effect on respiration and synthesis of protein, nucleic acids and cell wall of fungi. *J. Agric. Chem. Soc. Jap.* **42**, 633.

Scheffer, R. R. and Pringle, R. B. (1967) Pathogen-produced determinants of disease and their effects on host plants. *The dynamics of molecular constituents on plant-parasite interaction* (eds. C. J. Mirocha and I. Uritani) p. 217. St. Paul: Amer. Phytopathol. Soc.

Scherer, C. M. (1927) Tree injection for control of fungus diseases and insect pests. *Phytopathology* **17**, 51.

Schicke, P., Adlung, K. G. and Drandarevski, C. A. (1971) Results of several years of testing the systemic fungicide Cela W-524 against mildew and rusts of cereals. *Proc. 6th Br. Insectic. Fungic. Conf.*, p. 82.

Schicke, P. and Arndt, S. (1971) Die Bekämpfung von Zierpflanzenkrankheiten mit triforine. *23e Int. Symp. over Fytofarmacie en Fytiatrie Gent, 1971* (in press).

Schicke, P. and Veen, K. H. (1969) A new systemic, Cela-W 524 (N,N'-bis(1-formamide-2,2,2-trichloroethyl)piperazine) with action against powdery mildew, rust and apple scab. *Proc. 5th Br. Insectic. Fungic. Conf.* **2**, 569.

Schloesser, E. (1965) Sterols and the resistance of *Pythium* species to filipin. *Phytopathology* **55**, 1075.

Schlüter, K. and Weltzien, H. C. (1971) Ein Beitrag zur Wirkungsweise systemischer Fungizide auf *Erysiphe graminis. Meded. Fac. Landbouwwetensch. Rijksuniv. Gent* **36**, 1159.

Schmeling, B. von and Clark, M. (1970) Oxathiin-induced plant growth stimulation. *VII Int. Congr. Pl. Prot. 1970,* p. 227 (abs.).

Schmeling, B. von and Kulka, M. (1966) Systemic fungicidal activity of 1,4-oxathiin derivatives. *Science N.Y.* **152**, 659.

Schmeling, B. von, Kulka, M., Thiara, D. S. and Harrison, W. A. (1966) *Control of plant diseases*. U.S. patent 3 249 499.

Schönherr, J. and Bukovac, M. J. (1970a) Preferential polar pathways in the cuticle and their relationship to ectodesmata. *Planta* **92**, 189.

Schönherr, J. and Bukovac, M. J. (1970b) The nature of precipitates formed in the outer cell wall following fixation of leaf tissue with Gilson solution. *Planta* **92**, 202.

Schreiber, L. R. (1970) An evaluation of thiobendazole, 2-(4-thiazolyl) benzimidazole for the control of Dutch elm disease. *Pl. Dis. Reptr* **54**, 240.

Schroeder, W. T. and Provvidenti, R. (1969) Resistance to benomyl in powdery mildew of cucurbits. *Pl. Dis. Reptr* **53**, 271.

Schumann, G. von (1967) Stand und Entwicklung der Bekämpfung von Getreid Krankheiten durch Saatgutbehändlung. *Biol. Bund für Landungforst. 2. Pflanzenkrankh. Pfl. Path. Pflanzenschutz* **74**, 155.

Schwer, J. F. (1971) Private communication.

Schwinn, F. J. (1969) Azepines, a new class of fungicides. *Proc. 5th Br. Insectic. Fungic. Conf.* **2**, 584.

Scott, D. H., Draper, A. D. and Maas, J. L. (1970) Benomyl for control of powdery mildew on strawberry plants in the greenhouse. *Pl. Dis. Reptr* **54**, 362.

Scott, K. J. and Roberts, E. A. (1967) Control in bananas of black end rot caused by *Gloeosporium musarum*. *Aust. J. exp. Agric. Anim. Husb.* **7**, 283. (*Rev. appl. Mycol.* **46**, 3 147, 1967.)

Scott, K. J. and Roberts, E. A. (1970) Thiabendazole to reduce rotting in Packham's Triumph pears during storage and marketing. *Aust. J. exp. Agric. Anim. Husb.* **10**, 235. (*Rev. Pl. Path.* **49**, 3 381, 1970.)

Seberry, J. A. and Baldwin, R. A. (1968) Thiabendazole and 2-aminobutane as post-harvest fungicides for Citrus. *Aust. J. exp. Agric. Anim. Husb.* **8**, 440. (*Rev. appl. Mycol.* **48**, 471, 1969.)

Selling, H. A., Vonk, J. W. and Kaars Sijpesteijn, A. (1970) Transformation of the systemic fungicide methyl thiophanate into 2-benzimidazole carbamic acid methyl ester. *Chem. and Ind.*, p. 1625.

Sempio, C. (1936) Influenza di varie sostanze sul parassitamento: ruggine del Fagiola, ruggine e mal bianco del Frumento. *Riv. Patol. veg., Padova* **26**, 201.

Sempio, C. (1938) Influenza di alcune sostanze, date alle piantine per assorbimento, sullo svillupo della carie del Grano. (Nota preventiva). *Riv. Patol. veg., Padova* **28**, 399.

Seth, A. K. and Wareing, P. F. (1967) Hormone directed transport of metabolites and its possible role in plant senescence. *J. exp. Bot.* **18**, 65.

Shabi, E. (1971) Pear: scab *(Venturia pirina)*. *Fungicide–Nematicide Tests* **26**, 41.

Sharples, R. O. (1963) Therapeutic activity of a streptothricin-like antibiotic against apple mildew. *Nature, Lond.* **198**, 306.

Sharvelle, E. G. (1952) Systemic fungicides in 1951: British investigations. *Pl. Dis. Reptr* **36**, 35.

Shepherd, M. C. (1971) Report on Systemic Fungicide Symposium. *Nature, Lond.* **229**, 230.

Shillingford, C. A. (1970) Banana fruit rot control in Jamaica. *PANS* **16**, 69.

Siddiqui, M. Q. and Manners, J. G. (1971) Some effects of general yellow rust *(Puccinia striiformis)* infection on ^{14}C- assimilation, translocation and growth in a spring wheat. *J. exp. Bot.* **22**, 792.

Siegel, M. R. and Sisler, H. D. (1963) Inhibition of protein synthesis *in vitro* by cycloheximide. *Nature, Lond.* **200**, 675.

Siegel, M. R. and Sisler, H. D. (1964) Site of action of cycloheximide in cells of *Saccharomyces pastorianus*. I. Effect of the antibiotic on cellular metabolism. *Biochim. Biophys. Acta.* **87**, 70.

Siegel, M. R. and Sisler, H. D. (1965) Site of action of cycloheximide in cells of *Saccharomyces pastorianus. Biochim. Biophys. Acta.* **103**, 558.

Silverman, W. B. and Hart, H. (1954) Antibiotics tested for control of wheat stem rust. *Phytopathology* **44**, 506.

Simpson, W. R. and Fenwick, H. S. (1971) Suppression of corn head smut with carboxin seed treatments. *Pl. Dis. Reptr* **55**, 501.

Sims, J. J., Mee, H. and Erwin, D. C. (1969) Methyl 2-benzimidazole carbamate, a fungitoxic compound isolated from cotton plants treated with methyl 1-(butyl carbamoyl)-2-benzimidazole carbamate (benomyl). *Phytopathology* **59**, 1775.

Sinha, A. K. and Wood, R. K. S. (1964) Control of *Verticillium* wilt of tomato plants with cycocel ((2-chloroethyl) trimethyl ammonium chloride). *Nature, Lond.* **202**, 824.

Sinha, A. K. and Wood, R. K. S. (1967) The effect of growth substances on *Verticillium* wilt of tomato plants. *Ann. appl. Biol.* **60**, 117.

Sisler, H. D. (1969) Effect of fungicides on protein and nucleic acid synthesis. *Ann. Rev. Phytopath.* **7**, 311.

Sisler, H. D. (1971) Mode of action of benzimidazole fungicides. *Proc. 2nd Int. Congress of Pesticide Chemistry IUPAC, Tel-Aviv, 1971*, p. 323.

Sisler, H. D. and Siegel, M. R. (1967) Cycloheximide and other glutarimide antibiotics. *Antibiotics. Vol. I. Mechanisms of action* (eds. D. Gottlieb and P. D. Shaw), p. 283. Springer, New York.

Slade, P., Cavell, B. D., Hemmingway, R. J. and Sampson, M. J. (1971) Metabolism and mode of action of dimethirimol and ethirimol. *Proc. 2nd Intern. Congr. Pestic. Chem., Tel-Aviv*, p. 295.

Smale, B. C., Montgillion, M. D. and Pridham, T. G. (1961) Phleomycin, an antibiotic markedly effective for control of bean rust. *Pl. Dis. Reptr* **45**, 244.

Smalley, E. B. (1962) Prevention of Dutch elm disease by treatments with 2,3,6-trichlorophenyl acetic acid. *Phytopathology* **52**, 1090.

Smith, B. D. and Corke, A. T. K. (1966) Effect of (2-chloroethyl)trimethyl ammonium chloride on the eriophyid gall mite *Cecidophyopis ribis* Nal., and three fungus diseases of blackcurrant. *Nature, Lond.* **212**, 643.

Smith, D., Muscatine, L. and Lewis, D. (1966) Carbohydrate movement from autotrophs to heterotrophs in parasitic and mutualistic symbiosis. *Biol. Rev.* **44**, 17.

Smith, P. M. (1970) The integrated control of *Botrytis cinerea* on glasshouse tomatoes. *Ph.D. Thesis Reading University.*

Smith, P. M. (1971a) Primary evaluation of new fungicides. *A. Rep. Glasshouse Crops Res. Inst.* 1970, p. 131.

Smith, P. M. (1971b) Narcissus leaf scorch caused by *Stagonospora curtisii*. *A. Rep. Glasshouse Crops Res. Inst.* 1970, p. 141.

Smith, P. M. and Spencer, D. M. (1971) Evaluation of systemic fungicides for use in the glasshouse industry. *Pestic. Sci.* **2**, 201.

Smith, W. L. (1971) Control of brown rot and Rhizopus rot of inoculated peaches with hot water or hot chemical suspensions. *Pl. Dis. Reptr* **55**, 228.

Smoot, J. J. and Melvin, C. F. (1970) A comparison of postharvest fungicides for decay control of Florida oranges. *Proc. Fla St. hort. Soc.* **82**, 243. (*Rev. Pl. Path.* **50**, 2279, 1971.)

Snel, M. and Edgington, L. V. (1970) Uptake, translocation and decomposition of systemic oxathiin fungicides in bean. *Phytopathology* **60**, 1708.

Snel, M. B. and Fletcher, J. T. (1971) Benomyl and thiabendazole for the control of mushroom diseases. *Pl. Dis. Reptr* **55**, 120.

Snel, M., Schmeling, B. von and Edgington, L. V. (1970) Fungitoxicity and structure-activity relationships of some oxathiin and thiazole derivatives. *Phytopathology* **60**, 1164.

Snell, B. K. (1968) Pyrimidines: Part I. The acylation of 2-amino-4-hydroxypyrimidines. *J. chem. Soc.* (C) 2358.

Soel, Z. (1970) The systemic fungicidal effect of benzimidazole derivatives and thiophenate against *Cercospora* leaf spot of sugar beet. *Phytopathology* **60**, 1186.

Solel, Z. (1970a) The systemic fungicidal effect of benzimidazole derivatives and thiophanate against *Cercospora* leaf spot of sugar beet. *Phytopathology* **60**, 1186.

Solel, Z. (1970b) The performance of benzimidazole fungicides in the control of *Cercospora* leaf spot of sugar beet. *J. Am. Soc. Sug. beet Technol.* **16**, 93.

Solel, Z. (1971a) Vapour phase action of some foliar fungicides. *Pestic. Sci.* **2**, 126.

Solel, Z. (1971b) Effective control of *Cercospora* leaf spot of sugar beet by aerial spraying with benomyl. *Pl. Dis. Reptr* **55**, 408.

Somers, E. and Richmond, D. V. (1962) Translocation of captan by broad bean plants. *Nature, Lond.* **194**, 1194.

Spalding, D. H. (1970) Postharvest use of benomyl and thiabendazole to control blue mold rot development in pears. *Pl. Dis. Reptr* **54**, 655.

Spalding, D. H. and Hardenburg, R. E. (1971) Postharvest chemical treatments for control of blue mold of apples in storage. *Phytopathology* **61**, 1308.

Spalding, D. H., Vaught, H. C., Day, R. H. and Brown, G. A. (1969) Control of blue mold rot development in apples treated with heated and unheated fungicides. *Pl. Dis. Reptr* **53**, 738.

Spanner, D. C. (1958) The translocation of sugar in sieve tubes. *J. exp. Bot.* **9**, 332.

Spanner, D. C. (1970) The electro-osmotic theory of phloem transport in the light of recent measurements on *Heracleum* phloem. *J. exp. Bot.* **21**, 325.

Spencer, D. M. (1957) Studies on direct and systemic fungicidal action. *Ph.D. Thesis, University of London.*

Spencer, D. M. (1971) Unpublished results.

Spencer, D. M., Topps, J. H. and Wain, R. L. (1957) An antifungal substance from the tissues of *Vicia faba. Nature, Lond.* **179**, 651.

Spinks, G. T. (1913) Factors affecting susceptibility to disease in plants. *J. agric. Sci. Camb.* **5**, 231.

Springer, J. K. (1970) Peach: Valsa canker *(Valsa cincta). Fungicide–Nematicide Tests* **25**, 48.

Spurr, H. W., Jr. and Chancey, E. L. (1970) The chemical control of powdery mildew by fumigant redistribution. *Phytopathology* **60**, 1062.

Stalder, L. and Barben, E. (1970) Benomyl gegen Wurzelbräune der Cyclamen. *Gartenweld Hamb.* **70**, 289.

Staron, T. and Allard, C. (1964) Propriétés antifongiques du 2-(4′-thiazolyl) benzimidazole ou thiabendazole. *Phytiat.-Phytopharm.* **13**, 163.

Staron, T., Allard, C., Darpoux, H., Grabowski, H. and Kollman, A. (1966) Persistance du thiabendazole dans les plantes. Propriétés systémiques de ses sels et quelques données nouvelles sur son mode d'action. *Phytiat.-Phytopharm.* **15**, 129.

Stingl, H., Härtel, K. and Heubach, G. (1970) Hoe 2989, a new systemic fungicide to control various Basidiomycetes (rusts and smuts) as well as *Rhizoctonia. 7th Int. Congress Plant Protection,* p. 205 (abs.).

Stipes, R. J. and Oderwald, D. R. (1971) Dutch elm disease; control with soil injected fungicides and surfactants. *Phytopathology* **61**, 913 (abs.).

Stoddard, E. M. (1946) Soil applications of oxyquinolin benzoate for the control of foliage wilting in elms caused by *Graphium ulmi. Phytopathology* **36**, 682.

Stoddard, E. M. (1951a) Chemotherapeutic control of *Fusarium* wilt of carnations. *Phytopathology* **41**, 33.

Stoddard, E. M. (1951b) A chemotherapeutic control of strawberry red stele. *Phytopathology* **41**, 34.

Stoddard, E. M. (1952) Chemotherapeutic control of *Rhizoctonia* on greenhouse stock. *Phytopathology* **42**, 476.

Stoddard, E. M. (1954) Chemotherapeutic control of cucumber scab. *Phytopathology* **44**, 507.

Stoddard, E. M. (1957) A *Fusarium* rot of geraniums and its control. *Pl. Dis. Reptr* **41**, 536.

Stoddard, E. M. and Dimond, A. E. (1949) The chemotherapy of plant diseases. *Bot. Rev.* **15**, 345.

Stoddard, E. M. and Dimond, A. E. (1951) The chemotherapeutic control of *Fusarium* wilt of carnations. *Phytopathology* **41**, 337.

Stoddard, E. M. and Miller, P. M. (1962) Chemical control of water loss in growing plants. *Science* **137**, 224.

Stoker, G. L. and Dewey, W. G. (1970) Effect of loose smut infection and 'Vitavax' seed treatment on barley yields. *Utah Sci.* **31**, (2) 46.

Stokes, A. (1954) Uptake and translocation of griseofulvin by wheat seedlings. *Pl. Soil* **5**, 132.

Stott, K. and Jefferies, C. J. (1972) Fruit-set. *Rep. agric. hort. Res. Stn Bristol, 1971,* p. 149.

Stover, R. H. (1969) The effect of benomyl on *Mycosphaerella musicola. Pl. Dis. Reptr* **53**, 830.

Straib, W. (1941) Über die Wirkung organischer Verbindungen als Spritzmittel gegen Rostpilzinfektion. *Zentbl. Bakt. ParasitKde (Abt. II),* **103**, 74.

Strugger, S. (1949) *Praktikum der Zell- und Gewebephysiologie der Pflanze.* Berlin: Springer-Verlag.

Strugger, S. and Peveling, E. (1961) Die Elektronenmikroskopische Analyse der Extrafaszikulären Komponente des Transpirationsstromes mit Hilfe von Edelmetallsuspensoiden adäquater Dispersität. *Ber. dt. bot. Ges.* **74**, 300.

Stubbs, J. (1952) The evaluation of systemic fungicides by means of *Alternaria solani* on tomato. *Ann. appl. Biol.* **39**, 439.

Suhara, Y., Maeda, K., Umezawa, H. and Ohno, M. (1966) Chemical studies on Kasugamycin V. The structure of Kasugamycin. *Tetrahedron Letters* **12**, 1239.

Suit, R. F. and Ducharme, E. P. (1971) Cause and control of pink pitting on grapefruit. *Pl. Dis. Reptr* **55**, 923.

Sumi, H., Takahi, Y., Nakagami, K. and Kondo, Y. (1968) Controlling effect of pentachlorobenzyl alcohol on rice blast. *Ann. Phytopathol. Soc. Japan* **34**, 114.

Suzuki, S., Isono, K., Nagatsu, J., Mizutani, T., Kawashima, Y. and Mizuno, T. (1965) A new antibiotic, Polyoxin A. *J. Antibiotics (Tokyo) Ser. A.* **11**, 1.

Suzuki, S., Nakamura, G., Okuma, K. and Tomiyama, Y. (1958). Cellocidin, a new antibiotic. *J. Antibiotics (Tokyo) Ser. A.* **11**, 81.

Swarts, D. H., Jacobs, C. and Nel, F. G. (1969) The control of collar rot in bananas. *Fmg S. Afr.* **45**, 15, 17. (*Rev. appl. Mycol.* **48**, 3072, 1969.)

Szkolnik, M. (1969) Apple: scab *(Venturia inaequalis)*. Cherry (red tart): leaf spot *(Coccomyces hiemalis)*. *Fungicide–Nematicide Tests* **24**, 27.

Szkolnik, M. and Gilpatrick, J. D. (1969) Apparent resistance of *Venturia inaequalis* to dodine in New York apple orchards. *Pl. Dis. Reptr* **53**, 861.

Szkolnik, M. and Gilpatrick, J. D. (1971) Apple: scab *(Venturia inaequalis)*. *Fungicide–Nematicide Tests* **26**, 37.

ten Haken – see under H.

Tahori, A. S., Zeidler, G. and Halevy, A. H. (1965) Effect of some plant growth-retarding compounds on three fungal diseases and one viral disease. *Pl. Dis. Reptr* **49**, 775.

Talboys, P. W. (1962) Systemic movement of some vascular pathogens. *Trans. Brit. mycol. Soc.* **45**, 280.

Tamari, K., Ogasawara, N., Kaji, J. and Togashi, K. (1966) On the effect of a piricularin-detoxifying substance, ferulic acid, on tissue resistance of rice plants to blast fungal infection. *Ann. phytopathol. Soc. Japan* **32**, 186.

Tammes, P. M. L. and Die, J. van (1966) Studies on the phloem exudate from *Yucca flacida* Haw. IV. *Koninkl. Ned. Akad. Wetenschap. Proc.* C **69**, 656.

Tanaka, F., Tanaka, N., Youchara, H. and Umezawa, H. (1962) Studies on bacithracin, a new antibiotic from *B. megatherium*. I. Preparation and its properties. *J. Antibiotics (Tokyo)* **A 16**, 86.

Tanaka, N., Yamaguchi, H. and Umezawa, H. (1966) Mechanism of kasugamycin action on polypeptide synthesis. *J. Biochem.* (Tokyo) **60**, 429.

Tanton, T. W. and Crowdy, S. H. (1972a) Water pathways in higher plants. II. *J. exp. Bot.* **23**, 600.

Tanton, T. W. and Crowdy, S. H. (1972b) Water pathways in higher plants. III. *J. exp. Bot.* **23**, 619.

Tapley, R. G., Woodroffe, R. and Atkins, P. S. (1969) Field trials with benomyl on soft fruit in 1969. *Proc. 5th Br. Insectic. Fungic. Conf.* **1**, 130.

Tarjan, A. C. and Howard, F. L. (1953) Comparison of benzothiazole 1-2-thioglycolic acid derivatives with other chemicals for Dutch elm disease therapy. *Phytopathology* **43**, 486.

Taub, D., Kuo, C. H. and Wendler, N. L. (1962) Synthesis in the griseofulvin series: 7-fluoro-7-dechlorogriseofulvin, an active analogue. *Chem. and Ind.* 557.

Taylor, G. S. (1953) Control of tobacco blue mould by root application of zineb and ferbam. *Phytopathology* **43**, 486.

Taylor, G. S. (1970) Tobacco protected against fleck by benomyl and other fungicides. *Phytopathology* **60**, 578.

Taylor, J. (1971) A necrotic leaf blotch and fruit rot of apple caused by a strain of *Glomerella cingulata*. *Phytopathology* **61**, 221.

Tempel, A. and Kaars Sijpesteijn, A. (1967) Pentobarbital sodium salt, a systemic agent for control of powdery mildew of cucumber. *Nature, Lond.* **213**, 215.

Tempel, A., Meltzer, J. and van den Bos, B. G. (1968) Systemic fungicidal and insecticidal activities of 1- and 2-[bis(dimethylamido)phosphoryl]-3-alkyl-5-anilino-1,2,4-triazoles. *Neth. J. Pl. Path.* **74**, 133.

Thaine, R. (1962) A translocation hypothesis based on the structure of plant cytoplasm. *J. exp. Bot.* **13**, 152.

Thaine, R. (1969) Movement of sugars through plants by cytoplasmic pumping. *Nature, Lond.* **222**, 873.

Thanassoulopoulos, C. C., Giannopolitis, C. N. and Kitsos, G. T. (1970) Control of *Fusarium* wilt of tomato and watermelon with benomyl. *Pl. Dis. Reptr* **54**, 561.

Thanassoulopoulos, C. C., Giannopolitis, C. N. and Kitsos, G. T. (1971) Evaluation of sensitiveness and development of resistance of *Fusarium oxysporum* f. sp. *lycopersici* to benomyl. *Phyt. Zeitschr.* **70**, 114.

Thapliyal, P. N. (1970) Systemicity and fungicidal activity of three systemic fungicides in soybean. *Ph.D. Thesis, Univ. Illinois, Urbana,* 41 pp.

Thayer, P. L., Ford, D. H. and Hall, H. R. (1967) Bis-(*p*-chlorophenyl)-3-pyridine methanol (EL 241): a new fungicide for the control of powdery mildew. *Phytopathology* **57**, 833.

Thiollière, J. (1970) Prospects in the control of grey mould *(Botrytis cinerea)* in vineyards in France. *Summaries of Papers, VIIth Int. Cong. Plant Prot., Paris,* p. 256.

Thirumalachar, M. J. (1968) Antibiotics in the control of plant pathogens, *Advances in applied microbiology, Vol. 10* (eds. W. W. Umbreit and D. Perlman), p. 313.

Threlfall, R. J. (1968) The genetics and biochemistry of mutants of *Aspergillus nidulans* resistant to chlorinated nitrobenzenes. *J. Gen. Microb.* **52**, 35.

Ticknor, R. L. and Tukey, H. B. (1957) Evidence for the entry of mineral nutrients through the bark of fruit trees. *Proc. Am. Soc. hort. Sci.* **69**, 13.

Tillman, R. W. and Sisler, H. D. (1971) A chloroneb resistant mutant of *Ustilago maydis. Phytopathology* **61**, 914.

Tomlinson, J. A. and Webb, M. J. W. (1958) Control of turnip and cabbage mildew *(Erysiphe polygoni* D.C.*)* by zinc. *Nature, Lond.* **181**, 1352.

Tomlinson, J. A. and Webb, M. J. W. (1959) Powdery mildew diseases. *Rep. Nat. Veg. Res. Sta. for 1958,* p. 38.

Tomlinson, J. A. and Webb, M. J. W. (1960) Powdery mildew diseases. *Rep. Nat. Veg. Res. Sta. for 1959,* p. 43.

Topps, J. H. and Wain, R. L. (1957) Fungistatic properties of leaf exudates. *Nature, Lond.* **179**, 652.

Torgeson, D. C. (1967) (ed.) *Fungicides:* **1**, New York and London: Academic Press.

Tramier, R. and Antonini, C. (1971) Essai de lutte contre les trachiomycoses de l'oeillet Americain par le benomyl. *Annls. Phytopath.* **3** (in press).

Tsai, Y. P., Su, H. J., Ogawa, J. M. and Chen, S. S. (1968) Control of crown rot and finger spotting on banana hands with 1-(butylcarbamoyl)-2-benzimidazole carbamic acid, methyl ester (F 1991). *Pl. Prot. Bull. Taiwan* **10**, 35. (*Rev. Pl. Path.* **49**, 1101, 1970.)

Tsao, P. H., Leben, C. and Keitt, G. W. (1960) The partial purification and biological activity of an antifungal antibiotic produced by a strain of *Streptomyces griseus. Phytopathology* **50**, 169.

Tugwell, B. L. and Wicks, T. J. (1969) New chemicals for mould control in Citrus. *J. Agric. S. Aust.* **72**, 372. (*Rev. appl. Mycol.* **48**, 2979, 1969.)

Tukey, H. B., Jr. (1970) The leaching of substances from plants. *A. Rev. Pl. Physiol.* **21**, 305.

Tyree, M. T. and Fensom, D. S. (1970) Some experimental and theoretical observations concerning mass flow in the vascular bundles of *Heracleum. J. exp. Bot.* **21**, 304.

Uesigi, Y. (1970) Development of organo-phosphorus fungicides. *Japan Pestic. Inform.* **2**, 11.

Umezawa, H., Okami, Y., Hashimoto, T., Suhara, Y., Hamada, M. and Takeuchi, T. (1965) A new antibiotic, Kasugamycin. *J. Antibiotics (Tokyo) Ser. A.* **18**, 101.

Umgelter, H. (1969) Erfahrungen mit plantvax gegen Rostkrankheiten bei Zierpflanzen und Gemüsekulturen. *Gesunde Pfl.* **21**, 53.

van Andel – see under A.

van den Bos – see under B.

von Meyer – see under M.

etc.

Vaartaja, O. (1955) Chemotherapeutic action of dibenzothiophene against *Rhizoctonia solani. Bi-m. Progr. Rep. Div. For. Biol., Dep. Agric. Can.* **11**, 2.

Venkata Ram, D. S. (1969) Systemic control of *Exobasidium vexans* on tea with 1,4-oxathiin derivatives. *Phytopathology* **59**, 125.

Venn, K. I., Nair, V. M. G. and Kuntz, J. E. (1968) Effects of TCPA on oak sapwood formation and the incidence and development of oak wilt. *Phytopathology* **58**, 1071.

Ventura, E., Bourdin, J. and Berthier, G. (1970) Etude de l'efficacité de quelques fongicides systémiques nouveaux vis-à-vis des principaux parasites des semences de céréales. *VII Int. Congr. Pl. Prot. 1970* (abs.).

Ventura, E., Bourdin, J. and Piedallu, C. (1968) Activité fongicide de la carboxin vis-à-vis des principaux parasites des semences de céréales. *Phytiat.-Phytopharm.* **3**, 197.

Viala, P. (1893) *Les maladies de la vigne.* 3rd edn. Paris: Montpellier et Cie.

Vidali, A. and Cifferi, R. (1951) Esperienze di lotta contro l'oidio del Tabacco *(Erysiphe cichoracearum)* a mezzo di tiosulfato e sali di litio. *Tabacco* **55**, 95.

Visser, R., Shanmuganathan, H. and Savanayagam, J. V. (1962) The influence of sunshine and rain on tea blister blight, *Exobasidium vexans* Massee, in Ceylon. *The Tea Quart.* **33**, 34.

Vojvodić, D. (1970) Investigation on the influence of late autumn treatment on the formation of perithecia of *Venturia inaequalis* (Cke.) Wint. *Zašt. Bilja* **21**, 151. (*Rev. Pl. Path.* **50**, 2345, 1971.)

Volger, C. (1959) Über die Möglichkeit einer systemischen Wirkung von T.M.T.D. Präparaten. *Meded. LandbHoogesch. OpzoekStns. Gent* **24**, 837.

Vomvoyanni, V., Georgopoulos, S. G. and Kappas, A. (1971) Complexity of genetic control of biochemical processes in fungi as evidenced by studies on resistance to toxicants. *Radiation and radioisotopes for industrial microorganisms,* Intern. Atomic Energy Agency, Vienna, p. 233.

Vonk, J. W. and Kaars Sijpesteijn, A. (1971) Methyl benzimidazol-2-ylcarbamate, the fungitoxic principle of thiophanate-methyl. *Pest. Sci.* **2**, 160.

Waard, M. A. de (1971a) Germination of powdery mildew conidia *in vitro* on cellulose membranes. *Neth. J. Pl. Path.* **77**, 6.

Waard, M. A. de (1971b) Effect of systemic and non-systemic compounds on the *in vitro* germination of powdery mildew conidia. *Meded. Fac. Landbouwwetensch. Rijksuniv. Gent* **36**, 113.

Waard, M. A. de (1972) On the mode of action of the organophosphorus fungicide Hinosan. *Neth. J. Pl. Path.* **78**, 186.

Waffelaert, P. (1969) Nouvelles perspectives de lutte contre le Monilia et l'Oidium de l'abricotier. *Phytiat. Phytopharm.* **18**, 51. (*Rev. Pl. Path.* **49**, 1433, 1970.)

Waggoner, P. E. (1956) Chemotherapy of *Verticillium* wilt of potatoes in Connecticut 1955. *Am. Pot. J.* **33**, 223.

Waggoner, P. E. and Dimond, A. E. (1952) Effect of stunting agents, *Fusarium lycopersici,* and maleic hydrazide upon phosphorus distribution in tomato. *Phytopathology* **42**, 24.

Waggoner, P. E. and Dimond, A. E. (1955) Production and role of extracellular pectic enzymes of *Fusarium oxysporum* f. *lycopersici. Phytopathology* **45**, 79.

Waggoner, P. E. and Dimond, A. E. (1957) Altering disease resistance with ionizing radiation and growth substances. *Phytopathology* **47**, 125.

Wagner, F. (1940) Die Bedeutung der Kieselsäure für das Wachstum einiger Kulturpflanzen, ihren Nährstoffhaushalt und ihre Anfälligkeit gegen echte Mehltaupilze. *Phytopath. Z.* **12**, 427.

Wain, R. L. (1951) Plant growth-regulating and systemic fungicidal activity: the aryloxyalkylcarboxylic acids. *J. Sci. Fd Agric.* **2**, 101.

Wain, R. L. (1952) Systemic fungicides. *Ann. appl. Biol.* **39**, 429.

Wain, R. L. (1953) Systemic fungicides. *Meded. LandbHoogesch. OpzoekStns. Gent* **18**, 394.

Wain, R. L. (1959) Some chemical aspects of plant disease control. *R. Inst. Chem. Monograph No. 3.*

Wain, R. L. (1963) Some developments in research on fungicides. *Meded. Landb-Hoogesch. OpzoekStns. Gent* **28**, 516.

Wain, R. L. (1969) Naturally occurring fungicides. *Proc. Symp. Potentials in Crop Protection, New York, 1969,* p. 26.

Wain, R. L. and Carter, G. A. (1967) Uptake, translocation and transformations by higher plants. *Fungicides, Vol. 1* (ed. D. C. Torgeson), p. 561. New York: Academic Press.

Wain, R. L., Spencer, D. M. and Fawcett, C. H. (1961) Antifungal compounds in seedlings of *Vicia faba. S.C.I. Monograph No. 15,* p. 109.

Walker, A. T. and Smith, F. G. (1952) Effect of actidione on growth and respiration by *Myrothecium verrucaria. Proc. Soc. expt. Biol. Med.* **81**, 556.

Walker, J. C. and Stahmann, M. A. (1955) Chemical nature of disease resistance in plants. *A. Rev. Pl. Physiol.* **6**, 351.

Wallace, H. A. H. (1968) Co-operative seed treatment trials, 1967. *Can. Pl. Dis. Surv.* **48**, (3) 82.

Wallen, V. R. (1955) Control of stem-rust of wheat with antibiotics. I. Greenhouse and field tests. *Pl. Dis. Reptr* **39**, 124.

Wallen, V. R. (1958) Control of stem rust of wheat with antibiotics. II. Systemic activity and effectiveness of derivatives of cycloheximide. *Pl. Dis. Reptr* **42**, 363.

Wallen, V. R. and Millar, R. L. (1957) The systemic activity of cycloheximide in wheat seedlings. *Phytopathology* **47**, 291.

Wallen, V. R. and Skolko, A. J. (1950) Antibiotic XG as a seed treatment for the control of leaf and pod spot of peas caused by *Ascochyta pisi. Can. J. Res. Ser. C* **28**, 623.

Wallen, V. R. and Skolko, A. J. (1951) Activity of antibiotics against *Ascochyta pisi. Can. J. Bot.* **29**, 316.

Wang, M. C., Erwin, D. C., Sims, J. J., Keen, N. T. and Borum, D. E. (1971) Translocation of ^{14}C-labelled thiabendazole in cotton plants. *Pestic. Physiol. Biochem.* **1**, 188.

Wardlaw, C. W. (1961) *Banana Diseases including Plantains and Abaca.* London: Longman Group Ltd.

Wardlaw, I. F. (1968) The control and pattern of movement of carbohydrates in plants. *Bot. Rev.* **34**, 79.

Wark, M. C. and Chambers, T. C. (1965) Fine structure of the phloem of *Pisum sativum.* I. *Aust. J. Bot.* **13**, 171.

Webster, R. K., Ogawa, J. M. and Bose, E. (1970) Tolerance of *Botrytis cinerea* to 2,6-dichloro-4-nitroaniline. *Phytopathology* **60**, 1489.

Weigun Yang and Lung-Chi Wu (1971) Tolerance to PCNB and simulated pathogenic strength of *Sclerotium rolfsii. Memoirs Coll. Agric. Nat. Taiwan Univ.* **12**, 191.

Weindlmayr, J. (1963) Untersuchungen über eine Beeinflussung der Phytophthora-Anfälligkeit von Kartoffelpflanzen durch Gibberellin Behandlungen. *Z. PflKrankh. PflPath. PflSchutz.* **70**, 599.

Weinke, K. E., Lauber, J. J., Greenwald, B. W. and Preiser, F. A. (1969) Thiabendazole, a new systemic fungicide. *Proc. 5th Br. Insectic. Fungic. Conf.* **2**, 340.

Welch, A. W. (1968) Grape: black rot, bitter rot; *Guignardia bidwellii, Melanconium fuligineum. Fungicide–Nematicide Tests* **23**, 59.

Welch, N. C., Greathead, A. S., Hall, D. H. and Little, T. (1969) Brussels sprout ring spot control with fungicides. *Calif. Agric.* **23**, 17.

Wells, J. M. (1971) Heated wax emulsions with benomyl and 2,6-dichloro-4-nitroaniline for control of postharvest decay of peaches. *Phytopathology* **61**,, 916 (abs.).

Wells, J. M. and Gerdts, M. H. (1971) Pre- and post-harvest benomyl treatments for control of brown rot of nectarines in California. *Pl. Dis. Reptr* **55**, 69.

Wensley, R. N. and Huang, C. M. (1970) Control of *Fusarium* wilt of musk melon and other effects of benomyl soil drenches. *Can. J. Microbiol.* **16**, 615.

Western, J. H. (1971) *Diseases of crop plants.* London and Basingstoke: The Macmillan Press Ltd.

Wettstein, F. O., Noll, H. and Penman, S. (1964) Effect of cycloheximide on ribosomal aggregates engaged in protein synthesis *in vitro. Biochim. biophys. Acta* **87**, 525.

Wheeler, H. and Luke, H. H. (1963) Microbial toxins in plant disease. *A. Rev. Microbiol.* **17**, 223.

Whiffen, A. J. (1950) The activity *in vitro* of cycloheximide (acti-dione) against fungi pathogenic to plants. *Mycologia* **42**, 253.

White, G. A. (1971) A potent effect of 1,4-oxathiin systemic fungicides on succinate oxidation by a particulate preparation from *Ustilago maydis. Biochem. Biophys. Res. Comm.* **44**, 1212.

Whiteside, J. O. (1970) Effect of fungicides applied to citrus trees on perithecial development by the greasy spot fungus in detached leaves. *Pl. Dis. Reptr.* **54**, 865.

Wiggell, P. and Simpson, C. S. (1970) Apple mildew trial. IV. 1965–69. V. 1967–69. *Rep. Luddington exp. Hort. Stn for 1969*, p. 10.

Wilson, E. M. and Ark, P. A. (1958) The effect of chlorophyll on the phytotoxicity of acti-dione on bean. *Pl. Dis. Reptr* **42**, 1069.

Winner, C. (1966) Untersuchungen über parasitogene Schäden an Wurzeln der Zuckerrübe, insbesondere durch Aphanomyces, und über Möglichkeiten ihrer Verhütung. III. Versuche zur Hemmung eines Aphanomyces-Befalls durch spezifisch wirkende Chemotherapeutica. *Phytopathol. Z.* **57**, 310.

Winner, H. I. and Athar, M. A. (1970) Induction of resistance to polyene antibiotics in *Candida* species. *Abstracts International Congress for Microbiology 1970, Mexico,* p. 149.

Winter, A. G. and Willeke, L. (1951) Über die Aufnahme von Antibiotika durch höhere Pflanzen und ihre Stabilität in natürlichen Böden. *Naturwissenschaften* **38**, 457.

Witchalls, J. R. and Close, R. (1971) Control of eyespot lodging in wheat by benomyl. *Pl. Dis. Reptr* **55**, 45.

von Witsch, H. and Kasperlik, H. (1953) Infecktionshemmende Wirkung der Dichlorophenoxyessigsäure. *Naturwissenschaften* **40**, 294.

Wolfe, M. S. (1971) Fungicides and the fungus population problem. *Proc. 6th Insectic. Fungic. Conf. at Brighton, 1971,* **3**, 724.

Wood, R. K. S. (1960) Pectic and cellulolytic enzymes in plant diseases. *A. Rev. Pl. Physiol.* **11**, 299.

Wood, R. K. S. (1967) *Physiological plant pathology.* Oxford and Edinburgh: Blackwell Scientific Publications.

Woodcock, D. (1971) Chemotherapy of plant disease – progress and problems. *Chem. Brit.* **7**, 415.

Woodward, P. J. (1971a) Apple: mildew, new materials observation 1970. *Rep. Luddington exp. Hort. Stn for 1970,* p. 22.

Woodward, P. J. (1971b) Gooseberry: American gooseberry mildew and Botrytis control 1970. *Rep. Luddington exp. Hort. Stn for 1970,* p. 56.

Wormald, H. (1954) The brown rot diseases of fruit trees. *Min. Agric. Tech. Bull. no. 3,* London: HMSO.

Wortley, W. R. S. (1936) The effect of salts of lithium on the resistance of certain plants to disease. *J. R. agric. Soc.* **97**, 492.

Yamada, M. (1955) The sporulation-inhibitive effects of sulfadiazine on wheat leaf rust. I. Some experimental results with wheat seedlings. *Ann. Phytopath. Soc., Japan* **19**, 146.

Yamaguchi, S. and Crafts, A. S. (1957) Translocation of 2,4-D in *Zebrina pendula* is greatly affected by growth rate. *Pl. Physiol. Lancaster* **32**, Suppl. xlii.

Yarham, D. J., Bacon, E. T. G. and Hayward, C. F. (1971) The effect on mildew development of the widespread use of fungicide on winter barley. *Proc. 6th Br. Insectic. Fungic. Conf.,* p. 15.

Yarwood, C. E. (1947) Therapeutic treatments for bean rust. *Phytopathology* **37**, 24.

Yarwood, C. E. (1948) Therapeutic treatments for rusts. *Phytopathology* **38**, 542.

Yarwood, C. E. (1950) Effect of temperature on the fungicidal action of sulphur. *Phytopathology* **40**, 173.

Yarwood, C. E. (1959) Predisposition, *Plant Pathology Vol. 1* (eds. J. G. Horsfall and A. E. Dimond), p. 521. New York: Academic Press.

Yoshinaga, E. (1969) A systemic fungicide for rice blast control. *Proc. 5th Br. Insectic. Fungic. Conf.,* p. 593.

Yoshinaga, E., Uchida, T. and Iwakura, T. (1965) Studies on the rice blast control. IV. Rice blast control of kitazin by root treatment. *Ann. Phytopathol. Soc., Japan* **30**, 307 (abs.).

Zaracovitis, C. (1965) Activity of phenobarbitone in controlling marrow powdery mildew. *Nature, Lond.* **206**, 954.

Zaumeyer, W. J. (1957) Comparative protection of bean leaves from fungus infection by antibiotic treatments of upper and lower surfaces. *Phytopathology* **47**, 539.

Zaumeyer, W. J. (1958) Antibiotics in the control of plant diseases. *A. Rev. Microbiol.* **12**, 415.

Zaumeyer, W. J. and Wester, R. E. (1956) Control of several fungus diseases of beans and lima beans with antibiotics. *Phytopathology* **46**, 470.

Zehr, E. I. (1971a) Apple: scab *(Venturia inaequalis). Fungicide–Nematicide Tests* **26**, 40.

Zehr, E. I. (1971b) Grape: black rot *(Guignardia bidwellii). Fungicide–Nematicide Tests* **26**, 59.

Zelitch, I. (1961) Biochemical control of stomatal opening in leaves. *Proc. Nat. Acad. Sci., U.S.* **47**, 1423.

Zelitch, I. (1963) The control mechanisms of stomatal movement. *Stomata and water relations* (ed. I. Zelitch), p. 18. *Conn. Agric. Exp. Stn Bull.* no. 664, New Haven.

Zentmyer, G. A. (1942) Toxin formation and chemotherapy in relation to Dutch elm disease. *Phytopathology* **32**, 20.

Zentmyer, G. A. (1954) Chemotherapy for the control of *Phytophthora* root rot of avocado. *Phytopathology* **44**, 511.

Zentmyer, G. A. and Gilpatrick, J. D. (1960) Soil fungicides for prevention and therapy of *Phytophthora* root rot of avocado. *Phytopathology* **50**, 660.

Zentmyer, G. A. and Horsfall, J. G. (1943) Internal therapy with organic chemicals in treatment of vascular diseases. *Phytopathology* **33**, 16.

Zentmyer, G. A., Horsfall, J. G. and Wallace, P. P. (1946) Dutch elm disease and its chemotherapy. *Conn. Agric. Exp. Stn Bull.* no. 498, p. 1.

Zentmyer, G. A., Moje, W. and Mircetich, S. M. (1962) Ethionine as a chemotherapeutant. *Phytopathology* **52**, 34.

Zimmermann, M. H. (1957) Translocation of organic substances in trees. I. *Pl. Physiol. Lancaster* **32**, 288.

Zweep, W. van der (1961) The movement of labelled 2,4-D in young barley plants. *Weed Res.* **1**, 258.

Index

(Bold type denotes pages showing structural formulae)

acetate metabolism inhibited by oxathiins, 58
Actidione = cycloheximide *q.v.*
Alternaria longipes causing brown spot of tobacco, 125
Alternaria solani, use in bioassays, 15
3-amino-1,2,4-triazoles, **35**
antibiotics, 30. *See also* under individual compounds
antimycin A, *Venturia inaequalis* resistant to, 161
antisporulants, 126, 245, 252
Aphanomyces, 231
  cochlioides, sugar-beet root rot, 125
  euteiches, pea root rot, 231
apoplastic movement, 100
apple,
  canker, *Nectria galligena*, antisporulants, 241
  disease control, 238–43
  mildew, *Podosphaera leucotricha*, 239
    control with triarimol, 240
    effect of phenyl alanine, 119
  scab, *Venturia inaequalis*, control with triarimol, 240
  storage rots, *Gloeosporium* spp., control, 241
application methods,
  in glasshouses, 218
  on field crops, 180
apricot diseases, 244
Ascochyta,
  chrysanthemi, 216
  fabae, controlled by benomyl, 226
  pisi, controlled by benomyl, 226
Aspergillus nidulans resistance to benomyl, 162
assessment of systemic and therapeutic activity, 10
avocado root rot, 124
8-azaadenine, **77**
8-azaguanine, **77**
6-azauracil, 3,5-dioxo-2,3,4,5-tetrahydro-1,2,4-triazine, **77**
  Cladosporium cucumerinum resistant to, 161
  metabolic conversion in host tissues, 118
  mode of action, 118, 143
azepines, **78**

bacimethrin, **67**
banana
  diseases, 253

banana – *cont.*
  Penicillium rots, control by thiabendazole, 236
barley mildew, *Erysiphe graminis*, 186–194
barley smuts, 199
BAS 3170, 2-iodo-benzanilide,
  controls cereal rusts, 198
BAS 3191, 2,5-dimethylfuran-3-carboxylic acid anilide
  controls bunt, 201
  controls smuts, 200
BAS 3270, 2,5-dimethylfuran-3-carboxylic acid cyclohexamide
  controls bunt, 201
  controls smuts, 200
BAS 346 F, (MBC) methyl benzimidazol-2-yl carbamate
  mammalian toxicity, 90
bean
  rust, *Uromyces phaseoli*, control, 234
  seed-borne fungi controlled by benomyl, 226
  white mould, *Sclerotinia sclerotiorum*, controlled by benomyl, 234
Benlate T, 232
benomyl, methyl (1-butylcarbamoyl)-benzimidazol-2-yl-carbamate, **64**
  antifungal spectrum, 135
  decomposition to MBC, 64
  effect on fine structure of *Botrytis fabae*, 138
  effect on soil microflora, 223
  mammalian toxicity, 89
  non-fungicidal effects on hosts, 221
  post-harvest applications, 183
  resistance of fungi to, 162
  seed treatment against bunt, 200
  seed treatment against rice blast, 205
  seed treatment against smuts, 201
  sporulostatic effect, 254
  translocation, 111
  use against cereal, fruit and vegetable diseases – *see under* individual crops
benzanilides, 58
benzimidazoles
  non-fungicidal effects of, 221
biocidal effects, 153
biological control, 222
blackberry diseases, 247
black currant
  diseases, 246
  leaf spot, *Pseudopeziza ribis*, control, 245
blast, *see Pyricularia oryzae*
blasticidin S
  inhibits protein synthesis, **49**, 149, 167, 204
  mammalian toxicity, 88
  resistance of *Pyricularia oryzae* to, 162
blueberry diseases, 247
Botrytis cinerea
  grey mould and ghost spot of tomatoes, 209
  grey mould of black currants, 246
  grey mould of strawberries, 248
  on French bean pods, control, 234
  pear storage rot, 243
  resistance to benomyl, 159
  resistance to dicloran, 164
Botrytis fabae
  effect of benomyl on fine structure, 138
  effect of captan, 29
  use in bioassays, 14
Bremia lactucae, lettuce downy mildew, 213
brown rust of barley, *Puccinia hordei*, 194
Brussels sprouts ring-spot, *Mycosphaerella brassicicola*, control, 234
bunt, *Tilletia caries*
  control by benomyl and thiabendazole, 201
4-*n*-butyl-1,2,4-triazole (= RH 124), translocation in wheat, 112

caffeic acid products inactivate polygalacturonase, 127
calcium nutrition effects on cell-wall pectins, 124
captan used against *Botrytis fabae*, 14, 29

carbinols, 71
carboxanilides, 60
carboxin, 2,3-dihydro-6-methyl-5-phenylcarbamoyl-1,4-oxathiin, **56**
  barrier effect of bean seed testa, 177
  controls
    carnation rusts, 214
    Helminthosporium gramineum, 202
    Urocystis agropyri, 201
    Ustilago nuda and *U. tritici*, 200
  tested against *Sphacelotheca reiliana*, 205
  translocation, 111
carboxylic acid anilides, 54
carnation rusts, *Uromyces caryophyllinus* and *Puccinia arenaria*, 214
carnation wilts, 214
Carolate used against Dutch elm disease, 12
carrot liquorice rot, *Centrosperma acerina*, control by benomyl, 236
Cela W 524, triforine, N,N′-bis-(1-formamido-2,2,2-trichloroethyl)piperazine, **74**, 144
celery blights, *Septoria apiicola* and *Cercospora apii*, control by benomyl, 233
celery crown rot, *Centrospora acerina*, control by benomyl, 236
cellocidin, **53**
Centrospora acerina, rots of carrot and celery, control by benomyl, 236
Ceratocystis fagacearum, cause of oak wilt, 121
Ceratocystis paradoxa, pineapple fruit rot, 253
Ceratostomella (ceratocystis) ulmi, effect of vessel size on invasion by, 122
Cercospora apii, early blight of celery, control, 233
Cercospora beticola, sugar-beet leaf spot, control, 233
Cercosporella herpotrichoides, eyespot of cereals, control by benomyl, 203
cereal rusts, treatments for, historical survey, 17
cherry diseases, 244
chestnut blight, *Endothia parasitica*, 7
chitin, fungal, inhibition of synthesis, 53, 165
chloramphenicol, 109
chloraniformethan, 1-(3,4-dichloranilino)-1-formylamino-2,2,2-trichloroethane, **74**
  spray against cereal mildew, 191
chlorogenic acid inactivates polygalacturonase, 127
chloroneb, 1,4-dichloro-2,5 dimethoxybenzene, **83**
  mode of action, 145
  resistance of *Ustilago maydis* to, 162
chlorquinox, 5,6,7,8-tetrachloroquinoxaline spray against cereal mildew, 191
chrysanthemum
  control of *Ascochyta chrysanthemi* on cuttings, 216
  control of *Puccinia horiana* by carboxin, 216
citrus rots, *Penicillium* spp., control, 236
Cladosporium carpophilum, peach scab, 244
Cladosporium cucumerinum, cucumber scab
  control by reducing sugar level, 130
  resistance to 6-azuracil, 161
  use in bioassays, 16
  reduced by growth retardants, 125
Cladosporium fulvum, tomato leaf mould, 209
Coccomyces hiemalis, cherry leaf spot, 48, 244
cohnen
  controls *Pyricularia oryzae*, 204
Colletotrichum
  circinans, 2
  fragariae, strawberry anthracnose, 248
  lindemuthianum, bean anthracnose, control by benomyl, 226

'conventional' fungicides, 157
conversion, metabolic, of systemic fungicides, 118, 153
cotton boll rots controlled by benomyl, 183
Cronartium ribicola, black currant rust, 246
crotonanilide, **61**
cucumber
  black root rot, *Phomopsis sclerotioides*, 212
  powdery mildew, *Sphaerotheca fuliginea*, 211
  scab, *Cladosporium cucumerinum*, 125
  stem rot, *Mycosphaerella melonis*, 203
cuticle structure, 94
cyclamen
  black root rot, *Thielaviopsis basicola*, control by benzimidazoles, 216
  heart rot, *Botrytis cinerea*, benomyl treatment, 159
cycloheximide, β-(2-(3,5-dimethyl-2-oxocyclohexyl)-2-hydroxyethyl)glutarimide, 31, **47**
  inhibition of protein synthesis, 168
  mammalian toxicity, 88
  mode of action, 149
  wheat rust control, 20
cycocel, (2-chloroethyl)trimethylammonium chloride induces tylosis in tomato stems, 122
cytokinin-like effects of benzimidazole compounds, 221
Cytospora cincta, peach canker, 48, 244

2,4-D translocation, 108
dichlozoline, 162
DNA synthesis inhibited
  by chloroneb, 145
  by MBC, 138
detoxication of quintozene, 167
Dexon, sodium-*p*-dimethylaminobenzenediazo sulphonate, 11
dialkylphosphorothiolates, 39
diazinon, **67**
dicloran, resistance of *Botrytis cinerea* to, 164
diffusion of liquids
  into leaves, 100
  into roots, 99
dimethirimol, 5-*n*-butyl-2-dimethylamino-4-hydroxy-6-methyl pyrimidine, **68**
  resistance to, 158
  selectivity, 142
Diplocarpon rosae, rose black spot, 215
disruption of cell structure, 133
distribution of benomyl and dinocap on leaves, 181
dithiouracil, **77**
dodemorph, N-cyclododecyl-2,6-dimethylmorpholine, **74**
  controls rose powdery mildew, 215
dodine, resistance to, 158
Dutch elm disease, *Ceratocystis ulmi* = *Ceratostomella ulmi*, 11, 120
  effect of aminotrichlorophenylacetic acid, 121
  natural recovery, 120

EBC, ethyl benzimidazole-2-ylcarbamate, 137
ectodesmata, 94
elm disease, *see* Dutch elm disease
Elsinoe veneta, raspberry cane spot, 247
Endothia parasitica, chestnut blight, 7
Entomosporium maculatum, pear leaf spot, 243
enzymes, wall-digesting, produced by pathogens, 129
eradicant effects on barley mildew, 191
Erysiphe graminis, 187–94
  resistance to ethirimol, 162
Ethionine
  checks powdery mildews and pea root rot, 125
  effect on cell walls, 124
ethirimol, 5-*n*-butyl-2-ethylamino-4-hydroxy-6-methyl pyrimidine, **68**
  mammalian toxicity, 90

mode of action, 142
resistance of *Erysiphe graminis* to, 162
soil and spray application, 187–94
translocation, 111
Eutypa armeniacae, apricot dieback, 244
Exobasidium vaccinii, blueberry red leaf, 247
eyespot, *see Cercosporella herpotrichoides*

F 427, 2,3-dihydro-6-methyl-5-(2-diphenyl)carbamoyl-1,4-oxathiin, 140
ferulic acid detoxifies pyricularin, 130
5-fluorodeoxyuridine, anti-sporulant, 126
p-fluorophenylalanine, anti-sporulant, 126
foliar fungicides, application techniques, 180
free space, definition and estimates of, 96–8
French bean, grey mould, *Botrytis cinerea*, of pods, control by benomyl, 234
fuberidazole, 2-(2′-furyl)benzimidazole, **66**
cross-resistance with MBC, 135
furane compounds, **59**
Fusarium
culmorum and *graminearum*, control by thiabendazole, 202
nivale, 202
oxysporum
f.sp.*cubense*, spore migration in banana xylem, 181
f.sp.*melonis*, benomyl-resistant strains, 164
f.sp.*lycopersici*
production of polygalacturonase, 123, 128
tomato response to, 119
use in bioassays, 12
roseum, potato dry rot control by benomyl, 236
banana fruit rot, 253

G 676, 2,4-dimethyl-5-carboxanilidothiazole, 196
β-galactosidase synthesis in *Verticillium albo-atrum*, 128
genetic changes affecting fungal cell resistance, 163
geodin, **44**
ghost-spotting of tomato, 209
glasshouse acreages, 207
Gloeodes pomigena, apple sooty blotch, 242
Gloeosporium
fruit rots, control by benomyl, 183, 241
musarum, banana anthracnose, 253
Glomerella cingulata, apple necrotic leaf blotch, 242
glucosamine in fungal cell wall synthesis, 42
glucose spraying to control 'low-sugar' diseases, 130
gooseberry, American mildew, control by benomyl, 247
diseases, 246
granules
carboxin, 197
ethirimol, 188
kitazin P, 178, 204
grape diseases, 251
griseofulvin, 31, 42, **43**
growth-regulant effects on pathogens, 125
Guignardia bidwellii, grape black rot, 251
gums obstruct spore migration in xylem, 122
Gymnosporangium juniperi-virginianae, cedar-apple rust, 242

HOE 2873, *see* pyrazophos
haustorium formation inhibited by 6-azuracil, 143
Helminthosporium
gramineum, control by carboxin, 202
maydis, trials with carboxin, 205
solani, control, 227
vagans, melting-out disease of turf, control, 130
'high-sugar' diseases, 129
Hinozan, O-ethyl-S,S-diphenyl phosphorodithiolate, **40**, 204

host tissue modification, 117
hydroxypyrimidines, 66
hyperauxiny, 122

inactone, **67**
3-indole-acetic acid induces tylosis, 123
Inazin, O-ethyl-S-benzylphenyl phosphonothiolate, **40**, 204
inhibition
  of acetate metabolism, 58
  of biosynthesis, 131
  of cell multiplication, 66
  of DNA synthesis, 48, 138
  of fungal chitin synthesis, 53, 147
  of haustorium formation, 78, 143
  of oxidative phosphorylation, 82, 132
  of polynucleotide synthesis, 80
  of polyphenol oxidase, 80
  of protein synthesis, 50, 149
  of pyrimidine biosynthesis, 78
  of respiration, 66, 132
  of RNA synthesis, 58
  of tricarboxylic acid cycle, 58
injections into woody tissue, 182

Japan, use of systemic fungicides in, 48

kasugamycin, **53**, 150
  controls *Pyricularia oryzae*, 204
  mammalian toxicity, 88
  mode of action, 150
  resistance of *Pyricularia oryzae* to, 170
Kasumin, 205
kinetin, 6-furfurylaminopurine, tested against *Sphaerotheca fuliginea*, 22
kitazin, O,O-diethyl-S-benzyl phosphorothiolate, **39**
  application in irrigation water, 178, 204
  mode of action, 41
kitazin P, O,O-isopropyl-S-benzyl phosphorothiolate, **39**

Ladogal, sodium salt of *p*-aminobenzylsulphonoxymethylamide-N-D-glucoside sulphonic acid, 18
leaf disc tests, 22
leaf intake, 100
Leonardo da Vinci, 6
lethal synthesis, 168
lettuce
  downy mildew, *Bremia lactucae*, 213
  grey mould, *Botrytis cinerea*, 213
lithium salts
  injected against *Endothia parasitica*, 7
  used against *Erysiphe graminis*, 7
'low-sugar' diseases, 129

MBC, methyl benzimidazol-2-ylcarbamate, **64**
  action of, 138
  decomposition product of benomyl, 135
  decomposition product of thiophanate-methyl, 136
  inhibitor of DNA synthesis, 138
  translocation, 112
maize head smut, *Sphacelotheca reiliana*, tests with carboxin, 205
'mal secco' of orange seedlings, 250
maleic hydrazide
  increases cucumber scab, 125
  reduces *Alternaria longipes* in tobacco, 125
mammalian toxicities of systemic fungicides, 86–91
'masked fungicides', 14
mebenil, 2-toluanilide, **60**
  against yellow rust of wheat, 198
Melanconium fuligineum, grape bitter rot, 251
melting-out disease of turf, *Helminthosporium vagans*, control, 130
metabolic conversion, 153
L-methionine checks powdery mildew and root rots, 125
Microthyriella rubi, apple fly speck, 242
Milcurb = 10% w/v solution of dimethirimol formulated as hydrochloride
mildew, cereal, *see Erysiphe graminis*

mitochondria, energy production affected by fungicides, 165
Monilinia fructicola, brown rot of stone fruit, 244
Monilinia vaccinii-corymbsi, blueberry blossom blight, 247
morphactins, esters of 9-hydroxy-fluoene-9-carboxylic acid, 125
morpholine, 74
mushrooms
  benomyl treatment, 218
  cobweb disease, *Dactylium dendroides*, 217
  dry bubble disease, *Verticillium malthousei*, 217
  wet bubble disease, *Mycogone perniciosa*, 217
mutations, effect on resistance, 169
mycomycin, **54**
Mycosphaerella
  brassicicola, ringspot of Brussels sprouts, control, 234
  fijiensis, banana black leaf streak, 253
  melonis, cucumber stem rot, 203
  musicola, banana Sigatoka leaf spot, 253

NF 35, thiophanate, *q.v.*
NF 44, thiophanate methyl, *q.v.*
NF 48, 2-(3-methoxycarbonyl-2 thioureido) aniline, controls smuts, **80**, 201
α-naphthalene acetamide induces tylosis in tomato, 123
naphthaleneacetic acid, effect on cell-wall pectins, 124
naramycin B, isocycloheximide, 46
nectarine diseases, 244
Nectria galligena, apple canker, antisporulants, 241
negative forecasting, 184
nickel compounds used against wheat rusts, 20
Nigrosphaera sphaerica, banana squirter disease, 253
nonane, 4-phenyl-4-(3-pyridyl)-3-oxatricyclo-, **72**
nucleic acid metabolism affected by fungicides, 165

oak wilt, *Ceratocystis fagacearum*, altered susceptibility to, 121
oats
  covered smut, *Ustilago hordei* (= *U. kolleri*), 201
  loose smut, *Ustilago avenae*, 201
onion smut, *Urocystis cepulae*, control by benomyl, 229
onion white rot, *Sclerotium cepivorum*, control by benomyl, 229
Oospora pustulans, skinspot of potato, control, 227
orange
  'mal secco' of seedlings, 250
  mould (*Penicillium* spp.) control on fruit, 236
organophosphorus compounds, 34–42
  mammalian toxicity, 87
oxathiins, **56**
  mode of action, 142
  resistance to, 162
  selectivity, 140
oxazines, substituted, 74
oxycarboxin, **57**
  against cereal rusts, 196
  against flag smut, 202
ozone damage affected by benomyl, 117

paddy blast, *Pyricularia oryzae*, 39
parinol, bis-(*p*-chlorophenyl)-3-pyridine methanol, **72**
pathocidin, 67
pea *Ascochyta* diseases controlled by benomyl, 226
peach diseases, 244
pear diseases, 243
pectate lysases, 124
pectolytic enzymes, inactivation of, 127
Pellicularia sasakii tolerant to blastocidin, 150
Penicillium
  brevicompactum, benomyl-resistant strain, 160
  digitatum on banana and citrus, 236, 249
  expansum, apple blue mould, 242
  italicum on banana and citrus, 236, 249

pelargonium rust, *Puccinia pelargonii-zonalis*, 216
phenols and derivatives, 83
phenolic compounds inactivate pectolytic enzymes, 127
phenylalanine induces resistance to apple scab, 119
4-phenyl-4-(3-pyridyl)-3-oxatricyclononane, **72**
phenyl-thiourea, **80**
phloem, functions of, 103
phloridzin, phloretin, effects on resistance to apple scab, 119
Phomopsis sclerotioides, cucumber black root rot, 212
phosphorothiolates, dialkyl, **39**
Phragmidium mucronatum, rose rust, 215
Physalospora obtusa, apple frog-eye leaf spot, 242
phytoalexins, 118
Phytophthora cinnamomi, root rot of avocado, 124
pineapple diseases, 253
piperazine, N,N'-bis(1-formamido-2,2,2-trichloroethyl)-piperazine = triforine, **73**
Piricularia, see Pyricularia
Plasmodiophora brassicae, club root, control with benomyl, 230
plum diseases, 244
Podosphaera
  leucotricha, apple mildew, control with triarimol, 239
  oxycanthae, cherry mildew, 244
poinsettia black root rot, *Thielaviopsis basicola*, control, 216
polyenes, 53
polygalacturonases, 123
  inactivated by chlorogenic and caffeic acid products, 127
  produced by *Fusarium oxysporum* f.sp. *lycopersici*, 123
polyoxins, **51**, 147
  polyoxin D, mode of action, 148
polyphenol oxidase, 119
polyynes, 53
post-harvest treatments, 183, 235
potato tuber diseases control by benomyl, 227
powdery mildews, treatments for, historical survey, 20
procaine hydrochloride used against cucumber powdery mildew, 21
protectant fungicides, mode of action, 151
protein synthesis, inhibition of
  in *Pyricularia oryzae*, 50, 149
  in *Saccharomyces cerevisiae*, 168
protoplast membrane permeability, fungicide effect on, 165
Pseudopeziza ribis, black currant leaf spot, control by benomyl, 246
Puccinia
  arenaria, carnation rust, 214
  graminis, black rust, 196
  hordei, brown rust of barley, 194
  horiana, white rust of chrysanthemum, 216
  pelargonii-zonalis, pelargonium rust, 216
  pringsheimiana, gooseberry rust, 246
  recondita (= *triticina*), brown rust of wheat, 196
  striiformis (= *glumarum*), yellow rust, 196
purines
pyracarbolid, 2-methyl-5,6-dihydro-4H-pyran-3-carboxylic acid anilide, 187
  controls *Urocystis agropyri*, 201
  controls *Ustilago nuda*, 200
pyran, substituted, 61
pyrazolopyrimidines, phosphoric esters of, **39**
pyrazophos, 239, 241, 247
Pyrenochaeta lycopersici, tomato brown root rot, 211
Pyricularia oryzae, rice blast, control, 40, 50, 204
pyricularin, toxin produced by *P. oryzae*, 130
pyrimidines, 66, 76, 142

8-quinolinol sulphate
  for therapy of vascular wilt pathogens, 29
  for treating root rot of *Matthiola* sp., 11

used against *Stereum purpureum*, 7
quintozene = pentachloronitrobenzene, detoxication of, 167

raspberry diseases, 247
resistance, fungal
avoidance, 171
of *Botrytis cinerea*, 159
of *Penicillium brevicompactum*, 160
to
benomyl, 162
induced by u.v. radiation, 163
blasticidin S, 162
cadmium, 158
chlorinated nitrobenzenes, 157
chloroneb, 162
dimethirimol, 158
dodine, 158
ethirimol, 162
kasugamycin, 162
kitazin, 162
oxathiins, 162
thiophanates, 162
triarimol, 162
respiration: effect of thiabendazole, 66
reversal of fungitoxicity of hydroxypyrimidines, 71
RH 124, 4-*n*-butyl-1,2,4-triazole, 35
used against rusts, 196
Rhizoctonia solani on peas, control, 228
ribosomes, protein synthesis affected by fungicides, 165, 168
rice paddy blast, *Pyricularia oryzae*, 39, 50
control by kasugamycin, 204
rimocidin, **54**
RNA synthesis inhibited by oxathiins, 58
root uptake, 99
rose
black spot, *Diplocarpon rosae*, 215
powdery mildew, *Sphaerotheca pannosa*, 215
rust, *Phragmidium mucronatum*, 215
rufianic acid, 1,4-dioxyanthroquinone sulphonic acid, inactivates pectolytic enzymes, 14, 127
rusts, *see Puccinia* and cereal rusts

salicylanilide, **56**
scab
apple, *Venturia inaequalis*, 238
pear, *Venturia pirina*, 243
Sclerotinia
fructigena, apple brown rot, 242
laxa f. *mali*, apple blossom wilt, 242
sclerotiorum, white mould of bean, control by benomyl, 234
spp., stone-fruit rots, control by benomyl, 229
Sclerotium cepivorum, onion white rot, control by benomyl, 229
Sclex, 3-(3,5-dichlorophenyl)-5,5-dimethyl-2,4 oxazolidinedione, 243
Seed dressing
techniques, 22, 176
with ethirimol, 187
selection pressure, 172
selectivity, mechanisms of, 152
Septoria
apiicola, late blight of celery, control, 233
nodorum, glume blotch of wheat
controlled by benomyl spray, 199
controlled by thiabendazole, 202
smut, head, *Sphacelotheca reiliana*, trials with carboxin, 205
smut, flag, *Urocystis agropyri*, controlled by carboxin, 201
smuts, *Ustilago* spp.
control by
BAS 3191 and BAS 3270, 200
benomyl, 200
carboxin, 199
NF 48, 201
pyracarbolid, 200
soil application of ethirimol, 187
soybean, 229
Sphacelotheca reiliana, head smut of maize, trials with carboxin, 245
Sphaerotheca
fuliginea, cucumber powdery mildew
control, 212, 213
resistance to benomyl, 161
resistance to dimethirimol, 158
macularis, strawberry powdery mildew, 248

Sphaerotheca – *cont.*
mors-uvae, American gooseberry mildew, control by benomyl, 246
pannosa, rose powdery mildew, 215
spore migration in plant tissue, 121
sporulastatic effects, *see* antisporulants
spray application on cereals
ethirimol, 189
other systemic fungicides, 190
Stereum purpureum, silver-leaf of stone fruits, 7
Stigmina carpophila on peach, 244
stomatal closure barring entry of pathogens, 119
stone-fruit diseases, 244
strawberry
diseases, 248
wilt, *Verticillium dahliae*, control by benomyl, 249
streptomycin, **45**
streptothricin tested against powdery mildews, 32
streptovitacins, 48
sugar-beet
leaf spot, *Cercospora beticola*, control, 232
root rot, *Aphanomyces cochlioides*, 125
sugar levels in host tissues affect susceptibility, 129
sulphonamides, therapeutic activity, 18, 28
symplastic movement, 102

Taphrina deformans, peach leaf curl, 244, 245
terramycin, 46
thiabendazole, 2-(4′-thiazolyl)benzimidazole, controls, **64**
bunt, 201
Fusarium graminearum and *F. nivale*, 202
Septoria nodorum, 202
mode of action, 139
translocation, 111
Thielaviopsis basicola, black root rot of cyclamen, 216
thiophanate, 1,2-bis-(3-ethoxycarbonyl-2-thioureido)benzene, **79**
resistance to, 162
thiophanate-methyl, 1,2-bis-(3-methoxycarbonyl-2-thioureido) benzene, **79**
conversion to MBC, 136
spray against cereal mildew, 191
resistance to, 162
thiosemicarbazides, substituted, 80
thiourea-based compounds, 79
thiram, 226
L-threo-β-phenyl serine reduces cucumber scab, 32, 125
Tilletia caries and *T. contraversa*, wheat bunts, 201
time of attack, effects of on cereals, 191
times of spray applications, 183, 192
tobacco brown spot, *Alternaria longipes*, 125
2-toluanilide (mebenil), **60**
tomato
brown root rot, *Pyrenochaeta lycopersici*, 211
ghost spot and grey mould, *Botrytis cinerea*, 209
leaf mould, *Cladosporium fulvum*, 209
stem rot, *Didymella lycopersici*, 209
verticillium wilt, *Verticillium* spp., 210
wilt, *Fusarium oxysporum* f. *lycopersici*, varietal reactions, 119
toxin inactivation, 130
toyocamycin, 67
'training' of fungal mycelium, 23, 157
translaminar movement, 15, 233
translocation downwards, 74
triamiphos = Wepsyn 1-bis-(dimethylamido)phosphoryl-3-phenyl-5-aminotriazole-1,2,4, **35**
mode of action, 145
triarimol, α-(2,4-dichlorophenyl)-α-phenyl-5-pyrimidine methanol, **73**
used against
apple scab, 144, 240
cereal mildew, 191

1,2,4-triazoles, substituted, 35
tricarboxylic acid cycle inhibited by
carboxin, 58
2,3,6-trichlorophenoxyacetic acid
effect on
bean chocolate spot, 14
Dutch elm disease, 121
oak wilt, 121
trickle irrigation 219
tridemorph, N-tridecyl-2,6-dimethyl
morpholine, **74**, 144
spray against cereal mildew, 190
triforine = Cela W 524, N,N′-bis(1-
formamido-2,2,2-trichloro-
ethyl)piperazine, **74**
used against
carnation rusts, 215
cereal mildew, 196
cucumber powdery mildew, 212
rose black spot, 215
tubercidin, 66
tyloses obstructing spore migration,
121, 122

ultraviolet radiation induces resistance to benomyl, 163
Uncinula necator, grape powdery
mildew, 251
uptake of ethirimol, 188
uptake of oxathiins affects selectivity,
140
Urocystis
agropyri, flag smut of wheat,
control, 201
cepulae, onion smut, control by
benomyl, 229
Uromyces
caryophyllinus, carnation rust, 214
phaseoli, bean rust, control, 234
Ustilago
avenae, loose smut of oats, 201
hordei, covered smut of barley and
oats, 201
kolleri = *hordei* q.v.
nuda, loose smut of wheat, control
with BAS 3191, BAS 3270, 200
carboxin, 200
pyrocarbolid, 200
tritici = *nuda*, q.v.

vapour phase redistribution, 72
vascular pathogens, see also
Fusarium oxysporum, 11
Venturia
inaequalis, apple scab, 238
control with triarimol, 240
effect of phenylalanine, 119
resistance to antimycin A, 161
nashicola, 243
pirina, pear scab, 243
venturicidin, 32
Verticillium
albo-atrum, synthesis of β-galacto-
sidase, 128
cinerescens, carnation wilt, 214
dahliae, strawberry wilt, 248, 249
malthousei, dry bubble of mushrooms, 217
vessel group size, 120

Wepsyn, *see* triamiphos
wheat
bunt, *Tilletia caries*, 201
eyespot, *Cercosporella herpotrichoides*, 203
flag smut, *Urocystis agropyri*, 201
mildew, *Erysiphe graminis*, 195
rust, *Puccinia graminis* and *P.
striiformis*, 17, 196
wilt diseases, *see* vascular pathogens
winter barley, control of mildew, 193
wyerone, factor in natural plant
resistance, 32

zinc chloride used against *Erwinia
amylovora*, 7

B&T 9-502